J. Gil-Aluja

Fuzzy Sets in the Management of Uncertainty

Springer-Verlag Berlin Heidelberg GmbH

Studies in Fuzziness and Soft Computing, Volume 145

Editor-in-chief
Prof. Janusz Kacprzyk
Systems Research Institute
Polish Academy of Sciences
ul. Newelska 6
01-447 Warsaw
Poland
E-mail: kacprzyk@ibspan.waw.pl

Further volumes of this series can be found on our homepage: springeronline.com

Vol. 126. J.-L. Verdegay (Ed.)
Fuzzy Sets Based Heuristics for Optimization, 2003
ISBN 3-540-00551-X

Vol 127. L. Reznik, V. Kreinovich (Eds.)
Soft Computing in Measurement and Information Acquisition, 2003
ISBN 3-540-00246-4

Vol 128. J. Casillas, O. Cordón, F. Herrera, L. Magdalena (Eds.)
Interpretability Issues in Fuzzy Modeling, 2003
ISBN 3-540-02932-X

Vol 129. J. Casillas, O. Cordón, F. Herrera, L. Magdalena (Eds.)
Accuracy Improvements in Linguistic Fuzzy Modeling, 2003
ISBN 3-540-02933-8

Vol 130. P.S. Nair
Uncertainty in Multi-Source Databases, 2003
ISBN 3-540-03242-8

Vol 131. J.N. Mordeson, D.S. Malik, N. Kuroki
Fuzzy Semigroups, 2003
ISBN 3-540-03243-6

Vol 132. Y. Xu, D. Ruan, K. Qin, J. Liu
Lattice-Valued Logic, 2003
ISBN 3-540-40175-X

Vol. 133. Z.-Q. Liu, J. Cai, R. Buse
Handwriting Recognition, 2003
ISBN 3-540-40177-6

Vol 134. V.A. Niskanen
Soft Computing Methods in Human Sciences, 2004
ISBN 3-540-00466-1

Vol. 135. J.J. Buckley
Fuzzy Probabilities and Fuzzy Sets for Web Planning, 2004
ISBN 3-540-00473-4

Vol. 136. L. Wang (Ed.)
Soft Computing in Communications, 2004
ISBN 3-540-40575-5

Vol. 137. V. Loia, M. Nikravesh, L.A. Zadeh (Eds.)
Fuzzy Logic and the Internet, 2004
ISBN 3-540-20180-7

Vol. 138. S. Sirmakessis (Ed.)
Text Mining and its Applications, 2004
ISBN 3-540-20238-2

Vol. 139. M. Nikravesh, B. Azvine, I. Yager, L.A. Zadeh (Eds.)
Enhancing the Power of the Internet, 2004
ISBN 3-540-20237-4

Vol. 140. A. Abraham, L.C. Jain, B.J. van der Zwaag (Eds.)
Innovations in Intelligent Systems, 2004
ISBN 3-540-20265-X

Vol. 141. G.C. Onwubolu, B.V. Babu
New Optimzation Techniques in Engineering, 2004
ISBN 3-540-20167-X

Vol. 142. M. Nikravesh, L.A. Zadeh, V. Korotkikh (Eds.)
Fuzzy Partial Differential Equations and Relational Equations, 2004
ISBN 3-540-20322-2

Vol. 143. L. Rutkowski
New Soft Computing Techniques for System Modelling, Pattern Classification and Image Processing, 2004
ISBN 3-540-20584-5

Vol. 144. Z. Sun, G.R. Finnie
Intelligent Techniques in E-Commerce, 2004
ISBN 3-540-20518-7

Jaime Gil-Aluja

Fuzzy Sets in the Management of Uncertainty

Springer

Professor Dr. Jaime Gil-Aluja
Universitat de Barcelona
Dept. d'Economia i Organització, d'Empreses
Facultat de Ciències
Econòmiques i Empresarials
Avgda Diagonal 690
08034 Barcelona
Spain
E-mail: jv00000@teleline.es

DOI 10.1007/978-3-540-39699-4

Library of Congress Cataloging-in-Publication-Data

A catalog record for this book is available from the Library of Congress.
Bibliographic information published by Die Deutsche Bibliothek.
Die Deutsche Bibliothek lists this publication in the Deutsche Nationalbibliographie;
detailed bibliographic data is available in the Internet at http://dnb.ddb.de

Originally published by Springer-Verlag Berlin Heidelberg New York in 2004
MyCopy version of the original edition 2004

Cover design: E. Kirchner, Springer-Verlag, Heidelberg
Printed on acid free paper 62/3020/M - 5 4 3 2 1 0
www.springer.com/mycopy

PREFACE

On the threshold of the XXI century society has raised certain important questions concerning the government of states, but also business management. This type of problem pertains to a very wide variety of scientific disciplines. Engineers, physicists, economists, chemists, biologists, sociologists, etc. all occupy their place among those who are seeking a more just, fraternal and solidary world.

Traditionally, economic science has considered as one of its fundamental objectives, the finding of balances. We feel that there can be no economist, at whichever school of thought he or she may have acquired the knowledge, who does not base decision taking on thinking of arriving at a point of equilibrium, or else thinking about breaking down an already existing balance. But always with the idea of finding another which may be for him or her more favourable.

It should come as no surprise, then, that we are living times of a certain amount of confusion, when a reality, that is full of convulsions which make life unstable wants to be treated as it had been under situations of equilibrium, which are surrounded by stabilities. We have to get used to the idea that society, economy and business activity, as we have known them up to this point, have no possibility of surviving in the very immediate future and that many, but many changes will be inevitable.

We believe that consensus and instability are concepts that are increasingly more united and they will become so even more in the future. This is so much so that new tendencies in scientific research in the sphere of economy and business management are increasingly occupying a more fundamental position. Instability and volatility of financial markets, instability and fluctuation of prices, instability and precariousness of job offers, are all very clear examples.

Those who work in the creation of instruments for depositing them in the hands of those who must make decisions, are seeking, in the last resort, the well being of society, But in order to allow this task to be successful it is necessary to be aware, even in an imprecise manner, what is going to be the nucleus of economic-financial relations in the social systems. Without a doubt those who are capable of having a feeling for and can put limits to the profiles of the fuzzy world of the following decades, will be the winners of tomorrow.

But in a context of changes, such as we are living in the present, who is capable of foreseeing the outcome of events with the necessary precision of a prophet? Perhaps we should be content with less and make better use of that which we have available. From this statement, personal work and skill will allow us to attain success, at least for a part of those who deserve it.

It might be useful, in this context, to make a brief comment on the possibilities that are offered by the proposals that proceed from the laboratories where the new scientific findings are tested. In them, we can see that scientific activity is at a crossroads where the future of humanity is at stake. On the one hand, the geometrical conception of the universe, and on the other, the Darwinian conception. On the further hand, the lofty, monotonous and well known repetitive chants, renewed only in their forms. The imposition of certain pre-established beliefs from the splendid Newtonian dawn, where it was dreamed of reducing the functioning of the world to the predictability of meccano. On the other, the void of the unknown. The diverse murmuring and sometimes clashing notes that sound unconnected. The attraction of adventure. The invitiation to jump over a bottomless precipice, only guided by the hope of opening up new horizons. The response to the call of Bertrand Russell, of Lukasiewicz, Zadeh, Lorenz, Prigogine, Kaufmann. The rejection of the yoke of predestination and the proclamation of freedom of decision, which over and over again collides, with the wall of uncertainty.

Uncertainty. A word which was almost banned just a few decades ago. Subjected to mistreatment and object of confusion by those who should have been the faithful guardians of the "Holy Grail" of Science. Today, thanks to the original works of Zadeh, it has risen again to recapture a leading role that facilitates new directions to so many researchers desirous of providing replies to the questions raised in this day and age by a convulsed society.

It would be unthinkable not to accept the fact that social and economic systems are particularly sensitive to variations of the initial conditions, or of existing conditions during some moment of their activity. To put this in another way, it will be seen that when a disruption exceeds a certain level, future deviations lead to a process that is uncontrollable by the system itself, causing the birth of unsuspected new phenomena. The complexity of these systems makes it inevitable that they are understood and explained only by means of determinist laws, supported and developed with linear equations. What was required, and will still be required, is a large dose of imagination in order to break the links that keep us tied to the past, by placing in their place differential "non linear" equations that carry a large arsenal that is capable of describing uncertain situations.

Although it may be difficult to find the deepest roots of the view points with which researchers scrutinise the world, there is no doubt whatsoever that something very important is striving to emerge to the surface of scientific activity when the first essences of evolutionism, the rich inheritance of the XIX century, were being distilled.

In fact, in his fundamental work "The origin of species" published in 1859, Darwin considered that the fluctuations in biological species, thanks to the selection of the medium, give rise to an irreversible biological evolution. From the association between fluctuations (which assimilates the idea of randomness, we would say uncertainty) and irreversibility an auto-organisation of systems with increasing complexity, takes place.

On the other hand, Clausius, drew up in 1865 the " Law of increase in entropy" with the corresponding division between reversible or irreversible processes. This distinction becomes explicit in the second law that postulates the existence of a function, entropy, which in an isolated system increases when irreversible processes exist and remains constant in the presence of reversible processes. Therefore, entropy arrives at a maximum value when the system is reaching an equilibrium and the irreversible process ends.

Both in the case of Darwin and in that of Boltzmann, randomness (or if you like, uncertainty) and evolution are very closely related, but the result of their respective research leads to opposing conclusions. For Boltzmann probability reaches maximum when uniformity is being reached, whilst, for Darwin evolution leads to new auto-organised structures.

Contrary to these prospects, the prototype of classical physics and consequently the marginalized economic theory, is the mechanics of the movement, the description of the processes of a reversible and determinist nature, where the direction of time plays no part whatsoever, where no place exists either for uncertainty nor irreversibility. In short, economic and management systems constitute enormous robots.

It is very true to say that some phenomena that arise on the life of states, institutions and businesses can be perfectly described by means of determinist equations but, on the other hand, others include uncertain processes, or, in any event, stochastic processes. It could happen, that our very own existence, with all its complexity, is also to be found inscribed in the general laws right from the original time of the Big Bang. But science, having spent so much time seeking these generalities, symmetries and laws, has found the changeable, the temporary, and the complex.

Scholars within the field of economy and management are observing processes in which the transition from chaos to order take place, that is to say sequences directed towards an auto-organisation. The question that must be asked is how does this creation of structures, that is to say auto-organisation, take place. Well, given that the entropy of a system, if disrupted in such a manner that the state remains sufficiently close to equilibrium, the system responds by re-establishing the initial situation. We are then referring to a stable system. But if a state is far enough removed from equilibrium, it enters into a situation if instability relative to the disruption. This point is normally called the junction point. From this point new situations take place that can correspond to behaviours that are removed from the original situation. In this context determinist processes have no use for predicting which path will be selected between those existing at the junction. In many of these junctions a breakdown in symmetry occurs. It is the typical case in which there exists a "left hand" solution and a "right hand" solution, but it is impossible to know "ex ante" which of them will occur.

Economy, which has traditionally not considered this manner of conceiving the operation of economic-business phenomena and the thought that revolves around them, was initially structured based on a mechanistic mathematics between 1880 and 1914 with the General Equilibrium (Walras, Pareto, Cournot, Edgewort,). Faced by a reality the basic characteristic of which was a certain "indolence" in the behaviour of social life and economic relations, groups of scholars made an attempt to formalise the processes, which occurred within them. The classical mechanics of Lagrange was used, which gives a sense of seriousness, compared to what Perroux called the "laxity of the economic discourse". But on the other hand, human thought remains trapped by certain economic laws, parallel to the laws of nature, which prevents it exercising one of its most precious treasures: imagination. For this reason the automatisms of the mathematics of determinism were widely preached and prevailed, and still prevail today, in many spheres of scientific activity in economics and business management.

But the findings of Zadeh with his fuzzy sets have changed the whole panorama of research within the spheres of social sciences. And in the revolutionary inrush of new proposals, concepts so well rooted as profitability, economy, productivity, … expressed by cardinal functions, are loosing their particular attraction in favour of other notions such as relation, grouping, assignment and ordering, which now acquire a new sense. This displacement is fundamental, because it means the transfer of non-arithmetical elements, in traditional studies considered complementary, to the position of privilege they currently occupy.

We find ourselves then at a crossroads, which could be called Geometric-Darwinism, at the centre of which there is an argument that has raged for more than two thousand years. In fact, Aristotle (384-322 B.C.) pointed out that: "with regard to things present and past, propositions, whether positive or negative, are by necessity true or false. And of the propositions, which are contradictory, there must be one true and the other false". The thoughts of the Stoics to one of whose central figures, Crisipo de Soli (ª 281 – 208 B.C.) can be attributed the formulation of the so-called "excluded middle principle". The Epicureans, with their founder Epicurus (341 – 270 B.C.) at the head, replied to this principle with vigour, pointing out that it is only acceptable if there is not a third possibility "tertium non datur" (excluded middle).

More than twenty two centuries had to pass before Lukasiewicz, taking up the idea of the Epicureans again, was to point out that there existed propositions that were neither true not false, but indeterminate. This allowed him to draw up his "principle of value" (each proposition has a value of truth). Initially he assigned three values to truth: true (1), false (0), indeterminate (0.5), later generalising to n values, for n equal or greater than 2. In this way the path was opened up for so called multivalent logic.

On the occasion of the International SIGEF Congress in Buenos Aires, we made an attempt to place the Epicurean position within the new coordinates arising from the important work of Zadeh, drawing up the "principle of gradual simultaneity" (every proposition can at one and the same time be true and false, on the condition of assigning a degree of its truth and a degree of its falseness). Before and after this, a large number of scientists have placed, stone on stone, the foundations of what can become a new building of economic knowledge. From our sphere of study, many names have become milestones along this fruitful path and numerous research groups from universities on the five continents have accepted the challenge and are working on the different branches of the tree of science.

The work to which these few words are the preface, represents the last work of whom, during more than thirty years, has attempted, first with Professor Kaufmann and later, after his sad death, alone, to provide to those immersed in the intricate world of decisions, certain theoretical and technical elements that will allow for a more balanced struggle while faced with the uncertainties of today and tomorrow.

TABLE OF CONTENTS

INTRODUCTION

It is now just over a quarter of a century, in conjunction with Professor Arnold Kaufmann, that we commenced a journey full of satisfaction and also quite a few heartaches. The lack of understanding of many researchers and teachers towards a new path for tackling Knowledge, along with the resistance to change, which is so entrenched in so many universities, were some of the obstacles put in the way and which slowed down the illusion we had of contributing to a better world.

Step by step we managed to open up for the younger generations certain channels along which their thought could flow, always with the idea of working at all times within current reality. A mathematical engineer and an economist, very keen on mathematics, always found a point of convergence, which also, was not so vital in teacher-disciple relationships.

The passage of time marked the works carried out either individually or in common. Conferences, papers or communications at congresses and articles for publications or poly-copied notes, were the normal vehicle chosen for making our modest research known. The idea of the fuzzy sub-set, occupied as time went by, a space among enterprising intellectuals, unhappy with the avarice of the guardians of official knowledge. After the findings of L. Zadeh[1] in 1965 and of the book published by A. Kaufmann[2] in 1973, commenced a period of complicated divulgation in which mainly mathematicians, engineers or computer experts played a relevant role.

Each incorporation of the abundance of theoretical and technical elements, all the time with an ever-increasing flow, to the sphere of economy and business management would occur slowly, in different spheres, initially in an incoherent manner. It was in 1986 when the first book dedicated specifically to the introduction of the theory of fuzzy sub-sets to management[3] appeared. It was translated into several languages and several editions were published in Spanish. The last in 1992 was sold out much quicker than expected. We thought about the possibility of a further reprint, but felt that it would be more honest to redo this work, taking into account and incorporating new findings, which were now very abundant, and which researchers were arriving at. The death of Professor Kaufmann in 1994 cut this project short. There have been many friends and acquaintances, both in the university field as in the professional, who over all these years have reiterated their interest for a new version of this work. Recently, the President of the Reial Acadèmia de Doctors, Joe Casajuana, suggested the publication of a book in which there was a coincidence of the characteristics of scientific thoroughness and the possibility of reaching a very wide spectrum of readers. As always his reasoning was totally convincing. Perhaps the time had come for us to get to work and pick up the thread of that old project.

Fortunately, time does not pass in vain and from the outset of our work we were aware of the fact that the task we had taken on would have a different structure. From those initial, in a certain manner, rudimentary elements, on which we based ourselves at that time we were now faced with a truly scientific body, with a conceptualisation and methods open to very wide horizons. New theories or generalisations of those that already existed, have given rise to the elaboration of models and algorithms that are capable of adequately replying to problems that in that instance were not resolved, or not even tackled. The theory of fuzzy sub-sets has been and continues to be enriched on a daily basis and new theories accompany it or complete it. Today we are talking about a theory of uncertainty, with a principle, that of «gradual simultaneity», with numerical mathematics and non-numerical mathematics of uncertainty. It would appear then that it would be logical to write a book thinking more on the introduction to the theory of uncertainty in management rather than in the introduction to the theory of fuzzy sub-sets. This does not prevent us from maintaining (and we are very pleased to do so) those subjects of the 1986 book that continue to be current. When this occurs we will indicate the fact in the corresponding footnote. In the event that current knowledge allows for a better treatment than that done before, we will also make a mention in the bibliography. But our obligation is to offer the best of ourselves, and this at all times. Only in this case will the two mottos we inherited from our master become reality: «to be useful to others» and «to creat our own concurrence» .

A few, but very few, scientists working in this field of knowledge were conducting research, at that time, into the problems that uncertainty caused in economy and management. An association, SIGEF (International Society of Fuzzy Management and Economy), grouped together hundreds of scholars from all over the world and its publication Fuzzy Economic Review includes many of their works. Many books have seen the light of day on specific aspects of management in businesses and institutions[4], and at the different Congresses papers and communications are presented with contributions that are worthy of great consideration. It is our duty and also great satisfaction to be able to take advantage of these findings, even more so that very often when coming from members of our own research group, who one day were students and are now colleagues, and who today, are perhaps doing things better than us.

All of us are united in the feeling that the society in which we live can attain greater prosperity, if we are only able to understand it, know how to explain it and treat its problems in an adequate manner. For this certain theoretical and technical elements are necessary that are different from those taught to us by our predecessors. The world with its globalisation, states with their complex economical interrelationships and business and institutions with their ever increasing complex flows within and without their particular spheres, have undergone great change relative to times gone by. And that knowledge, which before was valid in a certain context, is no longer so, because it has undergone modifications. Therefore it was necessary to try new proposals, to make a kind of leap ..., which at that time appeared to be into the void. But now we can definitely say that it was all well worth the trouble. Before our eyes a barren field was opened up, but which by simple fertilisation and receiving all contributions, the return was more than a thousand to one.

The instruments traditionally used, based on a strict mechanism, do not capture realities, which are full of nuances. In this way, the «excluded middle principle», rule and guide to research during so many centuries, is no longer the only point of reference. And, above all else, within the field of social sciences, where the human being monopolises the prominent role, there is the pressing need for searching out new horizons. But for this, very solid foundations are required for supporting the whole weight of this modern building. Step by step new elements have arisen that scientists from all over the world have offered for use in the different branches of the tree of science. What was missing was the agglutinating principle. This role can be played by the already mentioned «principle of gradual simultaneity»[5] the flexibility and generality of which is given faith to by the multiple logic supported by the same.

From here arises the possibility of efficient treatment of ever changing realities and, for this reason, charged with a very high degree of uncertainty. The dogmatic alternative between truth and false is no longer valid in absolute terms. Even a stopped clock tells the truth twice a day. Grading in levels between true and false has become vital for the better knowledge of the phenomena taking place in society and our businesses.

Following the thread of our previous works and taking as an obligatory reference the 1986 book, we have written this text, mainly, but not exclusively, directed at those who are interested in getting to know and treating in a better manner our business and institutions of this new millennium, which has just begun to take its first steps. We have entered into a varied range of activities within the sphere of management, basically in those areas where decisions have to be taken in economic-financial relations. Therefore appearing are problems of planning, programming and control, and brought to light is the whole financial network in its three variations, raising funds, analysis and investment. Production processes are looked at from several angles and this vision provides new scenarios where man and machine acquire a special significance. Marketing products and/or services attracts an ever-increasing interest for large, medium and small companies. Here we have also paid special attention. And in the centre of our concern *imagination*, fountain of creativity, which not only is essential for the survival of production units and distribution but also for consuming units. For this reason it is necessary to be aware of the theoretical and technical elements which will be used in knowledge, explanation and treatment of all these questions.

In our first books[6] we included in the preliminary chapters an exposition of the basic instruments that would be used later on. We feel that today this is not essential, taking into account that they are sufficiently well known by our eventual readers. We decided, therefore, for the alternative of giving the necessary explanations when the concept, method or algorithm appears and we feel requires clarification or extension. We feel that with this we will be presenting a fluid text and wherever possible enjoyable. If as well it attracts the attention of the academic world, we will have attained our objective. In any event we hope our efforts will serve as homage to our master and friend, with whom we shared our concerns and our initial projects of service to those desiring a better world, without closing their eyes to uncertainty.

Jaime Gil-Aluja

REFERENCES

[1] ZADEH, L.: *Fuzzy Sets, information and Control*. June 8, 1965. pages 338-353

[2] KAUFMANN, A.: *Introduction à la theorie des sous-emsembles flous*. Vol 1. Publ. Masson. Paris 1973

[3] KAUFMANN, A. and GIL ALUJA, J.: *Introducción de la teoría de los subconjuntos borrosos a la gestión de las empresas*. Publ. Milladoiro. Santiago de Compostela 1986.

[4] GIL LAFUENTE, A.M.: *El análisis financiero en la incertidumbre*. Publ. Ariel. Barcelona 1990.

GIL LAFUENTE, A.M.: *Fundamentos de análisis financieros*. Publ Ariel. Barcelona 1993.

GIL LAFUENTE, J.: *Marketing para el nuevo milenio*. Publ. Pirámide. Madrid 1997.

[5] GIL ALUJA, J.: *Lances y desventuras del nuevo paradigma de la teoría de la decisión*. Proceedings of the III International Conference of SIGEF. Buenos Aires November 10-13, 1996.

[6] KAUFMANN, A. and GIL ALUJA, J.: *Introducción de la teoría de los subconjuntos borrosos a la gestión de las empresas*. Publ. Milladoiro. Santiago de Compostela 1986.

KAUFMANN, A. and GIL ALUJA, J.: *Técnicas operativas de gestión para el tratamiento de la incertidumbre*. Publ. Hispano-Europea. Barcelona 1987.

I UNCERTAINTY IN BUSINESS MANAGEMENT

Current problems in business management

At the commencement of this millennium society is faced with profound changes from which neither the economic system nor business activity can be excluded. Faced with such a changing reality, and as a result of the same, uncertain, businessmen are inundated with new problems the solution of which requires a high dose of imagination, and above all else, very solid training.

These new problems arise, on the one hand, as a consequence of the very rapid evolution of the social surroundings in which businesses move and on the other hand, by the continuous and accelerated technological development which frequently modifies the economic horizon. All indications lead us to think that this situation will not only be maintained but will accentuate over time.

In this context, business executives must make decisions and adopt strategies the economic and financial repercussions of which are not limited to the precise moment of taking the decision but extend, in many cases, over several economic periods. The difficulties of foreseeing and estimating values, inherent to the tasks of an executive, are all the time increasing as a consequence of the growing climate of uncertainty. The world of our predecessors evolved slowly, our world and the world awaiting in the immediate future is submerged in a system with very rapid interactions.

Traditional values, considered for generations as an immovable guide have disappeared. Virtues such as hard work, perseverance, patience, which before were highly considered and admired, have been left behind on a second level, leaving room for audacity, the competitive spirit, the «kingdom» of the image... The time is not too far-gone when there was sufficient time for thinking and deciding. Today events happen at such a speed that we are overcome.

On the other hand life in business is characterised by over-information, the treatment of which is made easy thanks to computers but is increasingly more difficult to use. Many times we are walking, without knowing too well where we are headed.

Economic life, in all its varied aspects, is submerged within this context and the decisions to be taken within this field are for ever getting more complex as a consequence of the «uncertainty» in the outcome of future events.

In this respect:

- Legal provisions are constantly changing the rules of the game.
- External economic influences day to day modify the expectations of men of action.
- Technology requires that we vary production and distribution systems with ever increasing speed.

These and many other elements require that management techniques change at a great speed in order to adapt to the new structure of the economic universe.

Machines, which have always been the best support for human activity, carry out increasingly more varied tasks, freeing man from the more routine and heavy tasks. In workshops automation has advanced as far as robots, and in administration calculators have given way to modern computers. It is not inconceivable that in the not too distant future the workman and executive will be limited to certain tasks where **imagination** is essential. Their powers will be competition and initiative.

It is not easy to change customs or to carry out the necessary adaptation to this new world. For many years the word **specialisation** has been talked about. In the future this concept will leave room for another: **adaptable qualification.** For this reason the idea of **permanent training** will become the norm. If specialisation becomes diluted, if machines substitute the workers in routine or heavy tasks, the destination of the proletariat is its disappearance to be substituted by the «robotariat».

Advances occurring and which will occur in computers and in robots are enormous. Computers will become the irreplaceable companions of men of action, since power can only be exercised if it is supported by information. What will be important is to know how to filter and use it, because the information exists and is all the time becoming more abundant.

In order to be able to face up to these changes, businesses are also modifying their productive structures and attempting to offer new products and new services, which sometimes are variations on those already existing and on other occasions are truly new. Only a few years ago, for example, banks offered a few limited financial products; sight savings accounts, one year savings, long term savings, current accounts and a very specific range of credits. These financial products have had a long life. Today the whole panorama has changed. Currently every day new financial products appear. The Banks and Savings Banks have intensified the publicity of certain products with apparent originality. But once they are out in the market, there immediately appears another financial institution that launches other products with the object of attracting potential clients. This is only a small sample of what can be seen on a daily basis.

From here it can be stated that «the facts of nature are uncertain; the economic, social, financial sphere of business changes without ceasing; the acts of man – because he is free and provided with imagination – like relationships between mankind – because these are not robots – are all the fundamental causes of uncertainty».[1] What has occurred then is a change, which constitutes a veritable revolution in the understanding of economic activity, and those carrying out research in this field are seeking new ways that are capable of treating these new situations. The models that traditionally appear in texts on management based on determinism and probability, although they continue to be valid fortunately in many circumstances, in their current state do not reflect this new world that is immersed in uncertainty adequately.

The time is not too far away when it was possible to make estimates on future values taking as a reference data from the past, because it could be stated that the outcome of events would be a prolongation of those that had already occurred. To believe that this continues to be so today is, in many cases, to deceive ourselves.

It is for this reason that scholars of economic problems in business are seeking new techniques, that are capable of grasping in all their complexity, the phenomena brought up by life itself, in order to be able to formalise the same and act upon them.

By the mere fact that in society, as an addition to the elements with a capacity for coexistence, certain processes of acceleration and deceleration are occurring that not always are going in the same direction, certain pressures of a diverse nature are taking place within it that, in some way, cause problems arising from the absence of a future platform with sufficient stability in order to be able to make estimates on values which could conveniently limit the outcome of events.

Nobody is capable of doubting that social groups, whatever their relative position, or their geographical location, are facing profound changes that will cause undoubted imbalances, the result of which it is difficult to foresee.

Neither the economic system in general, nor business in particular, is immune to this new framework in which they are carrying out their activities. It is for this very reason that their vitality, and even their survival will depend on their capacity to adapt.

Adapting to new phenomena means resolving not a few problems in the solution of which we university professors occupied in the teaching and research into subjects related to business activities cannot be oblivious to. We must face a changing reality and, consequently uncertain, which is going to require a new vision and a different manner of focusing the solution to the problems that are brought up.

Measurement and valuation of business phenomena

In management studies the use takes place of a determined group of techniques of measurement which aspire, following the concept laid down by Galileo Galilei «Measure what is measurable, and attempt to make measurable that which is not yet». Nevertheless, not all that has an interest in economic life in business can be measured, but that, only certain parts of the phenomena, facts and relations can currently be measured (that is say, objective numerical assignment) and there remains an important task to be performed in order for us to be able to widen the numerable field to all aspects of management. We cannot in all faith say if this is so essential.

Now, this reality should not stand in the way of consideration being given, in management studies, to those phenomena that constitute a problem and which are not susceptible to measurement and incorporation to operative techniques that are considered as classical. And this is so, because over latter years new possibilities of treatment have appeared which get around measurement by means of subjective estimating – and which are capable of being expressed numerically by means of what are known as «valuations».

The advance which at the time was signified by the use of the set of techniques, known under the generic denomination of «operational research» with its efficient application to complex problems, led to the development and widening of studies in the field of business management. Starting out from a few initial ideas, scientists and researchers have developed models based mainly on formal logic, always thinking about the possibilities of its application to the economic problems that occur in reality. The objectives attained were very important.

On the other hand development of data processing has made it possible for the success attained by researchers not to be reduced to mere intellectual speculation and today we can be surprised by the results that are arrived at in aspects such as order and the use of information and of the rational forecasts that are fundamental in the same.

But, as we have already mentioned, in this day and age economic activity of business takes place in an atmosphere of continual changes. In this context where decisions are taken with economic and financial repercussions that are not only limited to the time at which they are taken but also which extend, in many cases, over several years. The difficulties of forecasting and estimating, which are circumstantial to every executive, are all the time increasing as a consequence of the growing climate of uncertainty.

This approach to reality has taken place, on may occasions, by means of reasoning based on the concept of accuracy and frequently have been formalised by means of classical mathematical perceptions. The results have been certain models that constitute a "modified reality" in order for this to be adapted to certain mathematical knowledge, instead of all to the contrary, that is an adaptation of the models to real facts of life.

There is no doubt whatsoever that reality is precise in itself, but in the process of grasping it a more or less large part of its nuances are lost: the human brain grasps its surroundings in a simplified manner. And this is so because anything that is examined, however small, and take the example of the atom, is so complex that grasping it can only be done by means of a simplifying process. Even when looking at an object, our eyes capture it in a very gross manner, schematic, which will be refined as we use progressively more powerful lenses. But no microscope whatsoever exists that is capable of showing the object "just as it is" in reality.

The fact that formalisation normally means a restricted view obliges the researcher to elect between carrying out from the outset a selection of elements to be considered in order to be able latter to operate with precise instruments, or grasp the reality with all its inaccuracies and operate with this "fuzzy" information, even though knowing that the results will be given in a somewhat inaccurate manner. The decision is reduced to select between a precise model that reflects reality imperfectly, or a vague model that is perhaps more adequate to reality.

If to this we add that both the thought and action processes of humans are the result of a mixture of intuitions and logical rigor, their study, on been made by means of certain simplified schemes, cannot be at one and the same time strictly representative of reality and totally precise in its quantification. This has fed a tendency resistant to the use of mathematics for the study of the phenomena of social sciences. It is considered, all too frequently, that mathematics unnecessarily «complicate» facts and relation, which in reality are very simple.

In this context it has frequently been sustained that only that knowledge that is susceptible to measurement can be considered as scientific. If this were so, what a poor role would be played by Social Sciences. Situations are all the time becoming less in which precision allows for measuring with certainty and events are sufficiently well known or repetitive so that the notion of chance is associated to an idea of measurement through probabilities. Uncertainty appears with ever-increasing frequency, and is used in a subjective manner associating it to non probable facts, such as the concept of sensation and the notion of valuation, as much as with another of a fundamentally subjective nature.

Uncertainty is inherent to human thought and presides over much of its reasoning. In attempts to formalise social behaviour in general and in particular behaviour associated to the sphere of business, it is increasingly more necessary to introduce this reality, even though having to sacrifice oneself to this objective concept of measurement. When a phenomenon is not measurable then a valuation, ordering, assignment, grouping, relation, etc.. has to be made, this in itself being an important activity of the human brain. It is for this reason that the incorporation of these subjective concepts constitutes a significant advance to a better knowledge of the phenomena.

If a situation cannot be determined, but it can be stated that it is better than another situation, then we have passed on to a higher state of knowledge. But if it is stated that in the future an event is more «possible» than another, a fundamental field is being opened up in the prospects of reasoning and decision, since, with this subjective knowledge it can be practically submitted to the mechanisms of logic. Why is the distinction between inaccuracy and vagueness so fundamental? Vagueness, or fuzziness does not necessarily have to be inaccurate. In formal logic a thing is either true or false, but it cannot be both at one and the same time, as no nuances are permitted, but studies of fuzziness assign a fundamental importance to «degree» or level of grasping reality. Human thought is, in itself, full of nuances, full of vagueness.

Human beings need to consider, given that their brain is not a sequential machine in the manner of computers, certain ideas that are difficult to assign if only taking into account the whole or nothing. In fact, the brain feels the need to associate rigor with vagueness.

Due to the nature itself of things, either because of the thousands of ways which lead us to scientific progress, or the great variety of our thoughts, or the diversity of our opinions, makes it far more difficult today than it was yesterday to make forecasts, and increasingly more complicated to place ourselves within the safety of certainty. The world, seen from the perspective of both space and time, is increasingly more complicated because of the quantity of unforeseen events occurring that go to make up the framework of our surroundings.

We have crossed the threshold of the «age of communication» and all the information which scientific knowledge transports and divulges increasingly converts into combinatory that which can be grasped and introduced as the basis for decision. Even with the aid of the equipment for treating information, whatever its capacity, it becomes more difficult to remain in the sphere of rationality, because information and its divulging increases at an even greater pace than the treatment power of the equipment. And it would appear that this will continue for some time to come.. if humans do not tire of inventing and tire of this wonderful curiosity which has taken them to where they are today and will take them to an enthralling but... surely uncertain future.

Increasingly research into techniques for the treatment of problems within the sphere of uncertainty becomes more necessary. Treating formal, exact or even random data, that is probable data, is quite normal and convenient, because we always know where we are going, where we are proposing to go, if everything happens as it was conceived and expected. Treating uncertain data, accepting certain economic criteria without being sure of oneself, relying only on the will to comprehend almost constitutes an undertaking with irrationality.

What we can require from the models that we construct for representing general or special situations is that they be faithful to what we perceive. It is a bad thing to fool others but far worse to fool oneself; and much more so for those seeking the answers in the field of science. An involuntary and common way of deceiving is linked to semantic confusion, the sense we give or accept giving to words. In this way, some people confuse the subjective with the objective, others confuse uncertainty with probability. When we talk of chance, that is to say probability and are not capable of justifying or even vaguely specifying these probabilities, we are deceiving ourselves. It is almost certain that in all languages, in all dictionaries, the word chance appears as a synonym for uncertainty. And nevertheless there exists a difference between the contents of these two words; chance is linked to a measurement, that of probabilities, whilst, by mere definition, what is uncertain cannot be measured. Obviously the person who can measure or has to measure and does not do it cannot pretend to be a scientist, but when this is not possible one is left with the recourse of knowing and expressing its subjectivity in a better manner.

What we intend is to bring to light[2] that even without it being possible to measure in a formal way or by means of probability, we can also aspire to rational behaviour. Facts that are susceptible to true repetition belong to the sphere of nature, physics, chemistry, astronomy and even biology, but man introduces, apart from the uncertain facts of nature, those that stem from freedom, from the power of imagination.

It is true that in current circumstances, we economists feel uncomfortable, but we will feel even more so in the future. The theories of yesteryear are as obsolete as the machines that are exhibited in museums.

The theory of probabilities is supported on three centuries of perfecting, but does a theory of uncertainty exist?. It cannot be seen in what way it would be possible to define, to explain it.. Therefore, given that from the uncertain it is also possible to obtain information, this must be taken advantage of however weak it may be, even if is very little, in order to improve performance, making it more in agreement with a personal system of intentions. It is necessary to use in the best way possible all the information available in order to be able to construct mathematical models attempting to deceive ourselves as little as possible. For this the theory of errors may be used as well as that of confidence intervals, fuzzy numbers, fuzzy sub-sets and all the generalisations that have been proposed and which we have already used previously[3]. But we must insist on one point: if it can be measured then we should do so. And if it cannot be measured we should know in its absence to use what is perceivable. It can be stated that in the infinite field of uncertainty, it is nearly always possible to gather and use information, whilst maintaining efficiency and honesty.

Probability and uncertainty

Uncertainty and randomness are words, which are frequently used indistinctly, even in the scientific world. However, no man of science would ignore the existence of a difference between what is not measurable and what is measurable. Obviously, in every day language, that which is used for daily communication, this confusion is accepted without this fact causing too many upsets. But, when tackling the treatment of problems by means of a scientific method, the precaution has to be taken to make this differentiation. Mathematics for treating uncertainty and mathematics for treating randomness are not the same[4], but it is necessary to distinguish between one and the other at the same time as to be able to associate them correctly when this were to be necessary or useful.

The principal mathematical tool for the treatment of uncertainty is the theory of what is fuzzy and valuation with its many variations, while the theory of probabilities is the theory to be used relative to chance. Up to a certain date, the first were scarcely known or were unknown both by pure mathematicians as well as by specialists in applied mathematics. A hundred years ago the same occurred with the second option. Fortunately, as from a few years ago an epistemological change has taken place and today there is a better grasp of separating and associating what is measurable from what is not.

It can be stated that, with the object of maintaining survival and reproduction, all living things have carried out and are carrying out strategies, in an attempt to overcome extinction of the species, on occasions maintaining the same state, other by means of transformation. All living beings are permanently in a state of learning, either over hundreds of millions of years, or just for a mere instant. Thus instinct would be the result of a long learning period and intelligence of a short period. But what can be seen, in any event, is a learning period.

Man, just as the same as all other living beings, evolves in an uncertain environment. One of the objectives of all species is to lessen the effect of uncertainty. And for this it is necessary to collect repetitions, take them into consideration, and use them in order to decide.

The transition from uncertainty to probability and from probability to certainty is a procedure the study of which could be very interesting. In the first place it would be convenient to point out the difference existing between the words uncertainty and probability. For many people, among whom are highly evolved people and at times even intellectuals, these words mean almost the same abstract concept: Not much or perhaps nothing is known about what we are dealing with. Also even in the very best dictionaries the difference is not defined with clarity. And nevertheless uncertainty and chance do not correspond to one and the same level of information. Uncertainty posses no laws, probability does, known or not, but that exist by reason of hypothesis. Uncertainty is deficiently structured and when it is explained it is done in a subjective way. Chance on the other hand, is linked to the concept of probability, which in itself is a «measurement » on repeated observations in times and/or space. It is, therefore, a measurement on observed facts and constitutes an evaluation, which it is desired would be as objective as possible.

The frontier between what is objective and what is subjective is very vague. Considered as objective is that which is accepted without reticence by all, or by a group that is sufficiently important or powerful. Considered as subjective is that which refers to an individual or small group.

All too frequently it is stated that an event is probable when it should be said that it is possible. Probable is linked to measurement, possible with valuation, a lesser, weaker concept, but which can, in numerable cases, be more honest.

Therefore the words chance, random, stochastic should be used for those situations in which laws of probability are available and accepted. On the contrary it would be necessary to use words such as uncertain, uncertainty, vagueness, fuzziness. To put it in other words, there should be no confusion between situations that can be carried out and where measurements can be justified with those situations in which such measurements are not feasible or cannot be accepted.

Probability is structured uncertainty, measurable by means of logical and/or statistical reasoning. In times gone by chance was represented by σ τ χ ο ς, which represented the idea of a target to be attained, opportunity to guess, to foresee. Now the word «stochastic» is linked to all processes in which the different situations can be arrived at by means of probability. At first the notion of probability was intuitive (works by Pascal, Laplace), later a strict axiomatic was introduced by Borel and another by Kolmogorov.

The preparation of a decision, either simple or complex, is an organisational activity of the thought process in which intuition and logic are combined, and where, frequently, it is difficult to make a separation. Mathematics often serves as an aid to the mechanisms of logic, although at present we are capable of using them for stimulating the mind and sharpening intuition.

The models have their support, more often than not, in determinist or probabilistic theories. Determinist models take into consideration certain data, probabilistic models data that is statistically measurable or constructed from reasoning that permits allowing «a priori» the laws of probability. The novelty appears with the fact of knowing how to construct models in the event that neither the laws of probability nor the reasoning related to the same can be used.

These new models are constructed, in the majority of cases, using properties defined in the theory of fuzzy sub-sets and their infinite varieties. This theory constitutes an attempt, which is enjoying undoubted success (although controversial for those unaware of it) of scientifically vindicating subjectivity and vagueness. Instead of starting out from objective data, which unfortunately are unknown, this is done from subjective but accepted as reasonable data, and a logic is used which is qualified as maximum-minimum or «maxmin» in which all the hypotheses that are called «weak» (not measurable) are calculated by taking the maximum of the minimums (like the theory of the games in the sense of Von Neumann), which corresponds to the more prudent taking into account of the hypothesis. This way of acting can be compared with that corresponding to the hypothesis called «strong» (measurable) where the mathematically expected value is accepted by the sum of the products.

Now, in real problems, weak and strong data are found simultaneously, that is to say fuzzy data and random data. We know perfectly well how to associate this data of a different nature, although from not too far back, without wasting information. In this sense it should be pointed out that the maxmin logic is capable of undergoing modifications with the aid of different operators of the max and the min, provide there exists a justification due to the nature of the problem. What we are attempting to do is to make possible cohabitation in one and the same model of the fuzzy data and the random data (taking into account that formal data constitutes a special case of random data). Many ways have been discovered for associating uncertainty and probability. But it is always necessary to remember that uncertainty and chance are two very different concepts.

Why now has such an interest been awakened in uncertainty? Because it is increasingly more difficult, in spite of the progress made in processes and machines, to arrive at sequences of data that are sufficiently stable, stationary and even significant. The phenomena relative to human sciences are, at the present time, increasingly less foreseeable in a clear manner. On the other hand mathematical instruments are now available that are less formal, but no less strict (fuzziness does not exclude strictness).

We feel that what is important in a model, in the first place, is its fidelity. If what is real can only be grasped by means of sensations instead of events, are we not wasting information if we reject these sensations? Apart from this an event in human sciences is nothing less than a consensus of sensations. And even, as we have already pointed out, objective science is only objective by the mere fact that it has been grasped by the large majority, or by the most powerful group. No law whatsoever exists which one day or another, sooner or later, is not questioned. Another will be found which will contain its predecessor as a special case, or even will not contain it and then it will be revolutionary. The scientist should remain within modesty and treat in the best possible manner, in the current state of knowledge, that which is within reach.

This statement, which is valid for research activity, in general, is also valid for business phenomena. To stop and dwell on this reality inevitable means reconsidering the hypotheses on which the models are based, if it is desired that they allow for greater understanding of the operation of the sub-systems that make up the business world and constitute a valid path for decisions.

It is obvious that there exists a preference for the construction of those formal schemes that are based on precise and «safe» data. However this can be inoperative when reality does not allow for arriving at them. Therefore, if this information is not available it is necessary to resort to those numerical estimates of a subjective nature that we have called «valuations».

Following an inverse order of preference five types of models can be considered[5] relative to the different levels of information:

1.- Non determinist model with unknown situations.
2.- Non determinist model with known situations but events that cannot be evaluated (we know the possible situations, but we do not know how to assign

to them a subjective scale of values).

3.- Non determinist model with known situations and events that can be evaluated but not measured (we know the possible situations, we can assign a valuation to each event of each situation, taking into account that this valuation is not a measurement).

4.- Non determinist model with known situations and with measurable events (we know the possible situations and we know how to assign a probability to their occurrence).

5.- Determinist model (we know the situations and consider a hypothesis that the event of a specific situation is known).

In management studies, a modelling process is carried out in an attempt to go from level 1 to level 5. But the economic situation, which has already been described, forces researchers to stop at the lower levels. It is for this reason that resort has been made to the most general of theories that are capable of describing uncertain environments: the theory of fuzzy sub-sets.

It is common knowledge that the theory of fuzzy sub-sets forms a part of mathematics that is perfectly adapted to the treatment of both the subjective and the uncertain. It is an attempt to gather phenomena just as they occur in real life and to carry out their treatment without attempting to deform them in order to make them precise and certain.

Formalisation of the uncertain, taking fuzzy concepts as a starting point has given rise to a different manner of thinking which assembles the strictness of sequential reasoning and the richness of imagination, associating the sequential possibilities of the machine with the possibilities of the human brain.

The use of fuzzy schemes takes place, today, in practically all fields of study. It is already very consolidated in business management, but also in engineering, biology, medicine, geology, sociology, phonetics, and even in music, to mention just a few of them. Every problem that can be found within the sphere of uncertainty is susceptible to being treated by means of the theory of fuzzy sub-sets and its multiple variations, since as time goes by it is increasingly more feasible to introduce into formal schemes mechanisms of thought such as sensations and numerical and non numerical opinions.

For over 50 years now, a large number of mathematicians have become interested in multivalent logic, among these we can mention, RUSELL, LUKACIEWICZ, POST, etc., but it is 1965 when LOFTI A ZADEH published his first article on «Fuzzy sets»[6] and a few years have to go by for a certain amount of expansion to take place, since until the commencement of the 70's only a few papers on this subject were published.

At present, it is no longer possible to speak about a single logic, since there may exist as many logical developments as can be imagined. For this suffice it to establish certain axioms so that, from the same, propositions may fall into place correctly avoiding entering into any contradiction.

Therefore, today «fuzzy logic» is conceived in the same way as there was no problem at the time in conceiving Boolean logic. Furthermore, if for human-computer relations under the current situation it is essential to resort to binary logic, for relations between man and his fellow men the use of the theory of fuzzy sub-sets would appear to be more adequate.

The traditional theory of sets and Boolean algebra with their logic of belonging or not belonging, has allowed for the formalisation of certain situations brought up by reality, but other situations exist that were difficult to model by means of these theories.

In the day-to-day routine a multitude of examples appear, in which this unquestionable reality can be seen. Thus, when considering mankind as a whole and what is actually wanted is to define a sub-set of «young men», a problem immediately occurs, since the limits of such a sub-set are not very well defined. When from a distant hill a beach can be seen with a group of bathers and it is required to determine the sub-set formed by those already in the water, it is possible to point out exactly those that are in the water and those that are «in all surety» on the beach, but there will be some for which it will be necessary to establish a «degree or level of belonging», that is, that some will belong quite a bit whilst others will only belong slightly to said sub-set. With this it can be seen that we are operating with concepts that are defined in a vague manner, but which can be put into an order.

We have repeatedly pointed out that the economic, social and technological surroundings of business are much less foreseeable and are in a situation that is far unstable than in the immediate past. This reality has brought about the fact that, both from the macro-economic point of view as from a micro-economic perspective, new paths have been sought for studying the situations in which economic systems on the one hand and business on the other, are involved. It is for this reason that Economic Analysis for the study of the activity of the situations and Operational Research for the treatment of business problems have reached a highly accelerated rhythm

With the spread in the use of computers spectacular advances have been attained in the treatment of data which businesses could have available allowing them with this to improve management of the same. Thus, this aspect, absolutely fundamental, to a greater degree has been resolved, but another problem has arisen derived from the obtaining itself of the necessary data. Therefore the problem has been transfe-rred to the search for certain processes, which would allow for the introduction of certain primary information for its later treatment. Not even recourse to probabilis-tic models allows for advancing too far in certain fields of business activity.

Mathematicians and economists have been obliged to investigate in this field and have arrived at new schemes that allow for a far more complete consideration of reality, avoiding wherever possible their traditional deformation when resorting to numerical precision. In them, it is assumed that the taking of decisions is done in an environment in which the objectives to be attained, the limitations to which they are subjected and even the consequences for each on of the alternatives approached, appear in an vague manner.

It has been proven that in order to quantify this vagueness the techniques provided by the theory of probabilities are not sufficiently adequate, and to be more specific by the theories of decisions, because this would imply the fact of accepting that vague facts are equivalent to random facts. Within a new decision framework of business phenomena, vagueness is formalised by assigning to each situation a characteristic function in which there is a degree between absolute belonging and not belonging.

Thus, while the concept of probability is associated to randomness, the so-called membership function is associated to fuzziness. There exists a common point between probabilities and the membership function: both are included between zero and one. Nevertheless, since both concepts have a different origin, they also have different properties. "Uncertain numbers" do not have the same arithmetical rules as do "random variables" in the same way as the notion of "probabilistic entropy" should not be confused with the concept of "disorder in the field of uncertainty".

Works treating those problems arising from business management have a great variety of mathematical models the basis of which is to be found in determinism or randomness. But as and when is was seen that the requirements of these hypothesis meant that they gave place to a substantial distancing from the real world, the need was felt for introducing new schemes that were more flexible and adaptable. Technological advances, the diversity of markets, the multiplicity and variety of products has meant that the intuition of the businessman must be completed by ever increasingly complex models. The possibilities that fuzzy sub-sets offer for attacking problems of decision in the field of business action are so extensive that, without any doubt, they will enrich operative techniques for business management.

In traditional schemes, given the impossibility of precisely gathering the complex and uncertain reality of business, resort is made to an initial simplification in order to carry out later developments based on these simplified elements. The possible initial deviations accumulate and are amplified as the operative process advances. What is also lost is information from the very start and this is not recovered.

From a new perspective, what is proposed is to gather all the economic phenomena of business with all their vagueness and uncertainty, in order to carry out the pertinent developments *preserving the vagueness* (and also all the information) in order to let it «fall» as late as possible in the process. It is always possible (by loosing information) to reduce the uncertainty.

The possibilities of using this road are numerous[7] and go from long and short term forecasts, passing through investment selection, stock management, equipment renewal, personnel selection, creativity, right up to planning with hybrid data, and a very long etcetera. Nevertheless the enthusiasm for these models must not let us forget an unquestionable fact: traditional techniques cannot be put aside, but they are essential when the phenomena can be measured. But, when reality in business brings up a whole range of circumstances that are not measurable, then it is convenient to carry out an evaluation that is susceptible to being treated by means of criteria within the sphere of fuzziness.

Currently we are living in an age which requires a reality in the treatment of business management problems, and even though we feel that the time has not yet come to do without the genius and intuition of the businessman, the complexity of surroundings in which we are living causes the need for investigating new techniques so that, by means of the same, we can continue along the road to progress.

With these techniques an advance is being made towards the formal clarification of business attitudes, in this way increasing the coherence between the evolution of real processes and the schemes drawn up for their treatment. Models will be completed that traditionally have been used for resolving problems of decision for the businessman, with the appearance of a new nucleus of decision techniques that, without a doubt, are providing fruitful results for the solution of increasingly complex problems, occurring in the economic activity of businesses.

The principle of gradual simultaneity

These results, however modest, have constituted a stimulus for scholars of economy and business management who are attempting to channel their interests in order to be able to resolve the serious problems that the social, economic and business systems are bringing up[8]. Thus in scientific circles a large number of proposals have emerged that, in different senses, all converge in providing new treatments both for old problems and for those that arise from the complexity of economic-financial relations. Each of these proposals brings to light certain aspects that are intended to point out the new channels along which research activity should flow. Nevertheless in many of these what can be seen, in different forms, are old routines, which have been unable to generate new horizons that are capable of providing oxygen for the charged environment arising from contrasts between opinions.

Perhaps one the causes for so much disappointment is because of the influence of the «excluded middle principle», which constituted a rule and guide for scientists for more than 2000 years. We feel the time has come to announce a new principle, that of «gradual simultaneity», which can provide cover for a large amount of logical reasoning capable of creating concepts, establishing methods and drawing up models and algorithms, suitable for providing, at the very least, some of the hoped for answers. In spite of this, we are conscious of the fact of the limitations that may appear and will continue to appear as a consequence of certain works of which there is very little proven experience.

When the conglomerate of knowledge that forms what today we call economic science is analysed in certain depth, it will be seen that there are successive contributions in which appear elements, techniques and methods, which have been used previously in other fields of knowledge. The fact that our scientists may have delved into other sciences and taken advantage of apparent or real parallels for advancing in the process of acquiring knowledge of economic phenomena, should not cause any particular shame. Thus, in physics, for example, a mechanistic mathematics permitted describing a range of situations, which led to important discoveries known by all. The transfer of this mathematics to the field of social sciences in general and to economic and business management in particular, gave rise to a movement, marginality, the central figure of which is an individual called «homo oeconomicus» who only acted through stimuli of a «material» nature, who was devoid of any sentiment or any sensation that might lead him to carry out acts or take decisions in which there appeared the very slightest hint of subjectivity.

On the other hand, the transfer of techniques based on the Borel-Kolmogorof axioms to the filed of economy and management was, and continues to be, the starting out point for the flowering of models, the beauty of which and their formal attraction is undoubted. The possibilities of their use, nevertheless, would appear to be limited in a context such as the present that is characterised by change and, therefore, by uncertainty.

With the object of expressing the logical reasoning of human thought, resort has been made, in different spheres of science, to binary logic. In its origin there is a principle, that is the excluded middle principle, the formulation of which stretches back to Crispides and Aristotle. Its development and spreading about, however, takes place thanks to the form acquired by the work by George Boole, Laws of Thought, published in 1853. It is incorporated into economic sciences from diverse perspectives and very differentiated aspects, and attained spectacular results in very specific areas, such as computing, and in those aspects in which a relationship man-machines occurs. It must be recognised, nevertheless, that in relations occurring between humans and social groups, not everything is reduced to yes-no, true-false, black and white. In thought not everything is binary but nuances occur within the same, which gives rise to the fact that human logic is frequently vague, imprecise, ..., fuzzy.

It is generally accepted that the knowledge of things, of phenomena, takes place from models, that is to say, of nearly always schematic representations, which the brain process place in the memory in the form of images, sensations, numerical data, formulae, causality relations ... Scientific knowledge is mainly constructed from these models that are acquired by man by means of education and teaching, with the aid of mechanisms provided by logic. Now then, in the current situation of science, it is not possible to admit the existence of a single universal logic that is perfectly objective.

Aristotelian logic, used for so many centuries, is not a universal logic, but one amongst many others. For a great variety of models, as a consequence of a large variety of languages, there also exist a large variety of logics.

From this age-old logic stems an omnipresent principle, the excluded middle principle and of non-contradiction, which now forms a part of our mental patrimony. What we are saying here is that a proposition cannot at one and the same time be true and false, but that it is «always» either true or false. Nevertheless, in thought, between truth and falseness there exists an infinity of nuances. The need to find fluid and simple communication has accustomed us to the use of binary truth-values. The entry of computers constructed by binary elements has done nothing else but reinforce this tendency. In the field of electronics the binary system is the most convenient form of treatment. With 0 and 1 it is possible to express any number, concept, operation ... although the number of digits (position 0 or 1) can be enormous. Mathematicians call this «Boolean» algebra in honour of the creator of the binary rules of thought.

In this algebra two symbols are used: **V** which signifies true and **F** signifying false, although thcy can be substituted by another pair but they must explicitly correspond to **V** and **F**. Practically in a universal manner the pair {0, 1} have been selected, where 1 corresponds to true and 0 to false, in the same way that in the theory of sets belonging is expressed by 1 and not belonging by 0.

In the configuration of research activity, western science has constructed over centuries taking as one of the unquestionable supports the excluded middle principle, from which, as we have mentioned, a logical architecture takes form that is consolidated in the works of George Boole. A single logical operator allows for the linking of propositions by means of two «modes»: forwards with «modus ponens» and backwards with «modus tollens». The passage of logical reasoning, expressed in words, to its presentation by means of symbols has given rise to the development of the mathematics which in time has acquired different «forms», but always within a mechanism, both when following the path of certainty and when attacking chance. The formulation of binary mathematics clarifies what we have just stated. Based on this logical reasoning, expressed by means of symbols, specific techniques have been drawn up nearly always in the form of models and algorithms, the objective of which was and continues to be the provision of instruments capable of rendering aid in taking decisions. The use of these elements in the different spheres in which problems of decision arise constitutes the last step on this ladder, which researchers have climbed in order to provide the necessary assistance to those who are responsible for decisions and avoid in this way that they fall into inassimilable errors.

At the present time and, as a consequence of reasons that have been repeatedly expounded upon and others that we could add, it is not possible to base all research activity on the excluded middle principle, which has become limited for housing all the «explanations» of the complex realities and phenomena of our day and age. The formulation of a new principle therefore, becomes fundamental, if it is really desired to open up the gates to the new adventures proposed by the future.

A few years ago we commenced the task of defining a principle that was capable of harbouring the different logical operators, which arose, and which it was difficult to include within the excluded middle principle, since to do this was only possible by means of contrived means. What we also intended was that the new principle could generalise the excluded middle principle. From these considerations the principle of gradual simultaneity[9] was born. This principle can be expressed as follows:

«A PROPOSITION CAN BE AT ONE AND THE SAME TIME TRUE AND FALSE, ON THE CONDITION OF ASSIGNING A DEGREE TO ITS TRUTH AND A DEGREE TO ITS FALSENESS».

Let us move on, by means of certain examples[10], to bring to light the contents and significance of this statement, at the same time showing the relative position of the same, in relation to the traditional principle.

There are certain propositions for which normally there is no problem whatsoever in accepting the compliance of the excluded middle principle. Thus, when stating that «Peter belongs to the male sex», we conclude the *truth* of this proposition and the falseness of denying it. But certain other propositions can be made where compliance may not be quite so clear. In fact, propositions such as «Peter is tall» bring up certain problems given the relativity of the statement «tall». In order to resolve this, traditional studies normally establish a threshold (obviously subjective and arbitrary) as from which the truth of the proposition is assumed. If it is accepted that in this case the threshold is 1m. 80cm., those reaching or exceeding this height will be considered as tall people, and for those not attaining this height the falseness of the proposition will be assigned. What we arrive at in this way is to admit that a person with the height of 1m. 80cm. is *tall* while another person at 1m. 79cm is *not tall.* We feel that acceptance of the principle of gradual simultaneity allows for a good solution to this problem converting the excluded middle principle into a particular case. Let us take a look at this.

For greater ease we will continue with the same examples. Relative to the proposition concerning height we look for those measurements for which the truth and falseness of the proposition are fully complied with. The tallest human being in the world measures 2m 38 cm and the shortest (negative proposition of the latter) 0m 52 cm. From these two limits, in our case numerical, although they do not necessarily have to be so, an order is established from the truth (high) to the false (low). This scale can be framed by an interval, for example [0, 1], or by any other concept suitable for allowing an order. If for effects of greater simplicity and, in order not to become separated yet from the numerical field, the interval [0, 1] is accepted, we will assign a 1 to the truth and a 0 to falseness, and therefore the giant at 2m 38cm will be tall in a degree 1 and the dwarf of 0m 52 cm tall in a degree of 0. A person with a height of 1m 70cm could be considered tall to a degree of 0.3 for example. In this way it would be established that when the closer we are to the truth of the proposition the degree assigned will be closer to 1 and when we are removed from the truth the degree assigned will be neared to 0.

But when we formulate the falseness of the proposition, converting *tall* into *not tall*, or short, if you like, what happens is that the same actors play a different role, as they will be assigned a different degree. Thus, as short, the dwarf assumes the highest degree so is assigned a valuation of 1, whilst the giant is in no way short and will be assigned a degree of 0. The person with a height of 1m 70cm moves from a degree of 0.3 will now be short in a degree of 0.7 for example. In this way, any person can be tall and short at one and the same time, if we take the precaution of accompanying the qualification with a *degree* that can be expressed (as in our example) numerically or non numerically.

But also all those propositions to which the excluded middle principle can be suitably applied can also have the principle of gradual simultaneity applied. What happens, then, that of all the possible degrees only the extremes are applicable. In this way, with the reservations arising from the case, to the statement «Peter belongs to the male sex» a one in its degree of truth is assigned and to the statement «Peter does not belong to the male sex» a 0 in its degree of truth. Therefore, he belongs and does not belong to the male sex, but in a different degree.

This brief example allows, we hope, bringing to light the generality of the principle of gradual simultaneity, which is quite capable of including a great variety of logical developments. In fact, faced with the limitations existing in the formal logic arising from the strictness of the excluded middle principle, it can be seen that there is enormous flexibility and adaptability of the so called multivalent logics.

We have very briefly seen, how the human mind manifests itself by means of binary logic, and also how the mind carries out its activity in a much wider manner than in binary mechanisms.

In real problems, logical treatment, be it numerical or not, resorts to information of a very diverse nature. Some of this information is manifest by means of formal or probabilistic measurements, other information is simply the result of the opinion of experts and as such of a subjective nature. The estimation of the value of the truth of a statement cannot be limited to values 1 (true) or 0 (false). Language is full of adjectives and adverbs that position a statement between true and false. In certain cases we express an idea precisely, in others, in vague terms. The use of precise data and vague data is a constant of our coexistence, according to the level of knowledge and as a consequence of the inevitable subjectivity of our reasoning.

Therefore, the logic included in the principle of gradual simultaneity, in order to represent the real mechanisms of thought in a better way, accepting nevertheless binary logic as an important special case, takes into account the nuances existing between the states of truth and falseness. To put this in another way, a value of certainty can not only take the values 1 (true) or 0 (false) but also each one of the possible values existing between 1 and 0. This logic is called "multivalent logic" and in it the values of truth are normally subjective in the sense that application can vary from one person to another,

In short, reasoning, when we are talking of multivalent logic, means that a proposition is accepted with a "level of truth" by taking any number between 0 and 1, including the extremes.

The need to establish the corresponding semantics between the values in [0, 1] and the terms of the language that normally represent our thoughts, has lead to the setting up of different scales. Among these the one standing out is the one using the hendecagonal system (11 values), which could be as follows:

0	:	false
0.1	:	practically false
0.2	:	very false
0.3	:	quite false
0.4	:	more false than true
0.5	:	as false as true
0.6	:	more true than false

0.7	:	quite true
0.8	:	very true
0.9	:	practically true
1	:	true

There exists and it is possible to draw up an infinity of semantic scales. Among the simpler are the binary {0, 1}, the ternary {0, 0.5, 1}, the pentad {0, 0.25, 0.50, 0.75, 1}... It is normal to use scales with odd numbers of components with the object of, in every event, having a central value, which is 0.5. It is obvious that if greater nuances are required resort can be made to scales with a greater number of elements. For this it is sufficient to place intermediary numbers between the decimal points of the hendecagonal scale. What is brought to light here is the enormous possibilities that multivalent logic offers for covering the treatment of a wide range of problems which are introduced by the complex reality, drawing us closer to attaining that aspiration of a long time of reproducing the very rich nuances of human thought.

The idea of the fuzzy sub-set

The notion of the set constitutes a solid basis on which it is possible to build formal constructions of capital importance. The intuitive meaning of the word set is very simple. It is a "group of objects, either physical or mental, different one from the other, and which are very well defined". Taken as a "reference" a set can be considered as a nearly infinite number of sub-sets. A sub-set of a set does no necessarily contain all the "objects" (elements of the set) although, on certain occasions, it can be admitted that a set is a sub-set of itself. The set of reference is normally called "referential set".

Having established, colloquially more than scientifically, these basic elements, we will immediately move on to take a look at their possible significance in the sphere of social sciences. Therefore, for example, the group of people forming a potential market for the products of a business constitutes the *set* of possible purchasers. If, as from now, all that is studied has as a sole reference these possible purchasers, this set then will be called the "referential set". Let us assume that these potential consumers are Peter, John, Leslie, Sophia, Betty, Robert, Mary and Freddy. The referential set is represented by the capital letter E and is normally presented as follows:

$$E=\{Peter, John, Leslie, Sophia, Betty, Robert, Mary, Freddy\}$$

From this referential a large quantity of sub-sets can be established, each one of which represents a quality, characteristic, peculiarity, property, etc., of the persons (elements) that form the referential set. Therefore on stating the proposal "purchasers of a high-income level" a sub-set appears of the "rich" consumers:

Let this be:

$$r=\{Peter, Leslie, Susan, Betty, Robert, Freddy\}$$

Therefore Peter, Leslie, Susan, Betty, Robert, Freddy are consumers with a high income, while John, Sophia and Mary do not possess these characteristics.

A further proposal can be made: "consumers possessing real estate", and here another sub-set is established:

$$p=\{Peter, Leslie, Robert, Freddy\}$$

Peter, Leslie, Robert and Freddy own real estate and John, Sophia, Betty and Mary do not have this privilege.

As we reel off proposals successive sub-set will appear.

We have looked at a case in which the referential set was constituted by "real" objects (people). Let us now take a look at another in which the objects are "mental". For this it is sufficient to consider a group of "qualities" that are suitable for giving to a living being. In this sense we could take into account wealth, owning real estate, height, industriousness, beauty, chivalry and intelligence. Then it would be easy to consider a referential set such as:

E={wealth, property, height, industriousness, beauty, chivalry, intelligence}

Let us now take one person, in this case Peter who will possess some of these "virtues", for example, wealth, property, chivalry, intelligence and we will have another sub-set:

P={ wealth, property, chivalry, intelligence}

Another person, in this case John will have other virtues (and even it is possible, rara avis, all of them), for example, property, height, beauty and intelligence, and here again we have another sub-set:

J={ property, height, beauty, intelligence}

And so on we will arrive at as many sub-sets as there are people under consideration, in our case eight.

Both in the case of physical objects and mental objects the idea of the sub-set is associated to the fact of ***yes-no*** (possess-does not possess), ***truth-falseness*** (of the proposal). It is for this reason that it is easy to represent the sub-sets by assigning a value from *1* or *0* be it whether the *yes*, the *possessing*, the *truth* or the *no*, the does *not possess* or *the false*, respectively. A very normal manner of representing a sub-set is, relative to the referential of the physical objects of our example[11]:

For wealth:

	P	J	L	S	B	R	M	F
r =	1	0	1	0	1	1	0	1

For real estate ownership:

	P	J	L	S	B	R	M	F
p =	1	1	0	0	1	0	0	1

For Peter:

	r	p	e	l	b	c	i
P =	1	1	0	0	0	1	1

For John:

	r	p	e	l	b	c	i
J =	0	1	1	0	1	0	1

Up to this point we have not departed in the very least from the strictest binary orthodoxy, supported on the excluded middle principle. Some of the people are rich others not so, therefore they belong or not to the wealth sub-set. One or others are property owners or not and therefore, belong or do not belong in the property sub-set. The same can be said for the qualities, which are either possessed or not possessed by certain people, who are perfectly specified. Therefore there are no nuances possible. The mechanism appears as a fundamental element in the reasoning.

The passage from ordinary sub-sets to fuzzy sub-sets is easy, but fundamental. Now the belonging is no longer strict, nor radical, but contains nuances. A quality may be totally possessed, in a very high degree, or moderately, slightly, very slightly or even not possessed at all. In fact a proposal may also be true or false, very true, quite true. ... The incorporation of these nuances into the theory of sets constitutes the basic idea of the theory of fuzzy sub-sets. Instead of placing in each representative square of the sub-set either a 1, or a 0, any value between 1 and 0 can be placed, including these two extreme values. The values placed in the squares are called, by mathematicians, functional values that are characteristic of belonging, and which are normally represented by the Greek letter μ.

.If we consider the same referentials, in first place:

$$E = \{P, J, L, S, B, R, M. F\}$$

The fuzzy sub-set[12] for wealth could be.

	P	J	L	S	B	R	M	F
$\underset{\sim}{r}$ =	.9	.3	1	.2	.7	.8	0	.1

In it can be seen that Peter is very rich, John on the other hand poor, Leslie the richest, Sophia very poor, Betty is quite rich, Robert is rich, Mary is in abject poverty and Freddy extremely poor. These or other words, depending on the habits and customs will allow for a high degree of nuances and, therefore, of drawing close to a reality which is not assumable in ordinary sub-sets.

For real estate ownership we would do something similar:

	P	J	L	S	B	R	M	F
$\underset{\sim}{p}$ =	.7	.6	.2	.4	.8	.2	.3	.9

The interpretation of which could be something similar to the following: Peter is a person possessing quite a few properties, John somewhat less, Leslie is not an owner of real estate (a tiny apartment because she prefers to invest on the stock exchange), Sophia has a few properties (some are mortgaged) ... and so on successively we could make these interpretations or any others established previously.

One question must remain very clear right from the outset. "A fuzzy sub-set acts, at least in the sphere of social sciences, as a descriptor". The same applies when taking into account a referential of the qualities:

$$E = \{r, p, e, l, b, c, i\}$$

The "descriptor" fuzzy sub-set for Peter, could be as follows:

	r	p	e	l	b	c	i
$\underset{\sim}{P}$ =	.8	.7	.3	.2	.4	.7	.9

Having this fuzzy sub-set available allows for knowledge that is far fuller of nuances than when we limited ourselves to pointing out possession or not of these qualities or characteristics.. Thus in the Boolean sub-set it was said that Peter is rich, owns property, is not tall, is not industrious, is no beauty, is chivalrous and also intelligent. We feel that we are closer to the reality of our thoughts when we describe Peter by means of the fuzzy sub-set, which brings to light that he is very rich, possesses quite a few properties, is short rather than tall, not very industrious, not very good looking rather than good looking, is quite chivalrous and very intelligent. We can see that with a limited number of qualities and characteristics we arrive at quite a good profile of Peter, that is to say an adequate "description". The same occurs with John. His descriptor sub-set could be as follows

	r	p	e	l	b	c	i
$\underset{\sim}{J}$ =	.3	.6	.8	.4	.7	.2	.8

Also, on this occasion , we have certain knowledge about John by means of a fuzzy sub-set, who possesses certain characteristics that differentiate him from Peter, also described, as we have seen, by another fuzzy sub-set.

We can now ask ourselves if Peter and John are very different or, what amounts to the same, if they are very alike, quite alike or not very alike. If we compare characteristic by characteristic, we will see that Peter is rich, John quite poor. Both Peter and John are property owners, Peter somewhat more than John. Peter is quite short and John tall .... By means of this simple process we arrive at a good idea of how near Peter and John really are. However, if we take just one step more we will obtain a representative figure of the "proximity" there is between these two people. For this we resort to a mathematical concept called "*distance*". Many formulae have been drawn up to find the distance between two "*objects*". We will limit ourselves to just one that is very simple and practical called the "*Hamming distance*".

In order to arrive at the Hamming distance between the two fuzzy descriptor subsets of Peter and John, it is sufficient to arrive at the difference, in absolute values, between the level that one and the other possesses of each characteristic, in order later to carry out the sum of all of them. Therefore, the difference between the respective levels of wealth is $|.8 - .3| = .5$; the difference between the levels for property $|.7 - .6| = .1$; for height $|.3 - .8| = .5$; and so on successively. The consideration of the absolute value is justified since the distance is independent from whether the level of one or the other is greater. The sum of these differences (we repeat, all positive) gives us the total Hamming distance:

$$d = (\underset{\sim}{P}, \underset{\sim}{J}) = |.8-.3| + |.7-.6| + |.3-.8| + |.2-.4| + |.4-.7| + |.7-.2| + |.9-.8|$$
$$= .5 + .1 + .5 + .2 + .3 + .5 + .1 = 2.2$$

We now have a figure that provides us with the "distance" between Peter and John. Now, this value does not possess much representativeness since it is decidedly influenced by the number of characteristics that are taken into account. In fact, if in place of considering seven, as in our example, we were to have chosen twenty, for example, the figure arrived at would normally be greater, even though the differences were to be small. We would arrive, in this way, at the paradox that the greater Hamming distance would reflect two people who were closer (for the individually lower differences). For this reason, we normally use the so-called "relative

Hamming distance". In order to arrive at this suffice it to divide the total distance by the number of elements under consideration, in our case seven. The result being:

$$\delta(\underset{\sim}{P}, \underset{\sim}{J}) = \frac{1}{7} d(\underset{\sim}{P}, \underset{\sim}{J})$$

$$= \frac{1}{7} 2.2 = .314$$

Here then is the reason why Peter and John are not too distant between each other, as the relative Hamming distance is equal to 0.314.Normally when seeking proximity, the complement of the relative distance is calculated, that is one less the distance. In our case:

$$\pi(\underset{\sim}{P}, \underset{\sim}{J}) = 1 - \delta(\underset{\sim}{P}, \underset{\sim}{J})$$

$$= 1 - 0.314$$

$$= .686$$

The result shows us that Peter and John are quite close.

The coexistence of Olga and James

The use of this instrument is unquestionable. There are a number of occasions for which the Hamming distance allows us to resolve important problems. Merely as an illustration we are going to take stop at the threshold to the formation of a family.

Olga and James knew each other casually as they coincided on a trip. A few months later they decided to join their lives, in order to continue together, full of illusions and good intentions. From different origins and cultures, the notion floated over their heads that would they be able have a lasting relationship.

Having made this statement we must make a minimum consideration on what are the necessary and sufficient conditions in order for two human beings to be capable of living together. Simplified to the maximum in order not to make heavy reading, we are going to establish a limited number of personal characteristics that affect communication on a day to day basis between a couple. Let us not forget that happiness is not attained by the success of the great events, but is arrived at on enjoying the little things of every moment. Let us consider just a few of these:

A : Religious morality

B : Sexual activity

C : Sporting spirit

D : Cultural level

E : Tolerance

F : Personal dynamism

G : Open and communicative nature

H : Economic responsibility

It becomes obvious on making this list that we have no intention, in any way whatsoever, of the list being exhaustive and not even to include in these few elements, the most important conditioners for the coexistence of a couple.. There will be those who consider them adequate and for others, some of them will be superfluous and they would wish to add others that have not been taken into account. We are quite happy with all these ideas. In coexistence many factors that are very different in themselves are at play which change the characteristics from one country to another and from one time to another: the tradition of the place, morals of surrounding society, the example received from the respective families... and a very long etcetera. For our effects, we feel that there should be no inconvenience in experimentally accepting the eight characteristics we have used.

Below we move on to *discover* the two lovers by means of their respective fuzzy sets, taking as a referential these eight characteristics. For this it is necessary to estimate the level possessed of each one of them by Olga and James.

After some deep thought, we arrive at the conviction that it is possible to accept as descriptors the following fuzzy sub-sets:

	A	B	C	D	E	F	G	H
$\underset{\sim}{Olga}$ =	.6	.8	.4	.9	.7	.8	.6	.7

	A	B	C	D	E	F	G	H
$\underset{\sim}{James}$ =	.7	.7	.8	.6	.7	.6	.5	.9

Now we must see in some way if Olga and James are close or not close. For this we resort to a relative Hamming distance, which we are already aware of.

This is as follows:

$$\delta\,(\underset{\sim}{Olga},\underset{\sim}{James}) = \frac{1}{8}\ |.6-.7| + |.8-.7| + |.4-.8| + |.9-.6| + |.7-.7| + |.8-.6| + |.6-.5| + |.7-.9|$$

$$= \frac{1}{8}\ (.1+.1+.4+.3+0+.2+.1+.2)$$

$$= \frac{1}{8}\ 1.4 = 0.175$$

The result arrived at leads to a certain amount of optimism. In fact, the relative Hamming distance we have found is low but is *not zero*. And this is important. Excessive "*separation*" would make any dialogue between the two and life in common impossible as there would be continuous clashes. The idea, possessed by each of them of the most elemental things and at the same time basic, would be different, would generate misunderstanding and in the end lack of communication. At the very first difficulty, or introduction of a third party would cause a break up. At the other extreme, total and absolute identification, that is, no distance, would cause monotony and, in short, the lack of dialogue, as this would be unnecessary. The immediate consequences would be lack of communication and boredom. How many broken families are broken up by boredom? Fortunately this is not the case with Olga and James, who have a successful life ahead of them.

This very simple example allows us to make a technical comment, which can be of great use.

When the relative Hamming distance is found, by dividing the sum of the differences by the number of the same, what we arrive at is nothing less than the simple average of the distances. With this we implicitly accept that the *importance* of each characteristic, relative to the objective sought (the "*separation*" between Olga and James), is the same. And this is so because the operation carried out consists in multiplying each difference (in absolute values) by one and dividing the sum by the number ones. The "*weight*" given to each difference is the same (obviously we avoided this as we are dealing with a one and the product of any value by the unit is this same value).

But it could happen, and in reality does, that the importance of each characteristic, relative to coexistence, were to be different. To put this differently, it is necessary to assign a different weight to each one of the existing differences. When acting in this way, the result is a weighted value. Let us see what happens in the case of Olga and James. For this we will set, for each characteristic, a value between zero and one, in such a way that whenever a characteristic is more important the value set will be close to the unit. On the other hand, when less important then it will be further from one and therefore, closer to zero. Let us set, although quite arbitrarily, the following weights:

$$w_A = 0.6 \quad , \quad w_B = 0.9 \quad , \quad w_C = 0.4 \quad , \quad w_D = 0.7$$

$$w_E = 1 \quad , \quad w_F = 0.5 \quad , \quad w_G = 0.3 \quad , \quad w_H = 0.8$$

From these considerations we can find an index if distancing by multiplying each difference in absolute value (relative to each characteristic) by its respective weight (importance) and the result we will divide by the sum of the weights, which in our case is 5.2. We can see this below:

$$\eta\,(\underset{\sim}{\text{Olga}},\underset{\sim}{\text{James}}) = \frac{1}{5.2}(.1\text{x}.6 + .1\text{x}.9 + .4\text{x}.4 + .3\text{x}.7 + 0\text{x}1 + .2\text{x}.5 + .1\text{x}.3 + .2\text{x}.8)$$
$$= \frac{1}{5.2}(.06+.09+.16+.21+0+.10+.03+.16)$$
$$= \frac{.81}{5.2} = 0.156$$

With this new method it would appear that we have come even closer to reality. Olga and James are even closer, although sufficiently different in order to be able to talk and discuss, making time pass with a stimulating exercise of comprehension always seeking happiness.

Mathematicians would like to have a general expression of this index. In accordance with the nomenclature used up to now, we would arrive at:

$$\eta\,(\underset{\sim}{\text{Olga}}, \underset{\sim}{\text{James}}) = \frac{1}{n}\sum_{i=A}^{H} w_i \cdot |\mu_i^{Olga} - \mu_i^{James}|$$

Generalisation of this reasoning is immediate. In order for there to be coexistence between those of who form our society, it is necessary to exercise our freedom, banishing the greatest blight of our day and age which is mental torpidity, which prevents the exercise of the divine power called imagination. Freedom makes us different, from imagination arise ways for finding coincidences. Our most intimate desire is that all those opening this book harbour the desire of living in a society in which there are sufficient ingredients for true coexistence.

REFERENCES

[1] BARRE, Raymond: Prologue to the work by KAUFMANN. A. and GIL ALUJA, J.: Técnicas operativas de gestión para el tratamiento de la incertidumbre. Publ. Hispano Europea, Barcelona, 1987.

[2] KAUFMANN, A. and GIL ALUJA, J.: Técnicas operativas de gestión para el tratamiento de la incertidumbre. Publ. Hispano Europea, Barcelona, 1987.

[3] KAUFMANN, A. and GIL ALUJA, J.: Introducción a la teoría de los subconjuntos borrosos a la gestión de las empresas. Publ. Milladoiro, Santiago de Compostela 1986

[4] KAUFMANN, A. and GIL ALUJA, J.: *Las matemáticas del azar y la incertidumbre*, Publ. Ceura. Madrid 1990.

[5] KAUFMANN, A.: *Modèles mathématiques pour la stimimulation inventive*. Publ. , Albin Michel. Paris 1979, Pág. 53.

[6] ZADEH, L.: *Fuzzy Sets*. *Information and Control*, 8, Junio 1965, pág. 338-353.bis

[7] KAUFMANN, A. and GIL ALUJA, J.: *Introducción de la teoría de los subconjuntos borrosos a la gestión de las empresas*. Publ. . Milladoiro. Santiago de Compostela, 1986.

[8] GIL ALUJA, J.: *Elements for a theory of decision in uncertainty*. Kluwer Academic Publishers. Dordrecht, Boston, Londres, 1999, pág. 9 and siguientes.

[9] GIL ALUJA, J.: *Lances y desventuras del nuevo paradigma de la teoría de la decisión*. Proceedings from the III SIGEF Congress, Buenos Aires, November 10-13, 1996.

[10] GIL ALUJA, J.: *Elements for a theory of decision in uncertainty*. Kluwer Academic Publishers. Dordrecht, Boston, London 1999, page 12 and following.

[11] We have substituted the names of the people with their initials in capitals and the qualities also by their lower case initials.

[12] The fuzzy nature of the sub-set is indicated by means of a swung dash ,~, under the letter representing the same.

II. THE SETTING UP OF A BUSINESS

Viability of a business

Those who at one time have felt the temptation to start up a new business or institution will have considered a set of questions that nearly always derive into problems for which solutions must be found. It is quite obvious that in a work such as this we cannot nor must we carry out an exhaustive study. A few brief strokes of the paintbrush should be sufficient in order to introduce ourselves into the use of a method, which has given excellent results in several fields of financial, economic and commercial management.

One of the very first questions arises when thinking about the viability itself of the business that it is intended to set up. And this depends, among very many other factors, on the reception that the *"market"* gives to the products or services placed on sale. Given, then, a productive structure, that is to say, buildings, installations, equipment, estimating personnel at the different levels and specialities and knowledge of the goods that are the object of the activity, the first problem that occurs is relative to the calculation of the minimum quantity as from which the business will commence to generate profits. That is the amount that financiers call *"break even point"* or *"threshold of profitability"*. Once this figure has been estimated it must be determined whether the eventual clients will be able purchase, or failing that at what date will they be able to absorb this *"production"*. We shall limit ourselves here to two important subjects: arriving at the *"threshold of profitability"* and estimating the future period in which the new business will be attractive economically speaking. And all this framed within the idea of uncertainty that is inherent to our society.

For the development of a subject such as this, perhaps it would be useful to talk about a personal experience I had a short while ago. I have been friendly for a long time (from childhood, which is particularly good) with a prestigious surgeon, one of whose sons, Chris, was seeking to find a place for himself in economic activities and to forge himself a future in business. Clever and intelligent like few we know, he did not stand out precisely for his dedication to study, which cut off right from the start any career in teaching or research. On one of the many occasions we met, he expressed his concern by asking for advice. Throughout the conversation there was intermittent mention of the subject of sport, one of his passions. It was not illogical then that we should direct our ideas towards setting up a *"sporting business"*

From the outset this seemed like a good idea, not only from the point of view of the Chris's personal fulfilment, but also from the prospects seen in the geographical area in which this sporting business should be located (the coastal area of Mediterranean Spain). In fact, several indications convinced us of the spectacular demand for sports *"services"* (gymnasia for the middle and older aged population, tennis courts, golf courses…)

a) The first of these is becoming reality as a consequence of the proven fact that human life is lengthened day by day, but above all, the fact that increasingly, people getting older, do so with a far better quality of life. One cannot compare a present day *"old man"* of 60 with one of the same age of 50 years ago. All these old-young people, at the end of their active life (professionally speaking), or fully pensioned, are now seeking, and there will certainly be more who do so in the future, occupation for every day of the week. One cannot condemn them to be seated in front of the *"idiot box"* that is television. What better for them than to provide a good offer for maintaining them fit and morally healthy, by means of one or several suitable sporting activities.

b) Scientific advances and their technical applications make for the appearance of industrial equipment and computers that are increasingly improved which allow for the manufacture of products with the use of less human labour. This is the same as stating that productivity of labour in workplaces and offices will increase continually. In order to arrive at the same production levels, less presence of man will be required. If this tendency is suitably directed it will inevitably lead to a reduction in the number of weekly hours and, consequently in an increase in free time for leisure activities. As we are dealing with people of full working age, dedication of their free time to leisure activities and within this sport, would appear to be irrefutable. From here stems another growing segment of potential users of sporting activities.deporte, parece irrefutable. De ahí la existencia de otro segmento creciente de potenciales demandantes de actividades deportivas.

c) The industrialisation process, characterised by large human concentrations in closed spaces, meant an impoverishment of Southern European countries, with massive migration towards the North. Cold and rainy countries, accustomed to life closed in the homes, constituted a favourable environment for those work modes. But increasingly today repetitive, heavy and dangerous work is being done by machines. We have repeated over and over again that the era of the " *proletariat*" has ended up by being substituted by the era of the "*robotariat*". And more frequently those large human concentrations in modern industry are becoming less. Work from the home, although incipient, will become accelerated thanks to the facilities of data transmission networks. What is already occurring, but will happen much more in the future, is a migratory tendency from North to South. There can be no doubt whatsoever that work, above all intellectual work and that stemming from imagination and creativity is far less disagreeable when carried out bathed by the sun. This increase in the population with an economic potential and intellectual capacity, constitutes a segment with a high percentage of leisure time consumers and, consequently there is a demand for sporting activities.

These and other considerations, which could be made to complete or substitute those already mentioned, convinced us that the chosen activity was the correct one. But work had only just begun. The two fundamental questions remained latent: given a dimension for the business in accordance with Chris's possibilities, what was the minimum number of clients (problem of arriving at the "break even point" or "threshold of profitability") and knowing the threshold of profitability, in what future year would the population existing in the area of influence of the "sporting company" permit at least that number of clients. Let us now move on to study the first of these questions and later attack the second.

Determination of the minimum level of activity

To attempt to foresee the threshold of profitability requires having available information *relative to the future* on costs and income for levels of increasingly growing activity, that is to say as it is assumed the clientele will grow. But to talk of economic values (quantified data) situated in the future inevitably means delving into the mathematics of uncertainty and, consequently, the need to express these values by means of one (or several if required by circumstances) of the known uncertain numbers.

So as not to become lost in a plethora of details, we are going to, for now, provide an answer to both aspects. Let us commence by saying that a separation will be established between fixed costs (those that remain invariable whatever the number of clients, given the current structure of the business) and variable costs (those that depend on the number of clients, and increase proportionally relative to the same). The first will be represented by C_F and will include the cost related to premises, fixed installations, equipment, fixed apparatus, and the minimum permanent employees. For the second, it is considered prudent to foresee a cost per loyal client. This is represented by $C_v^{(m)}$ (average variable cost) and included in this figure are water consumption, electricity, sanitary products. Foreseeing income is very much easier, given the type of business under study, in which each client pays a fixed membership fee (pre-established) and makes complementary payments, consuming certain products, also paid for. These are indicated by $I^{(m)}$.

Having established the values to be considered C_F, $C_v^{(m)}$, $I^{(m)}$ we now have to select the uncertain numbers in which they are to be expressed. We decided for the simplest which is none other than the «confidence interval». Let us briefly recall that a confidence interval is comprised of two "levels" or "extremes". A lower extreme or level below which we believe that the represented figure does not exist and an upper level or extreme, above the value of which we will not arrive at the corresponding figure. Therefore, for example, on estimating fixed costs for the sporting business we say they are represented by a confidence interval of [30.000.000, 40.000.000], which indicates that fixed costs will be no less than 30 million nor in excess of 40 million monetary units, but we are totally ignorant as to where they will fit between these two extremes or limits. Normally in order to explain that a figure is represented by a confidence interval we use the symbol $\sim$ (broken slash) under the corresponding letter or letters. Therefore in our case we would write:

$$\underset{\sim}{C}_F = [30.000.000, 40.000.000]$$

We should remember that this fixed cost must be assumed, for example annually, whatever the number of clients (partners in the business who pay an annual membership fee).

For the other two values, we will follow the same path, making also expressing the forecasts in confidence intervals. In our case we accept the following:

$$\underset{\sim}{C}_V^{(m)} = [20.000, 24.000]$$

This interval brings to light the fact that the annual cost per client will be no less than 20.000 and no higher than 24.000 monetary units. Likewise it is foreseen that:

$$\underset{\sim}{I}^{(m)} = [100.000, 120.000]$$

result of adding to the annual fixed membership fee whatever could be consumed in products or services, a value fluctuating between zero (member with no additional services and 20.000 (members normally consuming complementary services).

We feel it unnecessary to insist on the importance of drawing up these forecasts very carefully.

In order to pass over from average variable costs (variable costs per client), $\underset{\sim}{C}_V^{(m)}$, to overall variable costs (variable costs for all clients) it is enough to multiply $\underset{\sim}{C}_V^{(m)}$ by the number of clients we foresee for any one year. The profitability threshold (an uncertain number to arrive at) will be the number of clients (members) necessary for arriving at neither profit nor loss. This is designated by $\underset{\sim}{N}$ and as this is an interval, will be formed by a lower extreme or limit n_1 and an upper extreme or limit n_2. Therefore we write:

$$\underset{\sim}{N} = [n_1, n_2]$$

In this way n_1 and n_2 are the values we are looking for and they form the "profitability threshold". In this figure neither profit nor loss should exist and also "total costs" should be equal to "total income". Only in this manner will total income less total costs be equal to zero.

If we call total costs $\underset{\sim}{C}_T$ and total income $\underset{\sim}{I}_T$ we can write::

$$\underset{\sim}{C}_T = \underset{\sim}{C}_F \,(+)\, \underset{\sim}{C}_V^{(m)} \,(\cdot)\, \underset{\sim}{N}$$

$$\underset{\sim}{I}_T = \underset{\sim}{I}^{(m)} \,(\cdot)\, \underset{\sim}{N}$$

The symbols (+) and (·) represent respectively the sum and product of confidence intervals. In our case we have:

$$\underset{\sim}{C}_T = [30.000.000,\ 40.000.000] \,(+)\, [20.000,\ 24.000] \,(\cdot)\, [n_1, n_2]$$

$$\underset{\sim}{I}_T = [100.000,\ 120.000] \,(\cdot)\, [n_1, n_2]$$

Given that in the profitability threshold total income is equal to total costs, that is to say:

$$\underset{\sim}{I}_T = \underset{\sim}{C}_T$$

We arrive at:

$$[30.000.000,\ 40.000.000] \,(+)\, [20.000,\ 24.000] \,(\cdot)\, [n_1, n_2] = [100.000,\ 120.000] \,(\cdot)\, [n_1, n_2]$$

Chris, in view of this equation with intervals, asks the logical question of how do we resolve an equation of this nature, and, in a previous step, in what way do we carry out the operations of addition, subtraction, multiplication and division of confidence intervals. The question was inevitable taking into account the fact that to resolve this equation in order to find the values of n_1 and n_2, extremes of interval $\underset{\sim}{N}$, means having a knowledge of the arithmetic of confidence intervals. Let us recall some of the more significant operations:

•*Addition*. The lower extremes are added together as are also the upper limits:

$$[a_1, a_2]\ (+)\ [b_1, b_2] = [a_1+b_1, a_2+b_2]$$

We illustrate this with an example:

$$[3, 4]\ (+)\ [6, 8] = [3+6, 4+8] = [9, 12]$$

• *Subtraction*. There are two types of differences, the so-called "normal difference" (this is the most often used) and the "Minkowski difference" (so called in honour of the Russian mathematician).

For the "normal difference" subtraction takes place by crossing over the two extremes of the intervals, that is: the *lower* extreme of the minuend less the *upper* extreme of the subtrahend (lower limit of the result) and the *upper* extreme of the minuend less the *lower* extreme of the subtrahend (upper limit of the result):

$$[a_1, a_2]\ (-)\ [b_1, b_2] = [a_1-b_2, a_2-b_1]$$

Again let us look at an example:

$$[6, 10]\ (-)\ [4, 7] = [6-7, 10-4] = [-1, 6]$$

For the "Minkowski difference" operation is even easier[1]. It is sufficient just to subtract the lower extremes from each other and the upper extremes from each other (just as in the case of addition, but with differences):

$$[a_1, a_2]\ (\overline{m})\ [b_1, b_2] = [a_1-b_1, a_2-b_2]$$

And using numerical data:

$$[6, 10]\ (\overline{m})\ [4, 7] = [6-4, 10-7] = [2, 3]$$

• *Product.* Arriving at a product for intervals requires multiplying each extreme- of an interval with the two extremes of the other. From the four products the least of these is chosen as the lower extreme and the highest as the upper extreme:

$$[a_1, a_2]\ (\cdot)\ [b_1, b_2] = [\text{Min}\ \{a_1{\cdot}b_1, a_1{\cdot}b_2, a_2{\cdot}b_1, a_2{\cdot}b_2\},$$
$$\text{Max}\ \{a_1{\cdot}b_1, a_1{\cdot}b_2, a_2{\cdot}b_1, a_2{\cdot}b_2\}]$$

Lo ilustramos con un ejemplo:

$$[3, 5]\ (\cdot)\ [10, 14] = [\text{Min}\ \{3{\cdot}10, 3{\cdot}14, 5{\cdot}10, 5{\cdot}14\},$$
$$\text{Max}\ \{3{\cdot}10, 3{\cdot}14, 5{\cdot}10, 5{\cdot}14\}]$$

$$= [\text{Min}\ \{30, 42, 50, 70\}, \text{Max}\ \{30, 42, 50, 70\}]$$

$$= [30, 70]$$

It can be easily seen that, as occurring in our case, *when all the values of the extremes of the intervals are positive*, operation is easier since it is sufficient to multiply the lower extremes by themselves (lower limit of the result) and the upper extremes by themselves (upper limit of the result):

$$[3, 5]\ (\cdot)\ [10, 14] = [3{\cdot}10, 5{\cdot}14] = [30, 70]$$

The result is obviously the same.

•*Quotient.* In the same way as the case with subtraction, here also there are two types of division. "Normal division" and the "Minkowski division"

"Normal division", as occurs with the product, requires doing four operations, each extreme of an interval with the two from the other. But now the operator is the quotient. The smallest figure is taken as the lower extreme and the largest as the upper limit:

$$[a_1, a_2] (:) [b_1, b_2] = [\text{Min} \{a_1/b_1, a_1/b_2, a_2/b_1, a_2/b_2\},$$
$$\text{Max} \{a_1/b_1, a_1/b_2, a_2/b_1, a_2/b_2\}]$$

Below is an example:

$$[8, 12] (:) [7, 9] = [\text{Min} \{8/7, 8/9, 12/7, 12/9\},$$
$$\text{Max} \{8/7, 8/9, 12/7, 12/9\}]$$

$$= [\text{Min} \{1.14, 0.89, 1.71, 1.33\},$$
$$\text{Max} \{1.14, 0.89, 1.71, 1.33\}]$$

$$= [0.89, 1.71]$$

When the values of all the extremes of the intervals are positive it is sufficient to divide by crossing over the extremes, that is, lower extreme of the dividend with the upper extreme of the divisor (lower extreme of the result) and the upper extreme of the dividend with the lower extreme of the divisor (upper extreme of the result):

$$[8, 12] (:) [7, 9] = [8/9, 12/7] = [0.89, 1.71]$$

Again we can see that the results coincide.

The "Minkowski division" (the result does not always culminate in an interval), takes place by arriving at the quotient between the lower extremes (lower limit) and the quotient between the upper extremes (upper limit):

$$[a_1, a_2] (\underset{m}{:}) [b_1, b_2] = [a_1/b_1, a_2/b_2]$$

If we put numbers to the extremes:

$$[8, 12] \ (\underset{m}{:}) \ [7, 9] = [8/7, 12/9] = [1.14, 1.33]$$

Chris did not think it was too complicated to Chris introduce these operators to the data he possessed. We did not want to go into further detail, which would have required a great deal of time and, perhaps, drift away from our objective which was essentially practical. On the other hand, he did not let the opportunity pass to bring up a couple of doubts he had. The first referred to the observations we made on the Minkowski subtraction and division, in the sense that, in some cases, the result does not appear in the form of a confidence interval. I gave him the following example:

$$[6, 10] \ (\overline{m}) \ [2, 8]$$

When we have 6-2=4 and 10-8=2, the result would be that the lower extreme would be greater than the upper extreme, which fact is contrary to the nature of intervals, in which the lower extreme must always be less or equal to the upper extreme. What should be done in this event?. There is only one alternative open to us, if the minuend and subtrahend intervals (data) are correct and immovable: convert the result into an exact number, "making the entropy fall" (add the differences and divide the sum by both) and continue the calculations using this exact number, even if it is in the form of an interval (every exact number is an interval in which the lower extreme is equal to the upper extreme). Thus in our example we would arrive at:

$$[6, 10] \ (\overline{m}) \ [2, 8] = (4+2)/2 = 3 = [3, 3]$$

The second question was in the form of a doubt. How can I know when to use, in subtraction and division, the normal operator and the Minkowski operator?. There is an example in which it is *always* necessary to resort to the Minkowsky operators in order *to resolve an equation.* In the remaining cases, the nature of the problem automatically determines which of the operators to use. The only precaution to be taken into account is when doing the subtraction or division with those extremes of one or the other interval that represent the "worst" situation that could occur and those others that represent the "best" of situations that could occur. I gave Chris an example which he considered very significant: if a product unit can be sold at

[100.000, 120.000] and the variable cost for this unit is [20.000, 24.000], the "worst" situation would be to sell the product very cheaply (100.000) and the cost be the highest (24.000), and the "best" situation is to sell expensively (120.000) with a low cost (20.000). It can be seen that the extremes must be "crossed over" and therefore the "normal subtraction" would be used:

$$[100.000,\ 120.000]\ (-)\ [20.000,\ 24.000] = [76.000,\ 100.000]$$

The profit per product unit before charging fixed assets would not be less than 76.000 nor higher than 100.000. to put this in simple term: the worst on the one hand and the best on the other.

We spent some time going over the arithmetic of confidence intervals, but it was worth it since by doing so, we were now ready to return to our immediate concerns. The first of these was to arrive at the values of the interval $\underset{\sim}{N}=[n_1,n_2]$ representing the number of clients (members) of the sporting enterprise who would pay membership fees every year.

Perhaps to make it rather more intuitive, we could have approached the equation of the equality of income-costs in some other way: as the first member the fixed costs and as the second member total profits before compensating for fixed costs (fixed assets), that is to say the profit per person multiplied by the number of people (members of the sports business). Therefore we write:

$$[30.000.000,\ 40.000.000] = [100.000,\ 120.000]\ (\cdot)\ [n_1,\ n_2]\ (-)\ [20.000,\ 24.000]\ (\cdot)\ [n_1,\ n_2]$$

And also:

$$[30.000.000,\ 40.000.000] = \big([100.000,\ 120.000]\ (-)\ [20.000,\ 24.000]\big)\ (\cdot)\ [n_1,\ n_2]$$

$$[30.000.000,\ 40.000.000] = [76.000,\ 100.000]\ (\cdot)\ [n_1,\ n_2]$$

And finally:

$$\begin{aligned}[n_1,\ n_2] &= [30.000.000,\ 40.000.000]\ (:)\ [76.000,\ 100.000] \\ &= [30.000.000/100.000,\ 40.000.000/76.000] \\ &= [300,\ 526]\end{aligned}$$

In this way, we arrived at the conclusion, although we will see provisional, that in order for us not to generate profit nor loss we would have to have a number of members-clients that in the very best of cases should not be any less than 300 and in the worst case it was not required that they exceed 526.

Chris remained very perplexed on seeing the extent of the resulting interval, in which the upper extreme was 175% that of the lower extreme, giving rise to very high uncertainty. What was the reason for this? Was there nothing to be done to diminish this uncertainty? The first of these questions allowed us to verify the coherence between the initial information and the result. In fact, a simple glance at the initial "data" brought to light the extent (uncertainty) of the interval relative to fixed costs, which Chris himself had estimated. Having located the problem, the "remedy" goes without saying: the estimate for "fixed assets" (in this case fixed costs and fixed assets coincided) had to be "refined". This was recognised by the interested party and a revision was made of the "numbers", finally arriving at the following interval for fixed costs:

$$\underset{\sim}{C}_F = [32.000.000, 36.000.000]$$

After this modification the following profitability threshold was arrived at:

$$\begin{aligned}[n_1, n_2] &= [32.000.000 / 100.000, 36.000.000 / 76.000] \\ &= [320, 474]\end{aligned}$$

With this very simple correction the extent of the interval, which we had arrived at, was reduced. Now the upper extreme was less than 48%. This result was acceptable.

This phase of the work was now finished. But had it had any influence on the mind of a young man accustomed to age-old mathematics? Yes...but. For even further consolidation of this scheme we proposed an alternative development to the reasoning we had used, starting out, obviously, from the same two ideas: "he profitability threshold takes place when total income is equal to total costs"; «when operating with intervals one must always link the best, on the one hand, with the worst, on the other». From the first we retain the equation:

$$[32.000.000, 36.000.000] (+) [20.000, 24.000] (\cdot) [n_1, n_2] = [100.000, 120.000] (\cdot) [n_1, n_2]$$

Of the second, given that the "worst" consist of the highest costs (upper limit of the first member) and the lowest income (lower limit of the second member), we construct the following equation:

$$36.000.000 + 24.000\ n = 100.000\ n$$

Where:

$$n = 36.000.000 : 76.000 = 474$$

Upper extreme of the resulting interval arrived at.

Likewise, taking into account that the "best" is given by the lowest costs (lower extreme of the interval of the first member) and the highest income (upper extreme of the interval of the second member) the equation is now:

$$32.000.000 + 20.000\ n = 120.000\ n$$

and therefore:

$$n = 32.000.000 : 100.000 = 320$$

extreme lower of the interval turned out obtained

This approach allows us to draw a graph, which will let us see what we have just commented on.

For this we draw a system of coordinates, in which the number of clients (members of the sporting company) is included as abscissa and the economic values (income and costs) as ordinates. In this system both total pessimistic income and optimistic income are represented as are total pessimistic and optimistic costs. We arrived at the following graph:

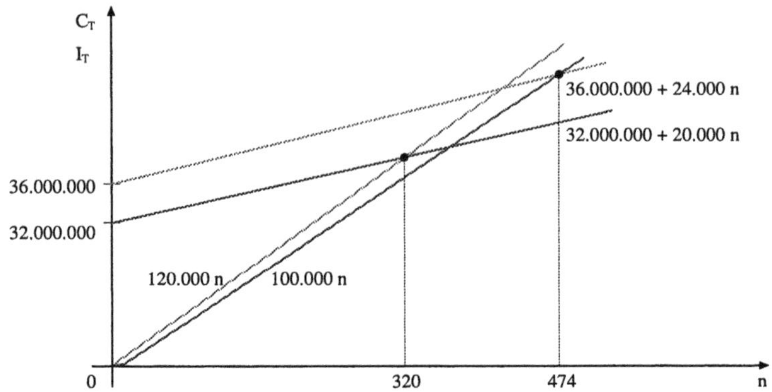

The intersection between the "pessimistic" strait lines and between the "optimistic" strait lines provide us, in abscissas, with the two limits of the confidence interval which show the minimum necessary number of clients (members) in order for the business not to show a loss.

We had now arrived at our first objective and Chris now possessed a basic element on which to base his hopes for success. But also, he had a very clear perception of why it was so necessary to count on so many members in order to cover fixed annual costs. He now felt secure in his position. Nevertheless, now there arose a further question and that was, on what current or future date would there be a sufficient influx of "potential" clients in the area where the sports business was to be set up of so that with those already existing, the "actual" clients would be [320, 474]?

Before replying to this question, it was necessary to know if the current population was sufficient placing, at least, the clients within the confidence interval we had found. And, if so, in which area would they be placed (towards the lower limit, that is the optimistic positions or in the upper limit and, therefore the pessimistic positions). We suggested for this that a survey be made. Fortunately two colleagues and friends, Joseph and James, were specialists in this field, with well-established and solid experience. We resorted to them. They did not take very long to deliver the results. They were conclusive: with a good launch of the new business a maximum of 350 member-clients could arrive at. But making a reasonable estimate (pessimistic) we could only hope for 290. With this very useful information, which we represented in the form of an interval [290, 350], we were able to arrive at the number of clients, which the migratory flow, which had already commenced, should contribute. Chris was quick to say that neither on this occasion was it necessary to use the Minkowski subtraction. In fact, the "best" of the possibilities shows the need for 320 members and a perfect "launching" would supply 350, which in the "most unfavourable" circumstances the threshold of profitability is located at 474 members, while a "bad" acceptance of the sports business would "only" attract 290 members. The necessary surplus was then:

$$\underset{\sim}{M} = [m_1, m_2]=[320, 474] (-) [290, 350]$$

$$=[320\text{-}350, 474\text{-}290] = [-30, 184]$$

We felt that this was particularly significant data. Nevertheless in a certain way it was confusing, because it was difficult to take a decision based on the statement according to which in the best of cases we exceeded the minimum required by 30, while in the worst of cases we were missing 184 members. It would not appear to be *normal* to put ourselves in one of the extreme situations of the interval because it is difficult (although possible) that things could go so well or so badly. I suggested that, *without abandoning the result* he should make the "entropy fall", that is to say, he should pass over from the interval to an exact number. The most suitable way of doing this is to arrive at the medium point of the former, by adding the two extremes and dividing the result by two. This we did immediately:

$$\overline{\underset{\sim}{M}} = (m_1 + m_2) : 2 = (-30 + 184) : 2 = 77$$

In short, the best representation of the interval [-30, 184] in the field of certainty was 77. This number of members is only an indication because in reality, the additional requirements could be located at any point of the confidence interval.

The next step to be taken was final: what was the necessary increase in population so that the new sporting business could increase its clients in the interval [-30, 184] or, if you like in 77 members? The survey done by Joseph and James was particularly useful but certain data which Chris had also helped greatlys and this referred to the percentages of immigrants who were keen on active sport over the total immigrant population, very much higher than the percentages corresponding to normal residents (at least in that particular geographical area). Also it was necessary to include the potential clients who would join other sports centres (clubs, associations, sporting companies). Having gathered together all the known data and available information, we estimated as the "key" figure somewhere around 1,400 new immigrants. In this way it was risky to commence activities immediately. But on the other hand, it could be particularly positive knowing *at what future moment* would this circumstance occur. The promoter of the project would have all this time available for suitably programming the task of preparation and start up of the new sporting business. From among the several techniques available for providing an answer to this question, we decided on the "Fuzzy Delphi" method.

Locating future events by the Delphi method

Chris had heard about the "Delphi method" but he only had a very superficial knowledge of the same. Therefore it was not possible to use the adaptation of this method to the field of fuzziness without first having very clear ideas. In order to arrive at this objective we mention a few pages of the first book written in conjunction with Professor Kaufmann, which we reproduce below[2].

Saturn gave Janus the power of knowing the past and the future. This is why he is represented with a double face: one looking to the past and the other to the future. This double face could symbolically represent the activity of those who have the mission of foreseeing the future.

Nowadays estimating takes place by means of models in which hypotheses are included, justified by reasoning and/or data on the phenomenon under study, and which are located in a past that is more or less remote. Obviously the range of models suitable for use can vary enormously depending on the relative field and objectives being sought.

With the theory of probabilities, statistical methods, analysis procedures of times series, stochastic models of evolution, etc...modern science is capable of providing instruments for analysis and estimating.

However, these instruments are limited in their possibilities of use in reality because they can only be used when information exists which is not, or is not yet, sufficiently structured for it to be integrated into the models. The problem arises when only having available information that permits the use of models for estimating. It should be taken into account that, given the economic context in which businesses find themselves, it is precisely this information that is insufficiently structured that abounds in reality.

Man today lives in a system in which countless information circulates and even considering the progress of estimating methods, the proportion of what remains to be foreseen starting out from intuition constantly increases, in spite of the means available, among which computers stand out.

In this respect, the more we resort to calculation and programming for treating structured information the more our imagination is going to require more estimates from us. In the pair information-expression, which forms a part of the nature itself of man, there can be coexistence of imbalances between the elements that make it up. The more information that is possessed, the more the need for expression. An imbalance would lead us to the robot man, as a consequence of the excess of information and lack of expression or, on the contrary to the wild man, due to the lack of information and excess of expression. It is for this reason that in the modern world intuition increasingly acquires more vital importance.

This intuition, which is not the fruit of programmable sequential reasoning by means of a machine, is vague, fuzzy, evolutionary and nevertheless constitutes the superiority of man over the means of treatment of information.

The establishment of an economic policy in a business has as one of its main supports the carrying out of estimates of long-term fundamental data. These reference points allow for the elaboration of lines of action aimed at attaining objectives, which are normally expressed numerically. Since these objectives are located in the future, "estimating" arises as a prime factor in business management.

Introducing oneself into the technical treatment of the future in businesses brings up some singular problems which commence from the very moment of attempting to define the fundamental concepts on which to base later developments. A representative example can be seen in the attempt to isolate the notion of estimating from other concepts such as predicting, research, etc... From a technical point of view prediction involves the possibility of certain known laws and logical chains. Foreseeing, on the other hand, is linked to the processes of estimating the bases of which, however, can be found principally in a past, which does not have a mechanical translation in the future. This is the reason why its most valid schemes belong to the field of randomness or the field of uncertainty.

The classical methods of estimating, which in a stationary economic system have their reason for existing, based on a mechanistic determinism have had to give way to others that were more capable of representing a reality with rapid changes and multiple accelerations of certain economic systems that are characterised by their continuous evolution. But even in these new methods the past plays a certain role, perhaps somewhat more discreet but at times no less important, since any process of subjective structuring or evaluation cannot be totally removed from the "experience" of the expert providing the same.

The risk that is run arises from the possibility that, in a conscious or unconscious manner, an excessive weight is granted to the incidence of factors that have acted in the past relative to their intervention in the future when this will have undergone substantial modifications both from the action of external elements and the intervention of human will.

This changing situation and the difficulties that exist in making a translation of the past to the future have given rise to the fact that a certain number of businesses have incorporated into their organic structures, groups of experts, whose objective consists in determining, by means of technical studies, how the business will be in a few years time, that is, in a "long term" future, by establishing their commercial, financial, organisational, personnel, technological, etc.. situation independently of the processes of transition from the current situation to the future.

In this same scheme it is determined which external elements for the business it is expected will constitute the surroundings in which it will carry on its activities. It is precisely these facts, which will condition its very survival, since a business cannot be considered as outside the sphere in which it acts. Among these elements we can mention legal conditions, the social structure, the population pyramid, technological development, etc...

What this would be then is a study of prospects in the sense of estimating the future from certain principles that can be accepted of a general nature, without considering the past. From this point of view, the study of prospects allows for a much wider horizon given its removal from current phenomena, giving way to the drawing up of certain schemes which starting out from the future return to today. From this then stems the important methodological change, which has occurred in the development of prospective methods and their possibilities in business management.

In this change it can be admitted that the quality of estimation by a group is higher than that of an individual person, on the condition that this group works suitably so that the principle of "two heads is worth more than one" becomes valid.

When there are several people issuing an opinion on a specific problem, the most significant parameter will be determined by the specific knowledge of each person on the subject in question. This is perhaps the subjectivist element, which is difficult to remove from the schemes.

The process that we are going to develop is formed by a set of stages in which the *group* appears as the central figure capable of fine tuning the estimates so that they become statistically more valid. The DELPHI method responds to three basic imperatives in the system of surveys which uses: anonymity, controlled retroaction and the statistical response of the group. Given the differentiation existing in mental habits in each country, and even in each business, it is convenient to adequately filter the DELPHI scheme with the object of adapting it to the requirements of each place and time. One of the characteristics of research work is that certain findings give birth to the study of other problems, which gives rise to the constant advance of scientific activity.

After reading this Chris remained thoughtful. Then, with an inquiring look on his face said: it seems interesting but.. how does one get off this theoretical cloud and arrive at an instrument which is valid for me and my needs to estimate when will I be able to start up my sports business? I feel that, even if we spend a certain amount of time on these basic aspects, they may be useful when the time comes for you to reason out the results. Just a little more patience and we will soon commence our specific task: a veneer of culture is not really excessively heavy baggage. Remember that the Delphi method is the result of the research of a group of American scientists from the "Rand Corporation" located in Santa Monica (California), among whom we should mention O. Helmer, E.S. Quade and N. Dalkey, whose results were published in the mid 60's [3].

These scientists proposed acting in the following manner:

a) Resort to a group of experts (people trained and experienced in the subject or subjects under study) in order for them to provide their opinions relative to the questions asked.

b) In a first stage the technicians from "Rand" conducting the programme, proposed to the experts that they state, by correspondence, the inventions, discoveries and interesting, desirable or feasible findings which might be carried out over the next fifty years.

c) Once the discoveries were selected for which it was desired to estimate their date of attainment, the experts were asked to state in which year of the future did they feel that this would take place or, also if they considered them definitely unrealisable.

d) The results established by the experts were included in a statistical study from which three values were extracted: the first quarto, the second quarto (the median) and the third quarto. In this process only those results where there was reasonable consensus were considered.

e) The experts were informed of these overall results, always by correspondence and their comparison with the personal opinion of the expert himself (not with that of the other individually) was made. Eventually there was a dialogue with those experts who had not given any opinion whatsoever and also with those whose opinion was removed from that of the majority. All of this with the object of arriving at new estimates, which it was hoped (it does not always occur in this way) would be closer together.

f) With this new information the statistics were drawn up again returning to phase d).

g) With the results obtained, phases e) and f) were repeated until the effects of re-adaptation of the experts was irrelevant.

As an anecdote it can be said that during the biennium 1964-1965 the "Rand Corporation" made an interesting test by inviting the participation in the experiment of 80 experts separated into 6 speciality groups. Initially 49 possible or desired discoveries were chosen. After multiple discussions this number was reduced to 30. The final result, after using the proposed method, can be seen today with a certain amount of blushing that is not exempt from admiration for the courage shown on attacking the subjects which at that time were so obscure. Here are some of the more shocking:

* Mining the bottom of the seas (1980, 1989, 2000)

* Economic production of synthetic proteins (1985, 1990, 2004)

* Substitution of organs in the human body (1967, 1972, 1982)

* Use of drugs to enhance intelligence (1984, 2012, 2030)

* Automatic direct translation of languages (1970, 1975, 1978)

* Control of hereditary deficiencies (1990, 2000, 2010)

* The use of telepathy as a means of communication (2021, never, never)

This brief outline had for Chris the attraction that is always exerted on youth by the unknown and by large projects. But after a short time, he did not take long to return to reality which for him was much more prosaic than saving humanity with unusual discoveries. His objective was now centred on trying to see when he should start up his sports business. And from here one doubt in particular among others had him worried: when the experts came up with a date in what terms did they do so, was it either in the absolute surety of its occurrence, or in terms of a probability? The response to this was provided by the Rand Corporation technicians on establishing that each one of the inventions would be estimated with a probability of $p \geq 0.5$ of its occurring on the indicated date. I was not surprised that Chris answered this reply by showing some doubt. The same would have happened to me at the time. For this reason I proposed to him that, in our case, we should adopt a method with a different basis to that established by the Rand Corporation.

The Fuzzy Delphi methods

It would appear to be clear that in the Delphi method two differentiable elements exist. On the one hand, the communication between the technicians treating the information and the experts providing the same and, on the other, the processes of estimating, analysis and grouping the data. We felt that it was convenient to maintain from the Delphi method the first of these and radically vary the second. We were moved by practical reasons but also there were theoretical aspects lying latent. One and the other could be expressed in the following terms:

1) Estimates in the medium and long term are difficult to fit into the field of randomness and, therefore, to talk about probabilities would appear to be out of place. On the other hand, the sphere of uncertainty would appear to be suitable for including and treating values that are charged with a high degree of subjectivity.

2) The grouping of the opinions of the experts constitutes a process that allows for a certain objectification. For this the use of statistical operators such as average and deviations are tolerated, even though other methods may be more adequate

3) The manner of putting the question to the experts must change radically. Instead of asking them for the date of a finding or event with a probability of 0.5, they should be asked about the nearest moment, the one they feel with the greatest intensity and the moment that is furthest removed (confidence triplet).

All this leads us to move from the sphere of randomness to that of uncertainty.

Having made these brief statements we are now in a position to commence the process, which should lead us to resolving Chris's problem. In the first place it became necessary to find a group of "experts" capable of giving us a valid opinion on the time at which, in the area of influence of the sports business, would occur the immigration of 1,400 people. After a lot of consultation, telephone calls, communications by fax and e-mail we selected 24 experts. We were conscious of the fact of how small this number was, but given the circumstances we had to be happy with what was available. Therefore we commenced the process with these twenty-four elements

Each expert was asked to provide a confidence triplet corresponding to the dates, which he considered (the closest date, the most likely date, and the furthest removed date). The result is shown in the following table:

Expert	YEARS		
	Closest date	Most likely date	Furthest removed date
1	2	4	5
2	3	5	7
3	2	3	5
4	4	4	4
5	1	3	4
6	5	8	10
7	3	4	6
8	2	3	4
9	1	4	4
10	3	5	6
11	5	7	8
12	4	6	7
13	2	3	4
14	1	3	5
15	6	10	12
16	3	4	6
17	2	5	6
18	4	5	5
19	3	4	6
20	2	2	3
21	1	3	4
22	10	12	15
23	5	7	8
24	6	8	10

Having gathered this information, we moved on to obtaining the "average confidence triplet". For this it was sufficient to add each one of the three columns and divide the result bythe number of experts, in our case 24. The result was the following:

$$\underset{\sim}{T}_1 = \left(\overset{m}{A}_1, \overset{m}{B}_1, \overset{m}{C}_1\right) = \left(\frac{80}{24}, \frac{122}{24}, \frac{154}{24}\right) = (3.33, 5.08, 6.42)$$

Thus in this first phase a result was arrived at according to which the desired number of immigrants would have been reached was in 3.33 years, the most likely date 5 years and the furthest removed date 6.42 years.

But this was only a very rough estimate. What was necessary was a complete analysis of the information received, prior to a new contact with the experts.

We commenced this analysis by seeking the differences between the average triplet and the triplet supplied by each one of the experts. We did this by looking for the differences in each one of the three columns. We then arrived at the following table:

Expert	YEARS			
	Closest date	Most likely date	Furthest removed date	
1	1.33	1.08	1.42	
2	0.33	0.08	-0.58	
3	1.33	2.08	1.42	
4	-0.67	1.08	2.42	
5	2.33	2.08	2.42	←
6	-1.67	-2.92	-3.58	←
7	0.33	1.08	0.42	
8	1.33	2.08	2.42	
9	2.33	1.08	2.42	
10	0.33	0.08	0.42	
11	-1.67	-1.92	-1.58	
12	-0.67	-0.92	-0.58	
13	1.33	2.08	2.42	
14	2.33	2.08	1.42	
15	-2.67	-4.92	-5.58	←
16	0.33	1.08	0.42	
17	1.33	0.08	0.42	
18	-0.67	0.08	1.42	
19	0.33	1.08	0.42	
20	1.33	3.08	3.42	←
21	2.33	2.08	2.42	←
22	-6.67	-6.92	-8.58	←
23	-1.67	-1.92	-1.58	
24	-2.67	-2.92	-3.58	←

A mere glance at the table was sufficient for us to see that some of the experts were a long way removed from the average opinion (see experts 15, 22 and 24) and others perhaps even further (see for example experts 5, 6, 20 and 21). We felt that we should have a brief conversation with the experts prior to sending them the average triplet. The result of this was positive except for experts 15 and 22 who remained unyielding and we felt that this was incorrect. As pleasantly as possible we finalised the process with them and from then on remained with 22 experts.

Once we had carried out this task we sent the overall result, triplet (3.33, 5.08, 6.42) to all the "remaining" experts with the request that they reconsider their estimates in view of the aggregate opinion of all the experts we had consulted. The new opinions are included in the table below:

Expert	YEARS		
	Closest date	Most likely date	Furthest removed date
1	3	4	5
2	3	5	6
3	2	4	5
4	4	4	5
5	2	4	5
6	4	6	7
7	3	4	6
8	3	3	5
9	2	4	5
10	3	5	6
11	4	5	6
12	3	5	6
13	3	4	5
14	2	4	5
16	3	4	6
17	3	5	6
18	4	5	5
19	3	4	6
20	3	3	4
21	2	4	5
23	4	6	6
24	4	6	7

Having calculated the average confidence triplet, we arrived at the following result:

$$\underset{\sim}{T}_2 = \left(\overset{m}{A}_2, \overset{m}{B}_2, \overset{m}{C}_2\right) = \left(\frac{67}{22}, \frac{98}{22}, \frac{122}{22}\right) = (3.05, 4.45, 5.55)$$

After this new sequence we arrived at the fact that the migratory affluence would take place between years three and five and a half, with a most likely date at nearly four and half years. This was good information for Chris.

In order to have a more precise idea, however, it was recommended that the average triplet should be converted into a single number. For this we advised him to give double importance to the maximum presumption over the extremes of the triplet. In this way what remained was:

$$\overline{\underset{\sim}{T}}_2 = \frac{3.05 + 2 \text{ x } 4.45 + 5.55}{4} = 4.38$$

Therefore according to this process it could be hoped that the business projected by Chris could operate without losses within approximately four years and three months.

The advisability of counter-expertise

The consultation could be repeated once again but we felt that this was unnecessary taking into account the results we had arrived at with the second opinions.

In order to be doubly convinced we thought that it was more convenient to make another type of analysis: That of "counter-expertise". This is something that is particularly simple. From the result we had arrived at, the extremes of the triplet are considered [3.05, 5.55], forming in this way a confidence interval. This information in conjunction with the most likely date 4.45 and the result of reducing the triplet to a single exact number 4.38, are the data to be supplied to *another* group of experts we will call "counter-experts". There position is to position themselves between 3.05 and 5.55, taking into account all the available information. That is to say, to see if their opinion was closer to 3.05 than to 5.55, in the centre of the interval or even at one of the extremes.

With the object of making this task easier it was proposed to them that their positioning should take place within the values [0, 1], following a hendecagonal scale similar to one we looked at in the previous chapter, but now with the following semantic correspondence: [4]

0	:	the value 3.05 is correct
.1	:	practically 3.05
.2	:	nearly 3.05
.3	:	near to 3.05
.4	:	closer to 3.05 than to 5.55
.5	:	as close to 3.05 as to 5.55
.6	:	closer to 5.55 than to 3.05
.7	:	near to 5.55
.8	:	nearly 5.55
.9	:	practically 5.55
1	:	the value 5.55 is correct

By combining efforts, we managed to get the agreement of 12 "counter-experts" who agreed to our plan. In a first conversation it became plain to us that they were not always capable of, or willing to set their position in an "exact number" in [0, 1] as, for example 0.6 or 0.7.

On many occasions they doubted between two or even three evaluations. It was therefore quite frequent to hear positions such as 0.6 or.0.7; 0.4 or 0.5,... We accepted this approach and admitted the possibility of position stated in confidence intervals. With this decision, it appeared that all were ready for emitting an opinion. These were as follows:

Counter-experts	Valuations
a	[.2, .4]
b	.2
c	[.1, .3]
d	[.3, .4]
e	.5
f	[.4, .5]
g	[.1, .3]
h	[.2, .4]
i	.3
j	[.3, .5]
k	.6
l	[.4, .5]

Once we had arrived at these opinions, we moved on to construct an "experton"[5]. For this we followed three stages:

1) Obtaining the number of times the experts gave the same value as the lower extreme and also the same value as the upper extreme. Here two experts considered 0.1 as the lower extreme; three came up with 0.2... As the upper extreme three provided us with 0.4; four 0.5,... We put this statistic together in a table as follows:

0		
.1	2	
.2	3	1
.3	3	3
.4	2	3
.5	1	4
.6	1	1
.7		
.8		
.9		
1		

statistic

A simple glance at this table brings to light the fact that the counter-experts are inclined for the lower numbers that is shorter time limits, for the migratory influx.

2) From this latter "statistic" the relative frequencies or normalised frequencies are arrived at. For this we divide each number of the preceding table by the total number of counter-experts, that is 12. In this way we arrive at the percentage of times that the same value has been estimated relative to the total. Here we arrive at:

0		
.1	17	
.2	.25	.08
.3	.25	.25
.4	.17	.25
.5	.08	.34
.6	.08	.08
.7		
.8		
.9		
1		

relative frequencies

As it could be in no other way, the sum of each of the two columns is equal to the unit. The values inside the table (to the left and to the right) represent, in "a certain way", the probability that the value of the lower or upper extreme be 0.1, 0.2,... 0.6.

3) Finally we are going to *accumulate* these relative frequencies commencing the accumulation from to bottom to the top (from 1 to 0, instead of doing it from 0 to 1). In this way we arrive at the complementary accumulated frequencies. These are:

$\underset{\sim}{a}=$		
0	1	1
.1	1	1
.2	.83	1
.3	.58	.92
.4	.33	.67
.5	.16	.42
.6	.08	.08
.7	0	0
.8	0	0
.9	0	0
1	0	0

Complementary accumulated frequencies

Nearly in an imperceptible manner we have entered into one of the more general theories of the arithmetic of uncertainty: the "experton theory". An experton (represented by the complementary accumulated frequencies) is an uncertain number that generalises random fuzzy numbers, fuzzy numbers and confidence intervals. It constitutes the *aggregation* of numerable opinions maintaining all the information and allowing for later operation either linear or non linear (which does not occur when aggregating by means of an average).

A superficial observation brings to light what was seen already with the initial opinions of the counter-experts: a concentration around 0.2 and 0.3 for the lower extreme and around 0.3, 0.4, 0.5 for the upper extreme. And this is important because if the "expected value" of this experton is arrived at (which is nothing else but an average) the information provided by it does not respond to the actual positioning of the "counter-experts". We are now going to verify this.

In order to find the expected value of an experton it is sufficient to add the values of the left-hand column, except for the value corresponding to 0, and to divide this by ten. The same is done with the right-hand column[6]. The result is as follows:

$$^{1}/_{10}\,(1+0.83+0.58+0.33+0.16+0.08) = {}^{2.98}/_{10} = 0.298$$

$$^{1}/_{10}\,(1+1+0.92+0.67+0.42+0.08) = {}^{4.09}/_{10} = 0.409$$

Therefore, the expected value of the experton is represented by the confidence interval:

$$[.298, .409]$$

This information "summarises" the experton by complementing the information provided by the same. It is customary to place the expected value below the experton:

	0	1	
	.1	1	
	.2	.83	1
	.3	.58	.92
	.4	.33	.67
$\underset{\sim}{a}$ =	.5	.16	.42
	.6	.08	.08
	.7	0	0
	.8	0	0
	.9	0	0
	1	0	0
		[.298 ,	.409]

After having developed this, Chris appeared to be a little lost with all this plethora of numbers. So a brief parenthesis was in order so as to summarise the path we had taken and to find out at what point we were.

From the opinion of several experts we had arrived at, by following the Fuzzy-Delphi method (aggregating opinions by means of the average), the estimate of the number of years neccssary for the migratory influx that would be capable of providing the minimum number of clients necessary for economic operation of the business. Several numbers represented this date: the triplet (3.05, 4.45, 5.55), the exact number 4.38.

In order to be surer and in an attempt to reduce the interval [3.05, 5.55] resort was made to the opinion of other experts whom we called counter-experts. A new technique, that of expertons, allowed for adequate aggregatin that positioned these counter-experts in the segment [0, 1] represented by 0 for the lower limit 3.05 and by 1 for the upper limit 5.55. What remained now was to convert the values in [0, 1] of the experton (and also its expected value) into moments in time and, therefore, in values within the field of positive real numbers. This could be attained by using an equation that we accept as linear, such as the following:

$$\underset{\sim}{A} = 3.05 \ (+) \ (5.55\text{-}3.05) \ (\cdot) \ \underset{\sim}{a}$$

By substituting the experton $\underset{\sim}{a}$ for its values we arrive at:

$\underset{\sim}{A} = 3.05 \ (+) \ 2.50 \ (\cdot)$

α	$\underset{\sim}{a}$		α	$\underset{\sim}{A}$	
0	1		0	5.55	
.1	1		.1	5.55	
.2	.83	1	.2	5.13	5.55
.3	.58	.92	.3	4.50	5.35
.4	.33	.67	.4	3.88	4.73
.5	.16	.42	.5	3.45	4.10
.6	.08	.08	.6	3.25	3.25
.7	0	0	.7	3.05	3.05
.8	0	0	.8	3.05	3.05
.9	0	0	.9	3.05	3.05
1	0	0	1	3.05	3.05
	[.298	.409]	=	[3.80	4.07]

Here then is the result in the form of an R^+-experton[7] in which it will be seen that the upper limit has been considerably reduced, as a consequence of the fact that the counter-experts have situated their opinions by reducing them more so than the expert opinions for the maximum time necessary for arriving at the pre-set migratory contingent.

If the expected value of the R^+-experton is taken as representative, that is to say the confidence interval [3.80, 4.07] it can be seen quite clearly that the time available to Chris for starting up his sports business was approximately four years. In this time he would have to set up the business under any of the legally existing forms, look for, if he so required, sources of necessary financing, make investments in installations, transport materials, etc., take on adequate personnel for efficient operation of his activity, prepare the marketing strategy.. and many other matters that constitute the tasks of a good promoter and better businessman, both qualities held by this dynamic young man, Chris. We wish him every success.

REFERENCES

1 In some cases the Minkowski difference does not lead to a solution in the form of a confidence interval

2 KAUFMANN, A. and GIL ALUJA, J.: Intoduccion a la teoría de subconjuntos borrosos a la gestión de las empresas. Publ. Milladoiro, Santiago de Compostela 1986, page 217-219

3 DALKEY, N., BROWN, B., COCHRAN, S.: *La prévision à long terme par la méthode Delphi*. Publ. Dunod. Paris 1972.

4 KAUFMANN, A. and GIL ALUJA, J.: *Técnicas especiales para la gestión de expertos*. Publ. Milladoiro, Santiago de Compostela 1993, page 263.

5 KAUFMANN, A. and GIL ALUJA, J.: *Técnicas de gestión de empresas. Previsiones, decisiones y estrategias*. Publ. Pirámide, Madrid 1992, pages 37-55.

6 KAUFMANN, A. and GIL ALUJA, J.: *Técnicas de gestión de empresas. Previsiones, decisiones y estrategias*. Publ. Pirámide, Madrid 1992, pages 242.

7 The denomination of the R± experton is due to the fact that it is an experton in R+, that is in the filed of positive real numbers.

III THE INVESTMENT PROCESS

Investment requirements in business

In order for a business to be able to carry out its economic activity it requires a series of basic elements among which stand out the buildings, business premises, installations, machinery, furnishings and other work instruments, all of which have a common characteristic: their duration (useful life) continues over several accounting periods. Their acquisition, upkeep and operation are the object of study by means of the so-called theory of investment.

Large industrial businesses devote large amounts of money to the purchase of "capital goods". And what is more, as a consequence of enormous technological advances, increasingly with greater frequency the participation of these capital goods in the production process is *percentage-wise* much higher. In days gone by, the "labour" contained in a product (time devoted by people to production) was much closer to the cost price of obtaining the same. Today machines do it all ... or nearly all of it. But machines have to be purchased and their acquisition got right, that is, take the optimum decision taking into consideration their characteristics and their *economic consequences*. The same occurs with land, offices, data processing equipment, industrial equipment, implements, tools, etc...

This tendency towards automation, characteristic of our day and age, would appear not to be remitting, but all to the contrary. Increasingly we will see how our office or workshop companions are being converted into computers or robots. At first it was the most repetitive, heavy or dangerous tasks. Later machines have occupied those areas in which calculation and memory was required. And tomorrow ... will some metal thing be able to substitute the human being in those acts where affection and imagination are essential? Where are the limits of the so-called "artificial intelligence"? In any event, the large concentrations of workers and employees are nearing their end. The "era of the proletariat" has given way to the "era of the robotariat".

For these and other reasons we are interested in the investment process. In large businesses, but also in medium and small businesses. In fact what would appear to be little for some is a lot for others. All, within their own dimension, must look for the way to "optimise" their resources, whether these amount to thousands of millions or in a few tens of thousands of monetary units. And this is where investment studies play their part in helping, where possible, to take the right decision.

Investment processes are present throughout the life of people and businesses. The setting up of a business, its expansion, renewal of its installations, a change in activity, are a few of the many examples that could be given. I remember, a few years ago the opportunity arose for verifying the reality of the usefulness that certain operating techniques can be as a support for taking investment decisions. Joe and Mary at that time carried on their business of selling wood that had been initiated by Joe's father. They frequently commented, at the usual Saturday coffee we had together, on the competition they had to face as a consequence of the setting up in their area of the activity of a large distribution and sales warehouse, on the one hand, and by the use of substitute materials in the construction of offices and homes, on the other. A time came when they felt that a change in direction of their business was necessary. But this meant for them, important *investments*, which required not only disposing of their savings but also resorting, at least for a time, to bank financing. As they were knowledgeable about the commercial activity and were aware of the "deficit" in the market of certain home and office furnishing and decorating products, they had decided to transform their purely commercial premises, into a manufacturing, assembly and sales centre for a determined range of furniture, ornaments and other decorating articles, such as curtains and carpets, among others.

The extension and transformation of the building, as well as the acquisition of the machinery and tools required certain almost immediate payments, but also arising, with a regular or an irregular rhythm, would be payments in each of the future periods, as a consequence of the operation of the new business. The question that was being asked was if the sales and corresponding collections would be sufficiently important for compensating these large economic sacrifices. What really worried

Joe and Mary was knowing if the flow of collections, overall, would be *higher* than the payment flow, also considered overall, over a certain interval of time, estimated as their economic horizon. Almost without knowing it, my dear friends had brought up the very genuine problem of investment selection. They had the kindness to invite me to a good meal and even better bottle of wine, as they said, to do "our numbers" and see if taking on this investment adventure was a good thing, or on the contrary not recommendable. As required by ritual I handed them a book1, neatly packed, which they duly opened and thanked me for, and then took a quick look at, after which you should have just seen their faces. They timidly asked me if it were not possible to give them a brief summary, in order for them to get an idea, although very superficial, on what is the theory of investment and to know how to use it.

We started with the term itself, "investment". In specialised works on the subject the word *investment* appears either as a reference to the *decision of investing* (I am going to make a good investment), or in order to talk about the invested asset (this work of art is a good investment). Therefore the very first thing we must do when we use the term investment is to specify if we are talking about an *act* (decision to invest) or an *object* (invested asset). As an example we could say that Joe and Mary are *the subjects* of the investment, each machine, building or installation that has to be purchased is an *object* of the investment, and the monetary consequences of the act of decision are an *outflow* of payments and an estimated *inflow of collections*. Now then there have been hundreds of books written on the subject and thousands of articles published on the same, all of which preferentially centre their study on the two currents: that of collections and that of payments, arising from the decision to invest in an investment object. Perhaps this is not all, but certainly a large part, which is known by the name of "investment selection".

The answer to this "speech" was only to be expected: in practice, what did this selection consist of?. To put it another way, what do we select?. Very approximately, as would be normal in the conversation we were holding we could say that the problem of selection (which is what we are interested in here) springs up in two different but connected, spheres. The first of these consists in determining, faced with an investment object, if in itself, *it is* or not convenient from an economical point of view. The second is to establish, among several investment objects, which is the best, obviously from an economical point of view. For now we will remain with these two cases.

Simple questions such as we have asked here, deserve certain simple criteria for their resolution. In a way we could say that an investment object is convenient from an economical point of view when the collection inflow *overall* is not less than the payment outflow, likewise considered overall. When it is possible to select between several investment objects the one to choose is the one where the difference between the collection and payments flows, both considered *overall* is the greatest. And we could have stopped here, enjoying, that is, the magnificent Priorato red wine. Joe and Mary were struck dumb. It cannot be that this is so simple. And in fact, it was not possible, but it was not so due merely due to a word: *overall.* But we would see this a bit further on. Now they had to tackle the task of *numerically estimating* the inflow or inflows of collections and the payment outflow or outflows.

During this phase of the work two aspects appear which are highly important. The first refers to the contents of each of the flows, that is, the concepts that make up the one and the other (cash payment for the "object", payments for yearly hours of work for its operation, payments for raw materials used,... cash sales of the products, delayed payments for the same,...) The second concerns the determination of the numbers in which the estimates are to be made (certain if confident of their strict occurrence; random when there are available valid statistics from the past; uncertain, in the event of data that is not very or not at all objective). Many questions that were put on the table that warranted, at least, some thinking about, We finished the very pleasant meeting, without forgetting that still requiring an answer was the "enigmatic" word *overall.*

Incorporation of interest rates

It was as a result of the following meeting, when Joe insisted in his desire for clarification of the meaning, which appeared fundamental, of the word *overall*. In the first instance, I clarified the fact that it was about a manner of expressing a customary behaviour, which, basically, forms a part of a sort of "natural law" known as *the law of sub-estimating future requirements.*

Having said this, I was once again aware of the fact, that I was allowing myself to be led by an excess of abstraction. And what Joe and Mary really wanted were very clear ideas, and ideas that were as close as possible to strictest reality. This then was the road we had to take. For this we commenced by saying: let us imagine a certain quantity of money, for example, a million monetary units. Do you prefer having them today or within two years? "Obviously today". Then if we generalise this idea a little we will reach the conclusion that as we move away in time the same amount of monetary means, is estimated less. And advancing even further, we could add that "a monetary mass located in different moments in time can be considered as if it was different goods" At one time as if it were oranges, at another apples, at even another butter, and further on chickpeas, only to mention edible products. Here, I think, we have arrived at the hub of our problem.

At the meal the other day, it appeared to me that when we established as a selection criterion that the collection inflow was no less than the payment outflow, Mary murmured: we add collections, add payments and we will immediately see if the "object" is economically valid. And now I ask, can you add apples, oranges, butter and chickpeas? I do not see how this could be done as we are dealing with heterogeneous products. Clearly, in a way, we do arrive at a *kind of addition*, but for this we must resort to their *value*, that is, to assigning a "price" to each product. The product of the price by its amount, homogenises in terms of value and, only then, is there any sense in doing the addition. The result of this addition allows us to say that we have bought (or sold, as the case may be) food products for a value of 3,000 monetary units. Therefore, we must do something similar in the case of the monetary mass to be collected or paid, located at different moments in time, as a consequence of the decision to invest. I expect you have now guessed that the role of the price will be occupied, in our case by the "interest rate".

If we place ourselves at the commencement of any year, 1 monetary unit today becomes in a year 1+0.03, when the annual interest rate is estimated at 3%. Theorists write this 1.03 as $1+i_1$, where i_1, is the interest rate for year 1. When the interest rate i_2 for the second year is estimated at 2%, then the monetary unit at the start will be equivalent to (1.03).(1.02) and this is written as $(1+i_1)(1+i_2)$. And so on successively, when the interest rate for year five is estimated at 4% we will have (1.03).(1.02). ... .(1.04) and this is expressed, in general terms as $(1+i_1).(1+i_2)$. ... $.(1+i_5)$. Notice that what we have stated signifies nothing else than locating a monetary unit from moment 0 in year 1, in 2, ... in 5 as can be seen from the following

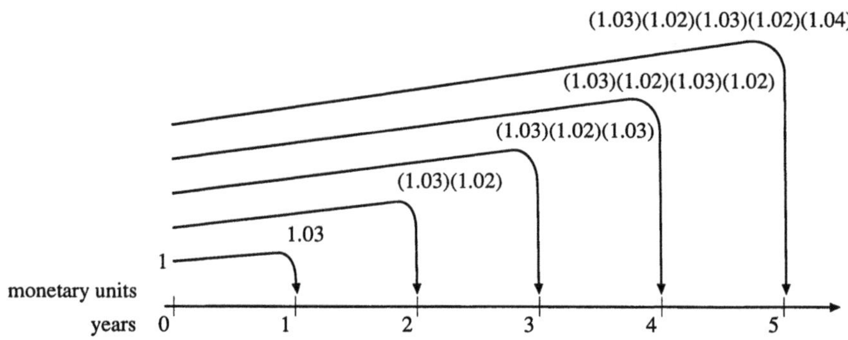

Now then, this approach has its double, in the sense of locating a monetary unit from year five in year 4, in 3, in 2 in 1 and at the initial moment 0. For this it is sufficient to use the simple *rule of three*. What we have then is 1 monetary unit from year 1 is equivalent to 1/(1+0.03) in year 0. A monetary unit from year 2 corresponds to 1/(1+0.03)(1+0.02) at source 0. And so on successively. One monetary unit from year five is the same thing as 1/(1.03).(1.02).(1.03).(1.02).(1.04). In general terms this is written as: $1/(1+i_1).(1+i_2)$. ... $.(1+i_5)$. This equivalence is shown in the following:

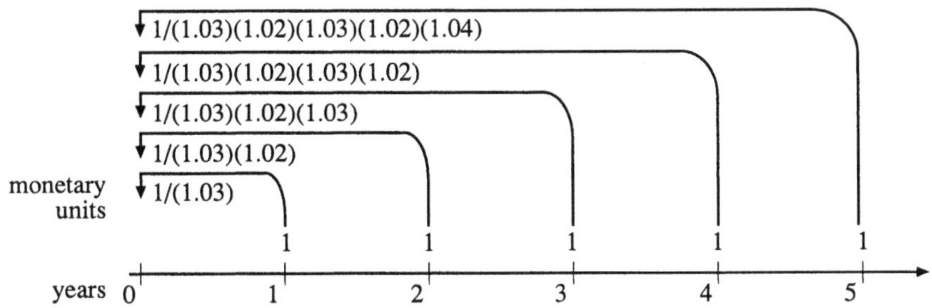

With this brief illustration we have now entered the *technique for updating income*. If instead of considering a monetary unit, what is wanted is to "transfer" from a future moment *t* to the initial moment 0 a mass of M units, for example, 30,000, it will be sufficient to multiply this money by a factor which is normally called *updating coefficient*. For our example, when *t* is equal to 1 year, 2 years, ... 5 years, we would have the following coefficients:

$$1/(1.03),\ 1/(1.03)\cdot(1.02),\ldots,\ 1/(1.03)\cdot(1.02)\cdot(1.03)\cdot(1.02)\cdot(1.04)$$

In a general way this is written as:

$$1/(1+i_1),\ 1/(1+i_1)\cdot(1+i_2),\ldots,\ 1/(1+i_1)\cdot(1+i_2)\cdot(1+i_3)\cdot(1+i_4)\cdot(1+i_5)$$

In theory, when it is considered that in the economic horizon there will be stability in the money markets, a simplifying hypothesis is adopted according to which the interest rate will be the *same* for all the periods (many people use this clause incorrectly when they are not capable or cannot make any valid estimates period by period). Therefore the above formulae are simplified and become as follows:

$$1/(1+i),\ 1/(1+i)^2,\ \ldots.\ ,\ 1/(1+i)^5$$

as:

$$i_1 = i_2 = \ldots. = i_5 = i$$

In an economic system such as we are living with, it is quite unthinkable that certain types of interest could be the same year after year. We find ourselves in a world submitted to great convulsions, which make economic and financial perspectives vary very rapidly, both for individuals as for businesses and institutions. For this reason estimating interest rates to be used in investment selection has become one of the problems that are most difficult to resolve.

And this ties in with the two aspects we pointed out at the end of the lunch the other day. Joe and Mary recalled that it was necessary to determine the *type of number* they should consider for representing the significant values of the investment process. Without a doubt that one of these values and not the least important is the interest rate.

Uncertain numbers

In the examples we have used up to now, we have started out from the presumption according to which it was possible to know with "certainty" what interest rates there were going to be in the future. For this we established for the first year 3%, 2% for year 2, ..., 4% for year 5. This is all very well in theory, said Joe, but for our specific problem it is extremely difficult, if we want to do things correctly, to set with any kind of certainty not only the interest rates there are going to be over the next few years, but also what payments will be made, and even more so, what we are going to receive from our sales (collections). How right he was! Fortunately today we have available, a large range of numbers that are capable of giving a response to the question posed by Joe and Mary. Only an *explanation*, even though very superficial, could calm our friends down. Were they ready for this "sacrifice"? I was only too willing to give them a brief description of uncertain numbers and, among them those most used. Schematically we made the following list:

a) **Confidence interval.** This is the simplest uncertain number, which "only" *puts limits* on the uncertainty. Let us take a look at an example, which we feel is useful. We can estimate the interest rate for the next year by saying that it will not be less than 2% nor greater than 4%. In this way we have established two *limits*. We are convinced that in reality the interest rate for the following year will be located somewhere between these two limits, although we know absolutely *nothing*. Nothing. Total lack of any knowledge. It is normal to write this as follows:

$$[0.02\ ,\ 0.04]$$

and in general theoretical representations:

$$[a_1\ ,\ a_2]$$

Now then, the interval, although uncertain, is a "number". Therefore it will be necessary to know how to do those operations common to all numbers. The most elementary were the ones we required for our objective, addition, subtraction, product and division.

I recalled that I found myself under a similar situation, when with Chris we had to resort to these operators. I referred my friends to the explanation given at that time[2].

With that my friends Joe and Mary found out how easy it was to operate with confidence intervals, if the information one has available only allows for limiting uncertainty[3]. But there are more to choose from. When greater information is available other uncertain numbers can be used, such as confidence triplets.

In fact, let us imagine that from a value, for example, interest rate, apart from estimating the extremes, one has information on the number, which it is felt has the greatest strength, with the maximum intensity, the *maximum presumption* of what is going to happen[4]. In this case three values will be available: Lower extreme or limit, maximum presumption, upper extreme or limit. We are then in the presence of a confidence triplet. In our case the interest rate for the next year, as well as estimating, for example, that it will not be less than 2% or higher than 4%, it is very strongly felt (maximum presumption) that it will be 2.7%. This then is written as:

$$(0.02,\ 0.027,\ 0.04)$$

and in general:

$$(a_1, a_2, a_3)$$

Operating with confidence triplets is very simple, if confidence interval arithmetic is known, since the same is done with the extremes and with the maximum presumptions which can be added, subtracted, multiplied and divided, according to which operation is required. Let us take a quick look at an example for each one of these operations:

$$(20, 21, 24)\ (+)\ (15, 17, 18) = (35, 38, 42)$$

$$(375, 400, 450)\ (-)\ (350, 400, 410) = (-35, 0, 100)$$

$$(375, 400, 450)\ (\overline{\text{m}})\ (350, 400, 410) = (25, 0, 40)$$

$$(20, 21, 22)\ (\cdot)\ (3000, 3100, 3200) = (60000, 65100, 70400)$$

$$(10, 20, 25)\ (:)\ (2, 3.5, 4) = (2.5, 5.71, 12.5)$$

$$(10, 20, 25)\ (\underset{\text{m}}{:})\ (2, 3.5, 4) = (5, 5.71, 6.25)$$

When talking of confidence triplets it is also normal to refer to another uncertain number, which is known by the name of extended triplet or "confidence quadruple".

The quadruple is differentiated from the triplet since the maximum presumption is not now a single number but an interval. It represents the case, which frequently appears, where the expert or person responsible for supplying this information is not convinced enough to give a single number of presumption and prefers to say: I feel that the value requested will be between such and such. In the oft repeated case of interest rates for the next year he would say: not less than 2% nor more than 4% and the maximum presumption would be between 2.5% and 3%. This then would be written as:

$$(0.02, [0.025, 0.03], 0.04)$$

and in the general presentation:

$$(a_1, [a_2, a_3], a_4)$$

Having reached this point, mention should be made of a matter, which, later on, will acquire special interest: the *levels of presumption*. Normally, presumption, or the belief in a value, if you like, although this is not quite correct, is presented by means of a "grading" in levels, from nil presumption to the maximum presumption. Likewise, the habit also exists of establishing a correspondence between a number and words normally used in every day language. This is what is known as the "semantic scale". The scale that is most used is the one based on the *hendecagonal system* (eleven positions). Let us take a look at a scale of this type in order for us to refer to the levels of presumption. The values will be taken from 0 to one by means of tenths, that is values in the segment [0, 1] as mathematicians would say:

1 : Maximum presumption

0.9 : a great deal of presumption

0.8 : a lot of presumption

0.7 : high degree of presumption

0.6 : quite a bit of presumption

0.5 : median presumption

0.4 : relative presumption

0.3 : low presumption

0.2 : little presumption

0.1 : very little presumption

0 : no presumption

Having accepted this scale, or even another which you could draw up yourselves, Joe and Mary, in accordance with your normal communication, we can now better understand the reason why confidence intervals as well as triplets and quadruples are often represented by a system of coordinates. In fact, if the estimated values are drawn with abscissas and the levels of presumption as ordinates, a confidence interval such as [0.02, 0-04] can be represented as follows:

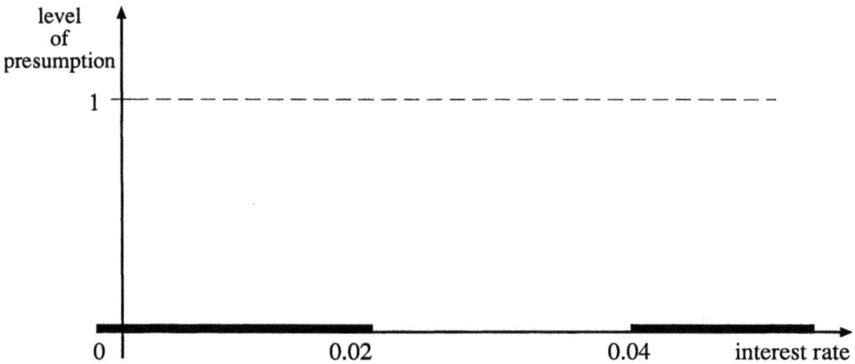

A triplet, such as for example (0.02, 0.027, 0.04) would be "explained" in the following figure:

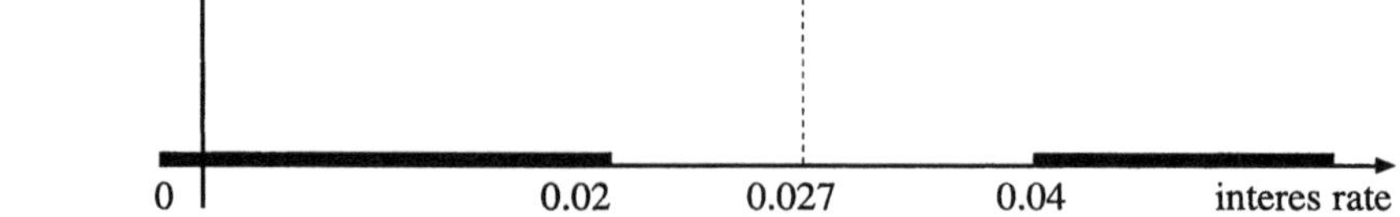

The quadruple (0.02, [0.025, 0.03], 0.04) could be represented as follows:

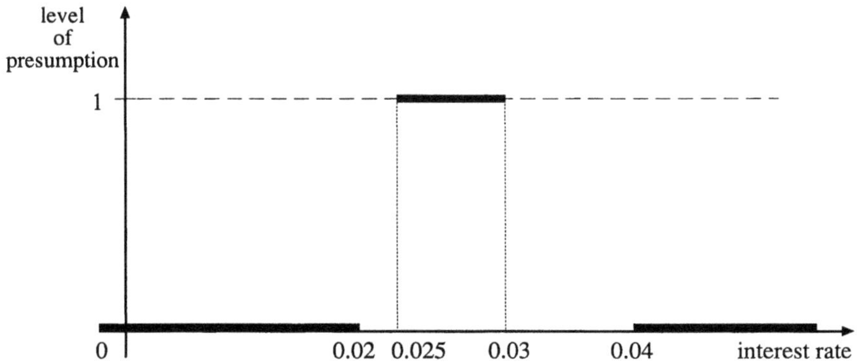

We can see that in the description of these three uncertain numbers we have only considered levels of presumption 0 (the minimum) and 1 (the maximum). But it is also possible to introduce intermediary levels. If as we move away from the maximum presumption towards the extremes the levels of presumption decrease (or they are maintained) we are faced with a fuzzy number.

a) ***Fuzzy Triangular Number.*** When it is possible to estimate all the levels of presumption between the extremes the fuzzy number arises. The following graph represents a fuzzy number:

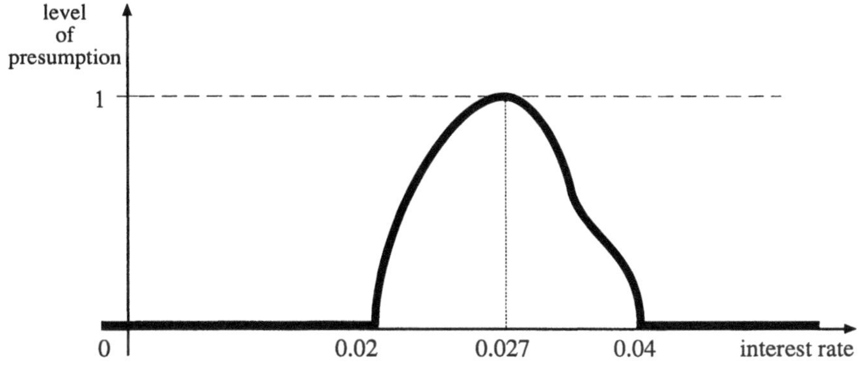

Dear friends do note that for each abscissa value between 0.02 and 0.04 there is a height that touches the curve. This height represents the level of presumption that the corresponding value possess, in our example of the interest rate.

Among fuzzy numbers there is one that by its simplicity and operation acquires special relevance. We are referring to a fuzzy triangular number (F.T.N.). The F.T.N.is characterised by the fact that from the maximum level of presumption the levels decrease proportionally, both to the left and to the right, even when they can decrease in a different proportion from one side to the other. It is precisely due to this fact that the geometric figure that it forms is that of a triangle and from there stems its name of fuzzy triangular number[5].

In the following figure we represent a fuzzy triangular number (F.T.N.)

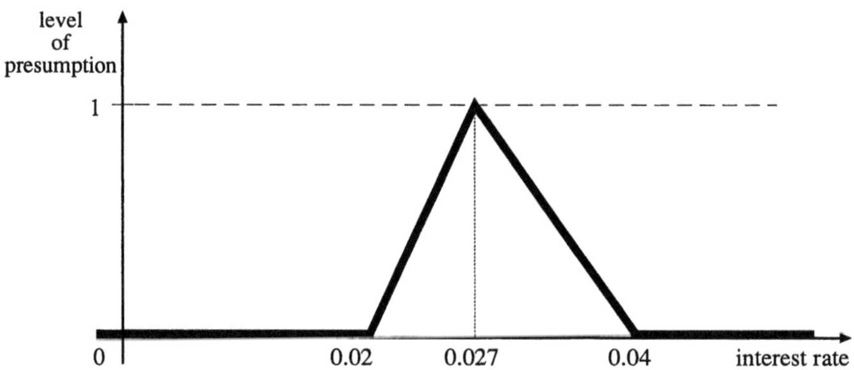

The F.T.N. can be expressed in different forms, the simplest and the one that is best adapted to the needs of non-experts is called the ternary form. If we designate the previous fuzzy triangular number by $\underset{\sim}{A}$ (capital letter with a swung dash below it), we can write:

$$\underset{\sim}{A} = (0.02, 0.027, 0.04)$$

Further on, when needed, we will take a look at operations with F.T.N.

b) ***Fuzzy Trapezoidal number.*** Another much sought after fuzzy number is the number that stems from the ideas of a confidence quadruple. In fact, if from the extremes of an interval of maximum presumption lines are drawn to the extremes of the quadruple we will arrive at the figure of a trapezoid. This is where the denomination for this fuzzy number F.Tp.N. comes from. Below we have shown the graphical representation of the fuzzy trapezoidal number (0.02, [0.025, 0.03], 0.04)

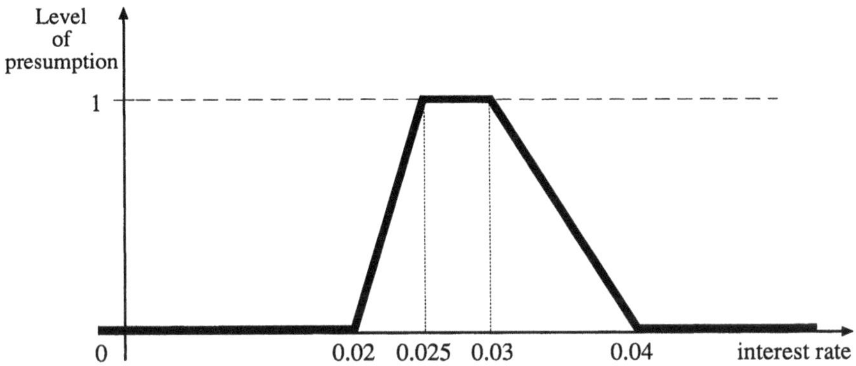

As in the case of the fuzzy triangular number, the F.Tp.N. shows us, between 0.02 and 0.04 the levels of presumption of each of the estimates for the interest rate by means of their respective heights of the trapezoid at each point. It goes without saying that for any values less than the unitary rate of 0.02 and higher than 0.04 presumption is nil. It is felt that these cannot occur.

Obviously other uncertain numbers could be shown, nevertheless I feel that we would greatly exceed our objective.

Joe and Mary now had in front of them a wide range of uncertain numbers from which to make their estimates and calculation relative to their investment project. There was even the possibility of taking different numbers for the different values, since, under certain conditions it is possible to operate with different types of uncertain numbers. They did not take long in expressing their preference for the fuzzy triangular number (F.T.N.).

Updating rates from fuzzy triangular numbers

Having taken this decision, with sound judgement, Mary suggested that we should commence using thes F.T.N. for estimating the interest rate, as they were quite clear about all we had stated on this phenomenon. After a number of consultations with experts they assumed the following interest rates for the following 5 years:

$$\underset{\sim}{i}_1 = (0.02, 0.03, 0.04)$$

$$\underset{\sim}{i}_2 = (0.015, 0.02, 0.03)$$

$$\underset{\sim}{i}_3 = (0.025, 0.03, 0.04)$$

$$\underset{\sim}{i}_4 = (0.01, 0.02, 0.03)$$

$$\underset{\sim}{i}_5 = (0.02, 0.04, 0.05)$$

Following the same path to the one already take in the sphere of certainty, the fuzzy updating rates could be arrived at as follows:

$$\frac{1}{1(+)\underset{\sim}{i}_1} = \frac{1}{1(+)(0.02, 0.03, 0.04)} = \frac{1}{(1.02, 1.03, 1.04)}$$

$$\frac{1}{1(+)\underset{\sim}{i}_2} = \frac{1}{1(+)(0.015, 0.02, 0.03)} = \frac{1}{(1.015, 1.02, 1.03)}$$

$$\frac{1}{1(+)\underset{\sim}{i}_3} = \frac{1}{1(+)(0.025, 0.03, 0.04)} = \frac{1}{(1.025, 1.03, 1.04)}$$

$$\frac{1}{1(+)\underset{\sim}{i}_4} = \frac{1}{1(+)(0.01, 0.02, 0.03)} = \frac{1}{(1.01, 1.02, 1.03)}$$

$$\frac{1}{1(+)\underset{\sim}{i}_5} = \frac{1}{1(+)(0.02, 0.04, 0.05)} = \frac{1}{(1.02, 1.04, 1.05)}$$

On reaching this point, the first important problem arose. Was it necessary to carry out the division between triangular fuzzy numbers. How? said Joe. In effect, in the denominators there are F.T.N. but in the numerator I can only see ones and, as far as I know 1 is an exact number. That is so, but it is also a F.T.N., specifically (1, 1, 1). Therefore for example the updating rate can also be written as:

$$\frac{1}{1(+)\underset{\sim}{i}_1} = \frac{(1, 1, 1)}{(1.02, 1.03, 1.04)}$$

Even with this clarification the question remained. How to divide a F.T.N.?. If they had been confidence triplets the answer would have been immediate:

$$\frac{(1, 1, 1)}{(1.02, 1.03, 1.04)} = \left(\frac{1}{1.04}, \frac{1}{1.03}, \frac{1}{1.02}\right)$$

but as they are fuzzy triangular numbers this reply was not correct. And what is worse still, is that in order to arrive at an adequate solution we have to resort to expressing the F.T.N. in a different form to the ternary, specifically in the form called "α-cuts". This is equivalent to saying that it was necessary to know how to pass from one to the other form. For this we opted for developing a very simple example. We started out from the fuzzy triangular number in the ternary form $\underset{\sim}{A}$=(3, 5, 9) the graphical representation of which is as follows:

Actually, although in the ternary form only three numbers are expressed (hence its name) which in our example are 3, 5, 9 the existence underlies of another three values that accompany the first. Thus when expressing the lower extreme, here with the number 3, in reality *the point* (3, 0) should be understood. When showing the maximum presumption, now 5, *the point* (5, 1) is implicit. Finally, when saying that the upper extreme is, as in our example 9, t*he point* must be considered (9, 0). The first numbers of these pairs represent the numbers in abscissas; the second the values in ordinates.

Having briefly said this, we are now in a situation to remember that *through two points one and only one straight line passes.* In our example a straight line passes between the points (3, 0) and (5, 1) (that of the left hand side of the triangle) and another straight line between points (5, 1) and (9, 0) (the right hand side of the triangle). If we use the known equation of the straight line passing through two points, these two equations would be respectively:

$$\alpha = \frac{x-3}{2}$$

$$\alpha = \frac{9-x}{4}$$

If instead of expressing the α in function of the x, we find the value of x, we would arrive at the following functions of α :

$$x=3+2\alpha$$
$$x=9-4\alpha$$

Do not forget Joe and Mary that the first of these equations describes the left hand side of the triangle and the second the right hand side. Therefore *if we form an interval* place the first of these functions as the lower extreme and the second as the upper extreme as follows:

$$[3 + 2\alpha, 9 - 4\alpha]$$

the function interval expresses the F.T.N in the form called α-cuts. In fact by giving values to α between 1 and 0 (for example, in tenths), that is to say by cutting the triangle by means of straight lines that are parallel to the axis of the abscissa and at different heights, we arrive at further confidence intervals. First we will take a look at it in the following graphical representation:

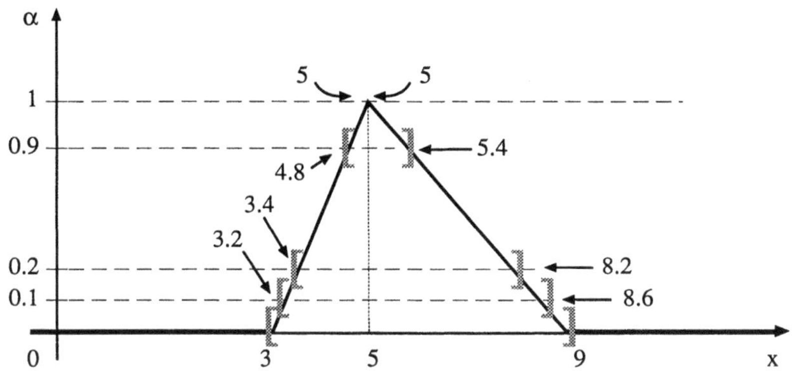

This is equivalent to giving values of between 1 and 0 to the α of the corresponding function interval. This then would read:

$$\alpha = 1 : [5, 5] = 5$$
$$\alpha = 0.9 : [4.8, 5.4]$$

....

$$\alpha = 0.2 : [3.4, 8.2]$$
$$\alpha = 0.1 : [3.2, 8.6]$$
$$\alpha = 0 : [3, 9]$$

After this, Joe dared to suggest: would there not be some simple and practical rule for passing over from the ternary form to the α-cuts form? We will now see if this is of any use.

If to the lower extreme of the ternary form the difference between the maximum presumption and this lower extreme is added, then this subtraction multiplied by α, we arrive at the lower extreme function in the form of α-cuts.

$$3 + (5 - 3)\,\alpha = 3 + 2\alpha$$

If to the upper extreme of the ternary form we subtract the difference betweens upper extreme and the maximum presumption, multiplying this difference by α, we arrive at the upper extreme function in the form of α-cuts.

$$9 - (9 - 5)\,\alpha = 9 - 4\alpha$$

It would appear to be quite simple to pass from one to the other form of expressing a F.T.N. and, nevertheless, with it not only is the division operation resolved but also any other that may be brought up. But always provided that the arithmetic of confidence intervals is a known factor. But you already know how to apply it adequately.

The only difference between operations with intervals and operations with F.T.N. can be found in the fact that numbers are added, subtracted, multiplied or divided in the first and functions in the second.

With what we have seen up to now I think that we have eliminated all the "technical" obstacles that could have cropped up in our work. Therefore let us now continue from where we left off, taking into account the difficulty, which arose at that time of the need of knowing how to divide the F.T.N. For this we will convert the quotients expressed in the ternary form into quotients shown in the form of α-cuts. In this way we arrive at:

$$\frac{1}{(1.02,\ 1.03,\ 1.04)} = \frac{[1,\ 1]}{[1.02+0.01\alpha,\ 1.04-0.01\alpha]} = \left[\frac{1}{1.04-0.01\alpha},\ \frac{1}{1.02+0.01\alpha}\right]$$

$$\frac{1}{(1.015,\ 1.02,\ 1.03)} = \frac{[1,\ 1]}{[1.015+0.005\alpha,\ 1.03-0.01\alpha]} = \left[\frac{1}{1.03-0.01\alpha},\ \frac{1}{1.015+0.005\alpha}\right]$$

$$\frac{1}{(1.025,\ 1.03,\ 1.04)} = \frac{[1,\ 1]}{[1.025+0.005\alpha,\ 1.04-0.01\alpha]} = \left[\frac{1}{1.04-0.01\alpha},\ \frac{1}{1.025+0.005\alpha}\right]$$

$$\frac{1}{(1.01,\ 1.02,\ 1.03)} = \frac{[1,\ 1]}{[1.01+0.01\alpha,\ 1.03-0.01\alpha]} = \left[\frac{1}{1.03-0.01\alpha},\ \frac{1}{1.01+0.01\alpha}\right]$$

$$\frac{1}{(1.02,\ 1.04,\ 1.05)} = \frac{[1,\ 1]}{[1.02+0.02\alpha,\ 1.05-0.01\alpha]} = \left[\frac{1}{1.05-0.01\alpha},\ \frac{1}{1.02+0.02\alpha}\right]$$

As you well know Joe and Mary, each one of these function intervals (fuzzy numbers) allows us to "transfer" any amount of money from the end of the corresponding year to the same time of the previous year. But what really interests us , is to "place" all the monetary masses at the same time and, if possible, as it is, place them at the initial moment (moment zero), even if only for the ease of calculation. The way of doing this is easy: Transfer from the last year to the previous year, accumulate by product the amount for that year and transfer to the previous year, and so on successively right up to the initial time. As, explained in this way it may seem complicated, we have clarified it with a drawing, before moving on to arriving at the *up-dating coefficients* for each year.

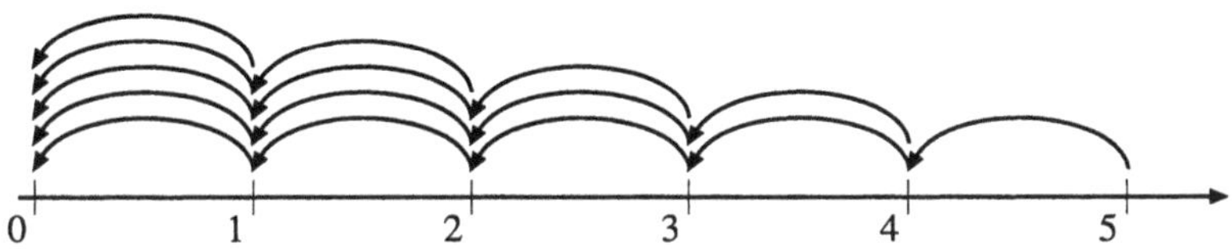

I think that we can now move on to express the uncertain up-dating coefficients relative to the case under study. From the end of the first year to the end of the fifth year, these are as follows:

$$\text{year 1}: \left[\frac{1}{1.04\text{-}0.01\alpha}, \frac{1}{1.02+0.01\alpha}\right]$$

$$\text{year 2}: \left[\frac{1}{1.04\text{-}0.01\alpha}, \frac{1}{1.02+0.01\alpha}\right](\cdot)\left[\frac{1}{1.03\text{-}0.01\alpha}, \frac{1}{1.015+0.005\alpha}\right]$$

$$\text{year 3}: \left[\frac{1}{1.04\text{-}0.01\alpha}, \frac{1}{1.02+0.01\alpha}\right](\cdot)\left[\frac{1}{1.03\text{-}0.01\alpha}, \frac{1}{1.015+0.005\alpha}\right](\cdot)$$
$$(\cdot)\left[\frac{1}{1.04\text{-}0.01\alpha}, \frac{1}{1.025+0.005\alpha}\right]$$

$$\text{year 4}: \left[\frac{1}{1.04\text{-}0.01\alpha}, \frac{1}{1.02+0.01\alpha}\right](\cdot)\left[\frac{1}{1.03\text{-}0.01\alpha}, \frac{1}{1.015+0.005\alpha}\right](\cdot)$$
$$(\cdot)\left[\frac{1}{1.04\text{-}0.01\alpha}, \frac{1}{1.025+0.005\alpha}\right](\cdot)\left[\frac{1}{1.03\text{-}0.01\alpha}, \frac{1}{1.01+0.01\alpha}\right]$$

$$\text{year 5}: \left[\frac{1}{1.04\text{-}0.01\alpha}, \frac{1}{1.02+0.01\alpha}\right](\cdot)\left[\frac{1}{1.03\text{-}0.01\alpha}, \frac{1}{1.015+0.005\alpha}\right](\cdot)$$
$$(\cdot)\left[\frac{1}{1.04\text{-}0.01\alpha}, \frac{1}{1.025+0.005\alpha}\right](\cdot)\left[\frac{1}{1.03\text{-}0.01\alpha}, \frac{1}{1.01+0.01\alpha}\right](\cdot)$$
$$(\cdot)\left[\frac{1}{1.05\text{-}0.01\alpha}, \frac{1}{1.02+0.02\alpha}\right]$$

Operations, relative to these coefficients, were in the final phase. In fact, we can see that, as from the second year, the only operator was the product from among the function intervals, which possessed the characteristic of all being positive. And we knew how to multiply them. It was sufficient to do so by the lower extremes between themselves and also the upper extremes between themselves. We did this and arrived at the following:

$$\text{year 1 } \left[\frac{1}{1.04\text{-}0.01\alpha}, \frac{1}{1.02\text{+}0.01\alpha}\right]$$

$$\text{year 2 : } \left[\frac{1}{1.04\text{-}0.01\alpha} \cdot \frac{1}{1.03\text{-}0.01\alpha}, \frac{1}{1.02\text{+}0.01\alpha} \cdot \frac{1}{1.015\text{+}0.005\alpha}\right]$$

$$\text{year 3 : } \left[\frac{1}{1.04\text{-}0.01\alpha} \cdot \frac{1}{1.03\text{-}0.01\alpha} \cdot \frac{1}{1.04\text{-}0.01\alpha}, \frac{1}{1.02\text{+}0.01\alpha} \cdot \frac{1}{1.015\text{+}0.005\alpha} \cdot \frac{1}{1.025\text{+}0.005\alpha}\right]$$

$$\text{year 4 : } \left[\frac{1}{1.04\text{-}0.01\alpha} \cdot \frac{1}{1.03\text{-}0.01\alpha} \cdot \frac{1}{1.04\text{-}0.01\alpha} \cdot \frac{1}{1.03\text{-}0.01\alpha}, \frac{1}{1.02\text{+}0.01\alpha} \cdot \frac{1}{1.015\text{+}0.005\alpha} \cdot \frac{1}{1.025\text{+}0.005\alpha} \cdot \frac{1}{1.01\text{+}0.01\alpha}\right]$$

$$\text{year 5 : } \left[\frac{1}{1.04\text{-}0.01\alpha} \cdot \frac{1}{1.03\text{-}0.01\alpha} \cdot \frac{1}{1.04\text{-}0.01\alpha} \cdot \frac{1}{1.03\text{-}0.01\alpha} \cdot \frac{1}{1.05\text{-}0.01\alpha}, \frac{1}{1.02\text{+}0.01\alpha} \cdot \frac{1}{1.015\text{+}0.005\alpha} \cdot \frac{1}{1.025\text{+}0.005\alpha} \cdot \frac{1}{1.01\text{+}0.01\alpha} \cdot \frac{1}{1.02\text{+}0.02\alpha}\right]$$

In this way, any mass of money, certain or uncertain estimated (by means of confidence intervals, triplets, quadruples, fuzzy numbers), located in any of the five periods of time can be "transferred" immediately by simply affecting the same (multiplying) by the uncertain up-dating coefficient corresponding to it according to the time of its occurrence. This is obviously valid both for the flow of collections as for payments both arising from the decision to invest.

Faced with this sequence of "formulae" Joe and Mary asked a question and made a complaint. The first was: the interest rates are expressed in F.T.N. and after their conversion into an up-dating coefficient, do they still continue to be fuzzy triangular numbers? No, and this is for the fundamental reason that the interest rates in order to convert them into up-dated rates have "undergone" *non-linear operations* as are the product and quotient. They accepted this answer and added not without reason, that these fuzzy numbers, even when not triangular they appear very much like this geometrical figure.

The complaint would have been made by any "businessman" having the patience for reading this work up to now: although the reasoning followed up to this point would appear to be correct on submitting it to certain executives these very abstract expressions ran the risk of being rejected and, suddenly, being branded as *theoretical and esoteric*. What was required was another kind of presentation. We opted for putting values to the a (carrying out the cuts) obviously in the segment of [0, 1]. We used, as is our custom, the 11 levels of the hendecagonal system, a = 1, 0.9, 0.8 ... 0.1, 0. Resorting to some tables would clarify the process we followed [6].

Up-dating coefficient for the first year:

α	$\frac{1}{1.04 - 0.01\alpha}$	$\frac{1}{1.02 + 0.01\alpha}$
1	.971	.971
.9	.970	.972
.8	.969	.973
.7	.968	.974
.6	.967	.975
.5	.966	.976
.4	.965	.977
.3	.964	.978
.2	.963	.978
.1	.962	.979
0	.962	.980

Up-dating coefficient for the secont year:

α	$\frac{1}{1.04-0.01\alpha}\cdot\frac{1}{1.03-0.01\alpha}$	$\frac{1}{1.02+0.01\alpha}\cdot\frac{1}{1.015+0.005\alpha}$
1	.952	.952
.9	.950	.953
.8	.948	.955
.7	.946	.956
.6	.944	.958
.5	.942	.959
.4	.941	.961
.3	.939	.962
.2	.937	.963
.1	.935	.964
0	.934	.966

Up-dating coefficient for the thirtd year:

α	$\frac{1}{1.04-0.01\alpha}\cdot\frac{1}{1.03-0.01\alpha}\cdot\frac{1}{1.04-0.01\alpha}$	$\frac{1}{1.02+0.01\alpha}\cdot\frac{1}{1.015+0.005\alpha}\cdot\frac{1}{1.025+0.005\alpha}$
1	.924	.924
.9	.922	.926
.8	.919	.928
.7	.916	.930
.6	.913	.932
.5	.910	.933
.4	.908	.936
.3	.905	.937
.2	.902	.939
.1	.899	.940
0	.898	.942

Up-datimg coefficient for the fourth year:

α	$\frac{1}{1.04-0.01\alpha}\cdot\frac{1}{1.03-0.01\alpha}\cdot\frac{1}{1.04-0.01\alpha}\cdot\frac{1}{1.03-0.01\alpha}$	$\frac{1}{1.02+0.01\alpha}\cdot\frac{1}{1.015+0.005\alpha}\cdot\frac{1}{1.025+0.005\alpha}\cdot\frac{1}{1.01+0.01\alpha}$
1	.906	.906
.9	.903	.909
.8	.899	.912
.7	.895	.914
.6	.892	.917
.5	.888	.919
.4	.885	.923
.3	.881	.925
.2	.877	.928
.1	.874	.930
0	.872	.933

Up-dating coefficient for the fifth year:

α	$\frac{1}{1.04-0.01\alpha}\cdot\frac{1}{1.03-0.01\alpha}\cdot\frac{1}{1.04-0.01\alpha}\cdot\frac{1}{1.03-0.01\alpha}\cdot\frac{1}{1.05-0.01\alpha}$	$\frac{1}{1.02+0.01\alpha}\cdot\frac{1}{1.015+0.005\alpha}\cdot\frac{1}{1.025+0.005\alpha}\cdot\frac{1}{1.01+0.01\alpha}\cdot\frac{1}{1.02+0.02\alpha}$
1	.871	.871
.9	.867	.876
.8	.863	.880
.7	.858	.884
.6	.854	.889
.5	.850	.892
.4	.846	.898
.3	.841	.902
.2	.837	.906
.1	.833	.910
0	.830	.915

Having arrived at the five fuzzy numbers that express the up-dating coefficients and before continuing with any later calculations, it may be useful, for Joe and Mary, to see that even when the geometrical forms are not that of a triangle they are very similar. This has frequently led to the fact that when this happens, the " *triangular approximations*" are considered and calculations are made, in a more simple manner, with confidence triplets. Having found the final result, the triplet is completed by joining the maximum presumption with the extremes, thus forming a F.T.N. In this way the *approximate* result is accepted as valid. As the calculations are not long I proposed to my friends that they obtain these *approximate up-dating rates.*

As we are dealing with *triplets* one should not (nor could) use any other form than the ternary and, in our case, as the estimated rates of interest were the ones already specified before, the successive confidence triplets representing the up-dating coefficients would be:

year 1 :

$$\frac{1}{1(+)\underset{\sim}{i}_1} = \frac{1}{(1.02, 1.03, 1.04)} = \left[\frac{1}{1.04}, \frac{1}{1.03}, \frac{1}{1.02}\right] = [.962, .971, .980]$$

year 2 :

$$\frac{1}{1(+)\underset{\sim}{i}_1}(\cdot)\frac{1}{1(+)\underset{\sim}{i}_2} = \frac{1}{(1.02, 1.03, 1.04)}(\cdot)\frac{1}{(1.015, 1.02, 1.03)} = \left[\frac{1}{1.04}, \frac{1}{1.03}, \frac{1}{1.02}\right](\cdot)\left[\frac{1}{1.03}, \frac{1}{1.02}, \frac{1}{1.015}\right] =$$

$$= \left[\frac{1}{1.04}(\cdot)\frac{1}{1.03}, \frac{1}{1.03}(\cdot)\frac{1}{1.02}, \frac{1}{1.02}(\cdot)\frac{1}{1.015}\right] = [.934, .952, .966]$$

year 3 :

$$\frac{1}{1(+)\underset{\sim}{i}_1}(\cdot)\frac{1}{1(+)\underset{\sim}{i}_2}(\cdot)\frac{1}{1(+)\underset{\sim}{i}_3} = \left[\frac{1}{1.04}(\cdot)\frac{1}{1.03}(\cdot)\frac{1}{1.04}, \frac{1}{1.03}(\cdot)\frac{1}{1.02}(\cdot)\frac{1}{1.03}, \frac{1}{1.02}(\cdot)\frac{1}{1.015}(\cdot)\frac{1}{1.025}\right] =$$

$$= [.898, .924, .942]$$

year 4 :

$$\frac{1}{1(+)\underset{\sim}{i}_1}(\cdot)\frac{1}{1(+)\underset{\sim}{i}_2}(\cdot)\frac{1}{1(+)\underset{\sim}{i}_3}(\cdot)\frac{1}{1(+)\underset{\sim}{i}_4} = \left[\frac{1}{1.04}(\cdot)\frac{1}{1.03}(\cdot)\frac{1}{1.04}(\cdot)\frac{1}{1.03}, \frac{1}{1.03}(\cdot)\frac{1}{1.02}(\cdot)\frac{1}{1.03}(\cdot)\frac{1}{1.02},\right.$$

$$\left., \frac{1}{1.02}(\cdot)\frac{1}{1.015}(\cdot)\frac{1}{1.025}(\cdot)\frac{1}{1.01}\right] = [.871, .906, .933]$$

year 5 :

$$\frac{1}{1(+)i_1}(\cdot)\frac{1}{1(+)i_2}(\cdot)\frac{1}{1(+)i_3}(\cdot)\frac{1}{1(+)i_4}(\cdot)\frac{1}{1(+)i_5}=\Big[\frac{1}{1.04}(\cdot)\frac{1}{1.03}(\cdot)\frac{1}{1.04}(\cdot)\frac{1}{1.03}(\cdot)\frac{1}{1.05},\frac{1}{1.03}(\cdot)\frac{1}{1.02}(\cdot)\frac{1}{1.03}(\cdot)\frac{1}{1.02}$$

$$,\frac{1}{1.02}(\cdot)\frac{1}{1.015}(\cdot)\frac{1}{1.025}(\cdot)\frac{1}{1.01}(\cdot)\frac{1}{1.02}\Big]=[.830,\ .871,\ .915]$$

A simple glance at these results brings to light the equivalence between the extremes and the maximum presumption of the *fuzzy numbers obtained* and the extremes and the maximum presumption of *the triplets*. If, in a graph, we join the extremes and the maximum presumption of the latter with straight lines, we will find certain triangular fuzzy numbers that constitute an *approximation* to the true results (fuzzy but no triangular numbers). From here stems the name of *triangular approximations*. For greater clarification below we have drawn the corresponding triangles in a system of coordinates:

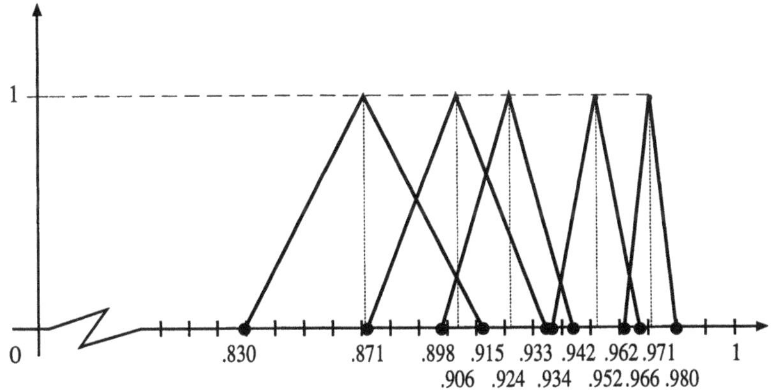

My friends remained thoughtful on seeing the five triangles. I instinctively felt that in their minds certain ideas were brewing, but when they thought about it they made two specific assertions: the first was relative to the fact that, as the F.T.N. became smaller the triangles moved towards the left; the second was that they saw an increase in the uncertainty (the bases of the triangles became larger) as the number of operations done with the uncertain numbers increased. I explained to them that both one and the other phenomenon did no more than automatically reflect aspects that are natural to ordinary routine. In fact, if instead of taking into account then F.T.N. they had taken into account numbers that are normally used, 7, 2, 4, 10.. and placed them along an axis, they would consider it as obvious that the 7 was on the left of the 10 because it is smaller, and the same would happen with the 4 and 2, each of which would be further to the left than the previous ones. The same occurs with the F.T.N.: the lower one to the left and the larger towards the right.

With regard to the increase in uncertainty when the calculations increase, this also is an aspect full of logic. When starting out with numerical information with a certain amount of uncertainty and to this information we add further more uncertain information, it would be quite normal for the accumulation to be reflected in the results obtained.

Estimating the results

Many other questions could have been asked due to the innate curiosity of Joe and Mary. But we had to make a cut-off somewhere. Maybe we had deviated a little too much from our objective. The result was that it was now urgent to make a decision in order to be able to continue with our study and answer the following question: should we continue the study with fuzzy numbers and then the result would be theoretically perfect; or was the use of confidence triplets that converted the result into triangular approximations, sufficient. In the first instance they decided to continue within the field of fuzzy numbers. In any event, as operating with triplets was easy and quick, they would use them later as a way of verifying the result. They now had to take into account the estimates for the other values of the investment process that is the payment and collections flows.

In this aspect, my friends had done some excellent work. They had taken the trouble to patiently accumulate the values representing each one of the cost components for each of the following five years, commencing with the initial "investment" as they called it. In this way they considered that for the new installations, adding machinery, installation, tools, transport, expansion and building transformation they were required to make a payment at commencement of the process of 120 million monetary units. In order to show this formally we decided to express it as follows:

$$\underset{\sim}{M} = (100, 120, 130) = [100 + 20\alpha, 130 - 10\alpha]$$

For each of the years, right up to the *economic horizon*, they took into account, wages and salaries, consumption of electricity, water, telephone, materials and raw materials to be used, office supplies, advertising,... All of this information was carefully added in order to provide the following annual F.T.N.:

$$\underset{\sim}{P}_1 = (50, 52, 55)$$

$$\underset{\sim}{P}_2 = (52, 56, 58)$$

$$\underset{\sim}{P}_3 = (52, 57, 60)$$

$$\underset{\sim}{P}_4 = (54, 60, 64)$$

$$\underset{\sim}{P}_5 = (56, 60, 68)$$

Each of these F.T.N. had to be multiplied by the fuzzy numbers describing the updating rates. For this it was necessary to transform them into α-cuts. Joe and Mary now knew how to do this:

$$\underset{\sim}{P}_1 = (50+2\alpha, 55-3\alpha)$$

$$\underset{\sim}{P}_2 = (52+4\alpha, 58-2\alpha)$$

$$\underset{\sim}{P}_3 = (52+5\alpha, 60-3\alpha)$$

$$\underset{\sim}{P}_4 = (54+6\alpha, 64-4\alpha)$$

$$\underset{\sim}{P}_5 = (56+4\alpha, 68-8\alpha)$$

In the same way they had sufficient experience and foresight in order to be able to estimate collections resulting from sales that they hoped for over the next five years. They also submitted the following numbers for our consideration in the F.T.N. form:

$$\underset{\sim}{C}_1 = (80, 85, 88)$$

$$\underset{\sim}{C}_2 = (84, 86, 90)$$

$$\underset{\sim}{C}_3 = (88, 90, 92)$$

$$\underset{\sim}{C}_4 = (90, 92, 95)$$

$$\underset{\sim}{C}_5 = (90, 94, 96)$$

For the same reason that existed for the case of payments, here it was also essential to express these F.T.N. in α-cuts. These are included below:

$$\underset{\sim}{C}_1 = [80 + 5\alpha, 88 - 3\alpha]$$

$$\underset{\sim}{C}_2 = [84 + 2\alpha, 90 - 4\alpha]$$

$$\underset{\sim}{C}_3 = [88 + 2\alpha, 92 - 2\alpha]$$

$$\underset{\sim}{C}_4 = [90 + 2\alpha, 95 - 3\alpha]$$

$$\underset{\sim}{C}_5 = [90 + 4\alpha, 96 - 2\alpha]$$

As we now had available estimates for payments and also estimates for collections, we could now find, by simple subtraction, the yearly "surpluses" which should, at the very least, compensate the initial payments arising from the investment which we have represented by M. By using the ternary form we arrive at:

$$\underset{\sim}{M}_1 = \underset{\sim}{C}_1 (-) \underset{\sim}{P}_1 = (80, 85, 88) (-) (50, 52, 55) = (25, 33, 38)$$

$$\underset{\sim}{M}_2 = \underset{\sim}{C}_2 (-) \underset{\sim}{P}_2 = (84, 86, 90) (-) (52, 56, 58) = (26, 30, 38)$$

$$\underset{\sim}{M}_3 = \underset{\sim}{C}_3 (-) \underset{\sim}{P}_3 = (88, 90, 92) (-) (52, 57, 60) = (28, 33, 40)$$

$$\underset{\sim}{M}_4 = \underset{\sim}{C}_4 (-) \underset{\sim}{P}_4 = (90, 92, 95) (-) (54, 60, 64) = (26, 32, 41)$$

$$\underset{\sim}{M}_5 = \underset{\sim}{C}_5 (-) \underset{\sim}{P}_5 = (90, 94, 96) (-) (56, 60, 68) = (22, 34, 40)$$

Repeating the same subtractions using the α-cut form, we arrive at the following results:

$$\underset{\sim}{M}_1 = [80 + 5\alpha, 88 - 3\alpha] (-) [50 + 2\alpha, 55 - 3\alpha] = [25 + 8\alpha, 38 - 5\alpha]$$

$$\underset{\sim}{M}_2 = [84 + 2\alpha, 90 - 4\alpha] (-) [52 + 4\alpha, 58 - 2\alpha] = [26 + 4\alpha, 38 - 8\alpha]$$

$$\underset{\sim}{M}_3 = [88 + 2\alpha, 92 - 2\alpha] (-) [52 + 5\alpha, 60 - 3\alpha] = [28 + 5\alpha, 40 - 7\alpha]$$

$$\underset{\sim}{M}_4 = [90 + 2\alpha, 95 - 3\alpha] (-) [54 + 6\alpha, 64 - 4\alpha] = [26 + 6\alpha, 41 - 9\alpha]$$

$$\underset{\sim}{M}_5 = [90 + 4\alpha, 96 - 2\alpha] (-) [56 + 4\alpha, 68 - 8\alpha] = [22 + 12\alpha, 40 - 6\alpha]$$

When Joe saw these results he took out his pencil and very quickly showed me on a piece of paper that the result we had arrived at was the same in either form. I told him that this should not surprise him if he took into account that the operation we had done, subtraction, was a *linear operation*. You must be conscious of the fact that if the calculation ended at this point, the transfer from the ternary form to the α-cuts form would have been superfluous. But since it was necessary to continue with this operation in which the product (non linear operator) plays a part, it was *now* essential that we have the α-cuts form available. What was true was that for the previous calculations it would have been sufficient to do the subtraction in the ternary form and then convert to the α-cuts form, which is what Joe did in a very short time.

We had now reached a situation where Joe and Mary had been capable not only of gathering all the necessary numerical information but this information has been transformed into the adequate form, that of α-cuts, in order to be able to carry out the further calculation correctly. These were now fairly simple. As we already agreed upon at our first meeting, all that was required then was to "transfer" the surpluses for each year to the origin of the investment project (moment zero) and subtract the initial payment (moment zero) and substrac the initial payment (which Joe called "initial" investiment). This operation then was now simple and intellible, particulary if we epressed the "surpless" for earch year by means of the hendecagonal system, as we have done below:

Surplus for year 1:

$\underset{\sim}{M}_1=$

α	$25 + 8\alpha$	$38-5\alpha$
1	33	33
.9	32.2	33.5
.8	31.4	34
.7	30.6	34.5
.6	29.8	35
.5	29	35.5
.4	28.2	36
.3	27.4	36.5
.2	26.6	37
.1	25.8	37.5
0	25	38

Surplus for year 2:

$\underset{\sim}{M}_2=$

α	$26 + 4\alpha$	$38 - 8\alpha$
1	30	30
.9	29.6	30.8
.8	29.2	31.6
.7	28.8	32.4
.6	28.4	33.2
.5	28	34
.4	27.6	34.8
.3	27.2	35.6
.2	26.8	36.4
.1	26.4	37.2
0	26	38

Surplus for year 3:

$\underset{\sim}{M}_3=$

α	$28 + 5\alpha$	$40 - 7\alpha$
1	33	33
.9	32.5	33.7
.8	32	34.4
.7	31.5	35.1
.6	31	35.8
.5	30.5	36.5
.4	30	37.2
.3	29.5	37.9
.2	29	38.6
.1	28.5	39.3
0	28	40

Surplus for year 4:

$\underset{\sim}{M}_4=$

α	$26 + 6\alpha$	$41 - 9\alpha$
1	32	32
.9	31.4	32.9
.8	30.8	33.8
.7	30.2	34.7
.6	29.6	35.6
.5	29	36.5
.4	28.4	37.4
.3	27.8	38.3
.2	27.2	39.2
.1	26.6	40.1
0	26	41

Surplus for year 5:

$\underset{\sim}{M}_5 =$

α	$22 + 12\alpha$	$40 - 6\alpha$
1	34	34
.9	32.8	34.6
.8	31.6	35.2
.7	30.4	35.8
.6	29.2	36.4
.5	28	37
.4	26.8	37.6
.3	25.6	38.2
.2	24.4	38.8
.1	23.2	39.4
0	22	40

Finaly we move on to the aklready mentioned "transfer" of the surpluses to the initial mement by calculating the *products* of the surpluses for each tear whit their respective up-dating retes. Mary then asked: how does the product of fuzzy nember occur? The answer was simple as we were dealing whith *positive numbers*. The ower extremes are multiplied by themselves and the upper extremes by themselves. And we do this for every level as follows:

For year 1:

α	$\underset{\sim}{M}_1$	
1	33	33
.9	32.2	33.5
.8	31.4	34
.7	30.6	34.5
.6	29.8	35
.5	29	35.5
.4	28.2	36
.3	27.4	36.5
.2	26.6	37
.1	25.8	37.5
0	25	38

$(\cdot)$

α	$\underset{\sim}{I}_1$	
1	.971	.971
.9	.970	.972
.8	.969	.973
.7	.968	.974
.6	.967	.975
.5	.966	.976
.4	.965	.977
.3	.964	.978
.2	.963	9.78
.1	.962	.979
0	.962	.980

$=$

α	$\underset{\sim}{M}_1 (\cdot) \underset{\sim}{I}_1$	
1	32.043	32.043
.9	31.234	32.562
.8	30.427	33.082
.7	29.621	33.603
.6	28.817	34.125
.5	28.014	34.648
.4	27.213	35.172
.3	26.414	35.697
.2	25.616	36.186
.1	24.820	36.713
0	24.050	37.240

For year 2:

α	M_2	
1	30	30
.9	29.6	30.8
.8	29.2	31.6
.7	28.8	32.4
.6	28.4	33.2
.5	28	34
.4	27.6	34.8
.3	27.2	35.6
.2	26.8	36.4
.1	26.4	37.2
0	26	38

(·)

α	I_2	
1	.952	.952
.9	.950	.953
.8	.948	.955
.7	.946	.956
.6	.944	.958
.5	.942	.959
.4	.941	.961
.3	.939	.962
.2	.937	.963
.1	.935	.964
0	.934	.966

=

α	$M_2 (\cdot) I_2$	
1	28.560	28.560
.9	28.120	29.352
.8	27.682	30.178
.7	27.245	30.974
.6	26.810	31.806
.5	26.376	32.606
.4	25.972	33.443
.3	25.541	34.247
.2	25.112	35.053
.1	24.684	35.861
0	24.284	36.708

For year 3:

α	M_3	
1	33	33
.9	32.5	33.7
.8	32	34.4
.7	31.5	35.1
.6	31	35.8
.5	30.5	36.5
.4	30	37.2
.3	29.5	37.9
.2	29	38.6
.1	28.5	39.3
0	28	40

(·)

α	I_3	
1	.924	.924
.9	.922	.926
.8	.919	.928
.7	.916	.930
.6	.913	.932
.5	.910	.933
.4	.908	.936
.3	.905	.937
.2	.902	.939
.1	.899	.940
0	.898	.942

=

α	$M_3 (\cdot) I_3$	
1	30.492	30.492
.9	29.965	31.206
.8	29.408	31.923
.7	28.854	32.643
.6	28.303	33.366
.5	27.755	34.055
.4	27.240	34.819
.3	26.698	35.512
.2	26.158	36.245
.1	25.622	36.942
0	25.144	37.680

For year 4:

α	M_4	
1	32	32
.9	31.4	32.9
.8	30.8	33.8
.7	30.2	34.7
.6	29.6	35.6
.5	29	36.5
.4	28.4	37.4
.3	27.8	38.3
.2	27.2	39.2
.1	26.6	40.1
0	26	41

(·)

α	I_4	
1	.906	.906
.9	.903	.909
.8	.899	.912
.7	.895	.914
.6	.892	.917
.5	.888	.919
.4	.885	.923
.3	.881	.925
.2	.877	.928
.1	.874	.930
0	.871	.933

=

α	$M_4 (\cdot) I_4$	
1	28.992	28.992
.9	28.354	29.906
.8	27.689	30.826
.7	27.029	31.716
.6	26.403	32.645
.5	25.752	33.544
.4	25.134	34.520
.3	24.492	35.428
.2	23.854	36.378
.1	23.248	37.293
0	22.646	38.253

For year 5:

α	$\underset{\sim}{M}_5$	
1	34	34
.9	32.8	34.6
.8	31.6	35.2
.7	30.4	35.8
.6	29.2	36.4
.5	28	37
.4	26.8	37.6
.3	25.6	38.2
.2	24.4	38.8
.1	23.2	39.4
0	22	40

(·)

α	$\underset{\sim}{I}_5$	
1	.871	.871
.9	.867	.876
.8	.863	.880
.7	.858	.884
.6	.854	.889
.5	.850	.892
.4	.846	.898
.3	.841	.902
.2	.837	.906
.1	.833	.910
0	.830	.915

=

α	$\underset{\sim}{M}_5 (\cdot) \underset{\sim}{I}_5$	
1	29.614	29.614
.9	28.438	30.310
.8	27.271	30.976
.7	26.083	31.647
.6	24.937	32.360
.5	23.800	33.004
.4	22.673	33.765
.3	21.530	34.456
.2	20.423	35.153
.1	19.326	35.854
0	18.260	36.600

All these calculations allowed us to have all the surpluses available just as if they had taken place at the initial moment of the investment process. Now yes, we could do the addition. Nothing easier, since each of the five fuzzy numbers was represented by eleven confidence intervals (α=1, α=0.9, ..., α=0.1, α=0). Joe and Mary helped and together "recited": add the lower extremes together and the upper extremes together. They quickly took up pencil and paper and with their pocket calculator arrived at the following result.

$$\underset{\sim}{M}_1 (\cdot) \underset{\sim}{I}_1 (+) \underset{\sim}{M}_2 (\cdot) \underset{\sim}{I}_2 (+) \underset{\sim}{M}_3 (\cdot) \underset{\sim}{I}_3 (+) \underset{\sim}{M}_4 (\cdot) \underset{\sim}{I}_4 (+) \underset{\sim}{M}_5 (\cdot) \underset{\sim}{I}_5 =$$

=

α	$\sum_{j=1}^{5} \underset{\sim}{M}_j (\cdot) \underset{\sim}{I}_j$	
1	149.701	149.701
.9	146.111	153.336
.8	142.477	156.985
.7	138.832	160.583
.6	135.270	164.302
.5	131.697	167.857
.4	128.232	171.719
.3	124.675	175.340
.2	121.163	179.015
.1	117.700	182.663
0	114.384	186.481

We now had the fuzzy number representing the surpluses, or as Joe liked to put it "up-dated earnings". All we had to do now was to subtract the *initial payments* (also up-dated) in order to find the profit or loss generated by the decision to invest, in the event, obviously that Joe and Mary were to set out on this adventure. We had now reached the crucial point:

α	$\sum_{j=1}^{n} \underset{\sim}{M}_j (\cdot) \underset{\sim}{I}_j$	
1	149.701	149.701
.9	146.111	153.336
.8	142.477	156.985
.7	138.832	160.583
.6	135.270	164.302
.5	131.697	167.857
.4	128.232	171.719
.3	124.675	175.340
.2	121.163	179.015
.1	117.700	182.663
0	114.384	186.481

(-)

α	$[100 + 20\alpha, 130 - 10\alpha]$	
1	120	120
.9	118	121
.8	116	122
.7	114	123
.6	112	124
.5	110	125
.4	108	126
.3	106	127
.2	104	128
.1	102	129
0	100	130

There was no need to insist, my friends had learned the lesson well, that in order to carry out a *normal subtraction* this had to be done by "crossing over the extremes" obviously level by level. This is how we arrived at the following result:

α	$\underset{\sim}{Y}$	
1	29.701	
.9	25.111	35.336
.8	20.477	41.985
.7	15.832	46.583
.6	11.270	52.302
.5	6.697	57.857
.4	2.232	63.719
.3	- 2.325	69.340
.2	- 6.837	75.015
.1	-11.300	80.663
0	-15.616	86.481

Here then is a representative table of this type of "accumulated and up-dated profit", which would be attained if and when Joe and Mary definitely accepted making the investment. Here was a very wide range of possible values, which went from a loss of 15,616 right up to a profit of 86,481. The maximum presumption was located at 29,701 monetary units of profit. Mathematicians like to express the process we have followed by means of a formula called "net up-dated value" or "capital value of an investment". It is as follows:

$$\underset{\sim}{V} = \sum_{j=1}^{5} \underset{\sim}{M}_j (\cdot) \underset{\sim}{I}_j (-) \underset{\sim}{M}$$

To be quite honest, Joe and Mary were not interested in academic treatises but they did, on the other hand, query whether it was possible to arrive at the same or similar result by a simpler operation. Inevitably the question arose as to what would happen in the event of using confidence triplets in place of the F.T.N. in order at the end to arrive at the triangular form. The easiest thing to do was to redo the calculations with the available data, but now with the triplets. What we did then is set out below:

$$\underset{\sim}{M}_1 (\cdot) \underset{\sim}{I}_1 = (25, 33, 38) (\cdot) (.962, .971, .980) = (24.050, 32.043, 37.240)$$

$$\underset{\sim}{M}_2 (\cdot) \underset{\sim}{I}_2 = (26, 30, 38) (\cdot) (.934, .952, .966) = (24.284, 28.560, 36.708)$$

$$\underset{\sim}{M}_3 (\cdot) \underset{\sim}{I}_3 = (28, 33, 40) (\cdot) (.898, .924, .942) = (25.144, 30.492, 37.680)$$

$$\underset{\sim}{M}_4 (\cdot) \underset{\sim}{I}_4 = (26, 32, 41) (\cdot) (.871, .906, .933) = (22.646, 28.992, 38.253)$$

$$\underset{\sim}{M}_5 (\cdot) \underset{\sim}{I}_5 = (22, 34, 40) (\cdot) (.830, .871, .915) = (18.260, 29.614, 36.600)$$

We did the addition of these five F.T.N. with the following result:

$$\sum_{j=1}^{5} \underset{\sim}{M}_j (\cdot) \underset{\sim}{I}_j = (114.384, 149.701, 186.481)$$

In order to find the "net up-dated value" expressed as a triplet, all that is needed is to deduct the so-called "initial investment" from this addition, as follows:

$$\underset{\sim}{Y} = \sum_{j=1}^{5} \underset{\sim}{M}_j (\cdot) \underset{\sim}{I}_j (-) \underset{\sim}{M} = (114.384, 149.701, 186.481) (-) (100, 120, 130)$$

$$= (-15.616, 29.701, 86.481)$$

On seeing this triplet, my friends became aware of the fact that there was a coincidence in the extremes and the maximum presumption of the fuzzy number $\underset{\sim}{Y}$ found by following the strictly technical path. Were so many calculations required as before? Was it sufficient just to use triplets? We would see immediately by transforming the triplet $\underset{\sim}{Y}$ into a F.T.N. which in all truth would be a triangular approximation of the fuzzy number $\underset{\sim}{Y}$. For this we were going to suppose that the triplet was a F.T.N. and we would pass it over to the form of α-cuts. The *new triangular approximation* would be:

$$\underset{\sim}{Y}' = [-15.616 + 45.317\alpha, 86.481 - 56.780\alpha]$$

which, represented by the hendecagonal form was the following:

α	$\underset{\sim}{Y}'$	
1	29.701	
.9	25.169	35.379
.8	20.638	41.057
.7	16.106	46.735
.6	11.574	52.413
.5	7.043	58.091
.4	2.511	63.769
.3	- 2.021	69.447
.2	- 6.553	75.125
.1	-11.084	80.803
0	-15.616	86.481

A comparison between the fuzzy number $\underset{\sim}{Y}$ and the triangular approximation $\underset{\sim}{Y}'$ would allow us to assert that there were no substantial differences in the different levels of α. These were limited to a few tenths in the worst cases. Faced by this circumstance the following questions would appear inevitable: Does it always occur in this way? Was it worth doing all those calculations with fuzzy numbers? These were two aspects of one and the same question. In fact, this near coincidence takes place when, as in this case, either one or the two simultaneous circumstances occur: when the interest rates are low and/or when the "economic horizon" for the investment is limited to only a few years. Joe and Mary immediately replied: "you knew this was going to happen and you have made us do a whole lot of calculations which appear superfluous to us". In fact this is true but... are you sure that this is your very last "investment"? Now I am quite sure that you will not be requiring me for your future adventures. Also with the two ways you will be much surer of what you are going to do and one process has served as verification for the other. They did not seem particularly convinced, but they accepted this in good faith.

So as to finish with the above scenario, prior to their decision I suggested to them that they represent the triangular approximation $\underset{\sim}{V}'$ in the graphical form. This is shown below:

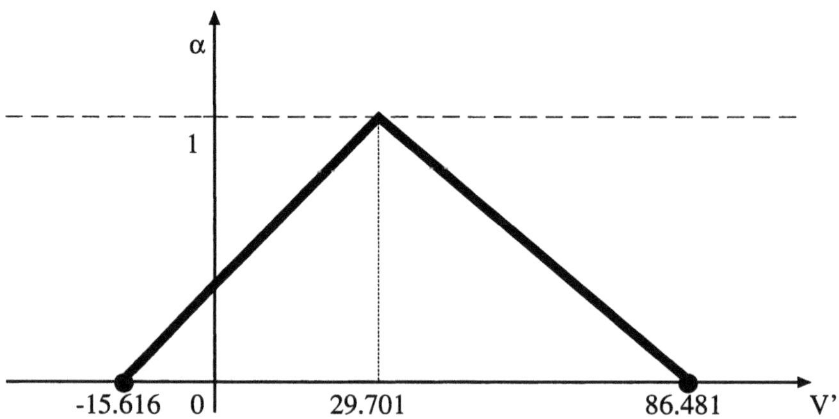

In view of the data obtained and of this graphical representation, certain reflections were required that would lead to a good decision.

One of the first characteristics of the result is that the lower extreme is negative (-15,616) which means the possibility of losses. If we look at the figure we will see that one part (although small) of the triangle is outside the first quadrant. It is precisely this part that corresponds to the negative results. All of this is true, as is also the fact that the maximum presumption and upper extreme are positive. Both points are within the first quadrant, giving rise to the fact that the greater part of the F.T.N. (triangular approximation) represents positive results. These simple facts encouraged Joe and Mary along the way towards a decision. But as we are all so used to numerical presentations I proposed calculating what we could call an "index of success" for the investment. We would arrive at this by obtaining the quotient between the area of the *part* of the triangle to be found in the first quadrant and the total area of the triangle and this would be designated by I_E.

In order to arrive at the numerator it is sufficient to subtract from the total area of the triangle the area of the small triangle that is outside the first quadrant, the base of which is 15,616 and its height, the result of the following equation:

$$-15.616 + [29.701 - (-15.616)]\,\alpha = 0$$

$$-15.616 = -45.317\,\alpha$$

$$\alpha = \frac{15.616}{45.317} = 0.345$$

Therefore the area of the small triangle is:

$$1/2\,(15.616 \times 0.345) = 2.694$$

and the area of the F.T.N.:

$$1/2\,[(86.481 + 15.616) \times 1\,] = 51.049$$

In this way we have arrived at the "index of success":

$$I_E = \frac{51.049 - 2.694}{51.049} = 0.947$$

There was no doubt about the success of the project, in spite of the small cloud which would only become a storm if events were to go really, really wrong. The possibilities of losses were very remote. To put this in another way they only came close to 5%. Therefore this risk could be taken. Also in the temperament of Joe and Mary pessimism hardly had a place. They were a happy and lively couple that always looked on the positive side of things. If the left hand side of the triangle representing the F.T.N. shows us pessimism and the right optimism, there is no doubt whatsoever that my friends are in the right hand side. This does not invalidate another of their characteristics: *prudence*. And this led them to carry out the process but... with some "modifications" in the initial investment making a sacrifice, that is, of greater effort in their personal work.

Sufficient time has gone by for us to make some verifications, obviously partial, of the estimates and calculations made in their day. To date the balance is highly positive. The activities of Joe and Mary's business continue with no ups and downs and the "surpluses" they obtain allow them to look to the future with smiles on their faces.

REFERENCES

1 GIL ALUJA, J.: *Investment in uncertainty*. Kluwer Academic Publishers. London, Boston, Dordrecht, 1999.

2 See pages 59-63 of this book.

3 For greater detail consult Kaufmann, A. and Gil Aluja, J.: *Las matemáticas del azar y de la incertidumbre*. publ. Ceura, Madrid 1990, pág. 21-34.

4 The maximum presumption does not imply "certainty". if this were so only this number would de taken and we would be operating in the field of determinism, with arithmetic known by all.

5 KAUFMANN, A. and GIL ALUJA, J.: *Técnicas operativas de gestión para el tratamiento de la incertidumbre*. publ. Hispano-Europea. Barcelona 1987, pág. 65-68.

6 KAUFMANN, A. and GIL ALUJA, J.: *Introducción de la teoría de los subconjuntos borrosos a la gestión de empresas*. publ. Milladoiro. Santiago de Compostela 1986, pág. 79-81.

IV. DRAWING UP A BUDGET

Considerations on budgeting activity

The word budget is present in nearly all spheres in which man appears as the active subject, although it is not always used for representing the same idea it intends to transmit. We quite often hear the phrase: *"I do not have a budget* for doing this trip".

This does not mean that the person emitting the sentence disposes of a formal state of chapters of estimated income and expenses to be made, but that mentally said person has made the necessary calculations and knows that there will not be enough monetary means for covering the costs of this leisure time activity.

He, or the head of the family, "in a certain way establishes a budget when determining how much income there is and how this income will be used for the different channels of expenses, such as food, child education, the home, recreation, purchase of durable goods such as (television sets, refrigerators, cars, etc.) and even when establishing that a certain residual portion will be saved. In a similar manner, countries also draw up their budgets of income and expenses in which appear, on the one hand the sources of financing, specifically taxes and other income, and on the other hand the application of these resources for covering the expenses of the different areas, such as defence, education, agriculture, industry, culture, etc". In businesses and institutions, the budgeting process constitutes one of the most important tasks from an economic-financial point of view. Estimating hopes for resources with which to cover the also estimated payments to be made for attaining objectives are the first steps towards adequately structuring policies and strategies.

"If initially the *budget* was conceived as an instrument directed at limiting the expenses that arise in businesses and institutions, today their objective is much wider since with it what is also intended is a more rational use of available resources for attending the conglomeration of their requirements"[1].

During our long experience of comparison between academic life and socio-economic reality, we have had several opportunities for participating in the drawing up and control of budgets. Furthermore, on graduating from university we were hired at the beginning of the 60's by the most important car manufacturing company in Spain in order to head up the Budget and Accounting Statistics Department. We already at that time had it quite clear in our minds that in order to draw up a budget it was necessary to have available certain strictly established but realistic updating rules, since the establishment of "ideal" budgets can lead to disillusionment, frustration and to a feeling of failure by those responsible for carrying out the functions leading to compliance of programmed activities. At the other extreme, when a budget is established without any ambition whatsoever, it provokes a lack of incentives and the inefficiency of control instruments which make the budget totally superfluous.

The techniques in fashion at the time are today quite unusable. Nearly always budgeting was done on estimates based on data recorded from previous periods which frequently hid inefficiencies taking into account the incorporation of previously existing expenses which were carried over to the following years, giving rise to unwanted payments.

As a novelty the so-called flexible budget made an appearance, in which at different levels of sales and therefore production, different volumes of expenses were charged. In this case a business or institution could authorise a different payment for each alternative level of invoicing. What it was in this case was an attempt to determine, from among the different alternative budgets, the one that should be considered for the corresponding period. The advantage on which those then in favour of flexible budgets supported themselves, was their capacity for being modified when circumstances changed.

Many years went by and with them a substantial change occurred in our labour activity, since we left the business world and became dedicated to academic life with particular dedication to research work. But destiny ensured that on the threshold of maturity we returned to once again find ourselves involved with budgetary matters.

Those who know me, even though superficially, are fully aware of my love of football, which I played in my youth and of which I have been an devoted spectator. I carry in my heart the Club of the city where I was born, which plays in the more modest categories. But from quite a few decades back it shares this place with a great club, the best in the world, which goes by the name of the capital of my country. We suffer with adversity and enjoy success. The ups and downs that are second nature to sport have not been the only ingredients that are the reason for the birth and development, within the sphere of this sports club, of certain sensibilities, that are difficult to explain and understand, by those who are removed from the life and history of this privileged corner of the peninsular.

In one of the more difficult moments in the life of the Club, there was a mutiny by the professional players, who were looking for the resignation of the Board of Directors and, at a press conference, advocated their taking over the Club management. There were several resignations in the different organs of the entity and the press aired a supposed financial "hole" of 3,000 million monetary units. Under these circumstances I received a call from the treasurer of the Club, a personal friend, William, proposing my incorporation as a member of the Economic-Statutory Committee of the entity. I declined his offer, above all taking into consideration the deplorable spectacle being offered by certain sectors of the social life of the city, who caused the Club damages of incalculable dimensions. Apart from this I considered that an invitation of this nature should come directly from the President. And this is what I said. Not two minutes had gone by when the telephone rang again. It was the head of the club, who after a very warm greeting persuaded me to attend on the next day a meeting prior to the setting up of the new Committee, obviously with no undertaking on my part.

What in principal should have been a first private contact became and official act, quasi public, due to the large number of the communications media present. In the blinking of an eye I found myself presiding the new Economic-Statutory Committee of the Club. Hand in hand with the President, Josep Lluis, we initiated a new era, which commenced with the outflow of some 3,500 million monetary units for renewing the players, constructing a team, the birth of what would become known as the "Dream Team". The extraordinary qualities of Josep Lluis, gifted with boundless imagination, an enviable capacity for work and a determination tested by a thousand obstacles, opened the doors to the reconstruction of a sporting entity, which is the pride of the citizens of our country, and the mirror in which other sporting institutions the world over see themselves. Few sincere people have any doubt that he was the finest president ever.

One of the first tasks that we tackled was to change the *philosophy* of the budgeting concept. To him we owe the idea that a budget relative to any season should be framed within a medium and long term planning process, which process was baptised with the suggestive name of *sporting project*. Jointly we established the fact that the Committee should meet monthly in order to analyse any eventual deviations existing between the budgeted values and actual values, which could be due to two types of causes: some external, as a consequence of elements having an incidence on the economic system in general, and others internal, due to the operation itself of our institution. This second aspect allows for greater control and constituted a healthy obsession of the President. His opinion, which we feel was quite relevant, that the budget should allow us to estimate, with sufficient notice, the changes required by the Club in order to meet objectives, and maintain good prospects for the future. The budget constitutes a synthesis of the values, where the Club *should be* manoeuvring, he was wont to say frequently and, also allows for better understanding of its operation, which in the long run should improve its capacity for auto-financing. By means of the budget an estimate of the balance sheet of the Club is arrived at for the next or future period or periods and an operating account with the estimated profit and loss statement.

This was his line of thought. All these cases had to be gathered up and with them be able to construct certain techniques that were adaptable to sporting entities. We knew that the tasks we had taken on for arriving at a budget meant certain additional work which frequently leads to an increase in internal coordination, since they call for the separation of responsibilities which, without these studies, are very often carried out by more than one decision centre in one and the same club. We should then *put our minds to the task*, and search among those methods that were best adapted to our ambitions. We decided on trying out a new technique, at that time in constant expansion, called Zero Base Budgeting (Z.B.B.). We looked for some texts on the subject and in them it was stated that one of greater problems of normal budgets arises from the fact of having their roots in preceding situations and their estimates constituting a projection of past data. In this way what are gathered together are all the deficiencies and inefficiencies that existed in previous periods. Zero Base Budgeting attempts to eliminate, among other things, this inconvenience. For this it establishes, as a principle, the need to justify any amount "budgeted", refraining from taking as a support the budgets from previous years and, for this, "starting out from zero".

We learnt that the birth of this technique took place in 1970 by the hand of Peter PYHRR, but it was developed a few years later, a five year period being necessary before it underwent any expansion. The primary objectives intended to be reached by means of Z.B.B. do not differ substantially from the aims sought after from classical budgeting, that is, to establish a planning process, arrive at a better assignment of resources, carry expenses at an adequate level and, obviously, improve the decision taking activity of businesses [2] and institutions. This is what we wanted for our Club.

We were discouraged when we saw that the belief was unanimous that in order for a budget drawn up according to traditional models to be viable, it had to comply with the following conditions:

1. that all the activities of the Club were equally important for normal operation of the same.

2. that in the future we should follow the same degree of efficiency as in previous times; and

3. that the assignment of resources had been done in the past in an adequate manner.

It was obvious that, at that time, sporting entities moved in a context in which changes took place very rapidly and the future could not be seen as a simple projection of the past. From this we reached the conclusion that the budget drawn up in accordance with classical schemes could not be accepted for the reality of our Club and were on the way to accepting the basic premises of Z.B.B. And this was so, because the principles on which Z.B.B. is based could reflect the realities of those time far better. What we wanted was a distribution of the resources that could be generated, in accordance with *previously set objectives*, separating our institutions into several Decision Centres, each one of which would have a certain amount of autonomy. On the other hand we were conscious of the fact of the limitation of the financial resources and this would oblige us to "point out a system for grading the requirements" of each decision centre. In other words, we were faced with the need to determine the relative importance of each of the objective of each centre, faced with those established for the remaining centres. The problem was how were we to instrument priorities. For this it was considered convenient that each Decision Centre should justify all and each one of the monetary masses it requested. And this could only be attained in the first place by *identifying the activities*, in second place *putting a value to them* and in third place establishing for them an *order of priority.*

With these elements it appeared we were ready to pass on to "finding a model for" the process. But the insight of the President made him see the necessity for certain complementary reflections. A man who was used to business fighting sustained that in the contents of each specific proposal there should be reasoning that was sufficiently clear in order for a decision to be taken with guarantees, consisting I accepting or not the level of the proposed activity., that is, to include or not the payments required by the corresponding budget. Josep Lluís considered as necessary elements, although not exclusive, the following: the description of each activity and its position within the general objectives of the Club; the expenses involved expressed by means of the monetary payments this would mean; as well as sporting and economic results that it was hoped would be attained by carrying out the same. Prudence cautioned that it was also convenient to determine what were to consequences of not including the studies activity at the specified level among the activities foreseen for the Club during the time covered by the budget, which coincide with a sporting season.

We were now increasingly more convinced of the advantages of using Z.B.B. One of the most significant advantages stems from the fact that each one of the specific proposals could be drawn up for different levels of activity and therefore for different levels of expenses. This characteristic was the determining factor for us, because it was perfectly suitable for the sporting realities of the Club, characterised by the fact that a team of a certain speciality, for example football, can be formed by different players for each position on the field at different levels of quality. With the object of not getting off the beaten track due to abstract ideas, we agreed on considering a minimum basis of activity, one maximum and the others intermediary, always taking into account that at each level we would comply with the objectives, also in a determined degree. The activity of the Club would be separated into certain Decision Centres which could make their proposals at several levels of activity. All of this led us to think in the possibility of using one of the fuzzy numbers. But at this time this appeared to be utopian, given the incipient situation in which we found ourselves. It was obvious that we should commence this new challenge without separating ourselves from the rules of normally accepted texts.

The Zero Base Budgeting (Z.B.B.)

The first important decision which had to be taken referred to the establishment of a number, obviously reduced, of *Decision Centres* with a certain capacity for action, certain powers and responsibilities. They would be provided with certain monetary means and they would be required to produce results. The figures assigned could not be exceeded and the amount of payments should be related to the results attained. Heading each Decision Centre would be a *President* who would be given adequate responsibility. At the time of putting names and providing contents to each *Decision Centre* which would be popularly known as "sections", Josep Lluis was very clear. The *sections* would be the following:

A. Football; B. Basketball; C. Handball; D. Roller skate hockey; E. Other sections. In this latter Decision Centre would found all those non professional activities among which were: figure skating, rugby, athletics...

It was not difficult to find a responsible person for each one of these sections, since already doing this task to the satisfaction of the President there were excellent people all of them successful in their professional lives. Thus the following were appointed:

> *John* as president of "section" A, *football*
>
> *William* as president of "section" B, *basketball*
>
> *George* as president of "section" C, *handball*
>
> *Edward* as president of "section" D, *roller skate hockey*
>
> *Joseph* as president of E, *other sections*

Once they had taken up their respective offices (and responsibilities, we must not forget), they were given all the information on their respective Decisions Centres, and they were put in contact with the administrative services of the Club, so that they could gather from the latter, if required, all the necessary data for them to be able to carry out an efficient task. Particularly it was insisted upon that they study, in great depth, the levels at which they could situate their section, taking into account the importance and significance of the Club in the national and international sporting context. As a first approximation, they should establish a reduced number of *level*s, each of which would mean certain expenses, and parallel to this, results. The greater the level of payments it was assumed there would be greater sporting success. Obviously with unlimited resources, it was assumed that each section would win everything. But as this was utopian, some or all the activities had to be limited to intermediary levels, in accordance with the financial means available or *susceptible* to being attained. Having clarified this basic idea these five presidents were called to a meeting at which the Vice-Presidents and Club Manager were also present.

The day came for this important meeting. Each one of the section presidents had made a situation analysis of his team in relation to all their adversaries in all competitions, a search had been made in the different spheres with the object of looking at the possibility of taking on new players, as well as an analysis of the actual situation of the current components of the teams After the initial statements a definition was made for each of the Decision Centres of certain expense levels, for each level of which all and every one of the items comprising the same were duly justified. As it could not be in any other way, the greater the level, the more success was hoped for. After several discussions on the subject, commentaries and adjustments a consensus was reached establishing the following *levels* for each decision centre:

A = Football : A_0, A_1, A_2, A_3

The objectives to be attained and the monetary eran, requirements were, respectively:

A_0 : Win H.M The King's Cup	6,000 m.u.
A_1 : Win the Spanish Championship	8,000 m.u.
A_2: Win both championschips	9,000 m.u.
A_3 : Win the "Champions League"	1,000 m.u.

B = Basketball: : B_0, B_1, B_2

These corresponded as follows:

B_0 : Win H.M. The King's Cup	1,000 m.u.
B_1 : Win the ACB League	1,500 m.u.
B_2 : European Championship	3,000 m.u.

C = Handball : C_0, C_1

Objectives and the necessary resources would be established to attain those listed below:

C_0 : National Championship	500 m.u.
C_1 :European championship	1,200 m.u.

D = Roller skate hockey : D_0, D_1

What was sought and for this it was necessary to have available:

D_0 :	Win the national championship	400 m.u.
D_1 :	Win the European championship	800 m.u.

E = Other sections : E_0, E_1, E_2

E_0 :	Grass roots football	800 m.u.
E_1 :	Grass roots football and Four non professional sports	1,000 m.u.
E_2 :	Grass roots football anb Twelve non profesional sports	1,500 m.u.

With this we had the first basic information, where it could be seen that the levels with a zero sub-index were the minimum essential levels for the existence of the corresponding Decision Centre, since below these levels it could not operate. Budgets indicated by sub-indices 1, 2, 3, contained improvements, obviously justified.

Having finalised this information stage we commenced the hardest phase in which the priorities of the Club had to be *put in an order*, either by giving priority or relegating one or other section at each one of the levels. In this we faithfully followed the methodology of the Z.B.B. one of the first steps of which consists in a *sequential selection*, commencing by the decision centre at the lowest level, always represented by a sub-index zero. On quite a few occasions Josep Lluis had to assume the responsibility of the final decision, as a result of the discrepancies that arose. He always led the sessions with sensitivity and prudence. For greater clarity below we have shown all the sequecens by measn of a graphical representation:

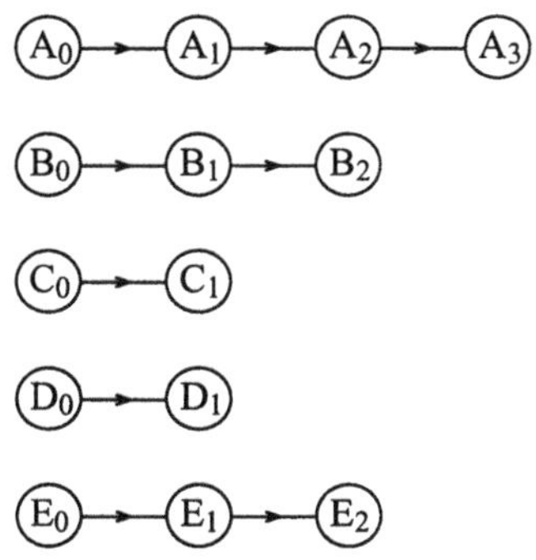

Initial graph

Below are summarised the development of each one of the decisions agreed upon at the meeting. There was no doubt among those present in highlighting the priority of football. That is, if only a minimum amount of resources were available that only allowed starting up one "section", then that section would be football. We express this by marking the A_0 vertex with a cross. In this case the budget of the Club will coincide with the budget of Decision Centre A, football, at level zero, that is to say 6,000 monetary units.

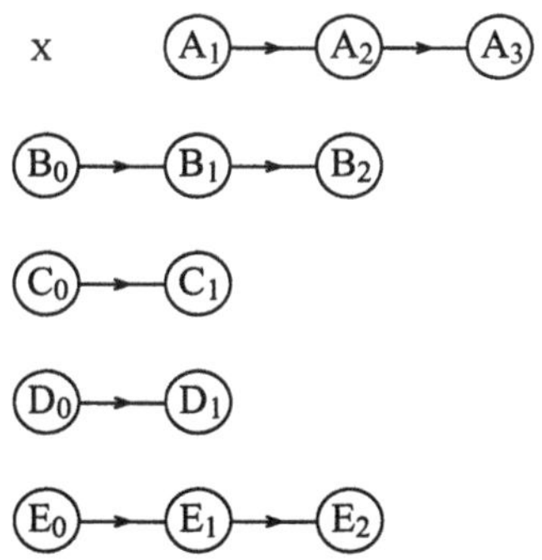

A_0 has been selected

Afterwards, the possibility of having a greater quantity of financial means available was considered. The alternative consisted in strengthening the football section, moving on to higher level A_1, with a cost of 8,000 monetary units, or to continue with level A_0 and start up another section. It was agreed to move on to the higher level in football. The Club budget then reached 8,000 monetary units, which is shown below:

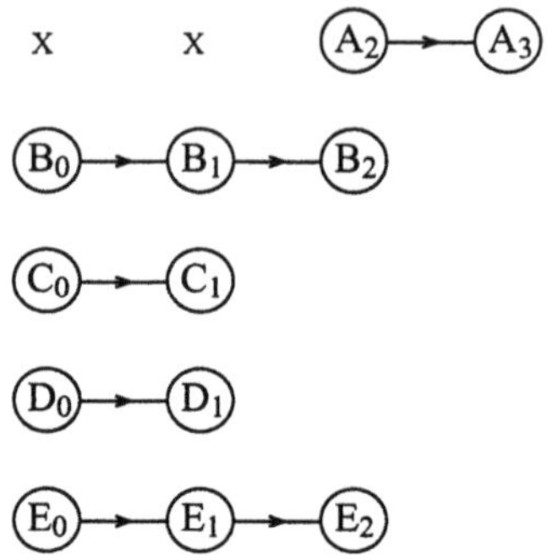

A_1 has been selected

In the next phase it was decided to start up the basketball section obviously at level zero. The budget for the Club would then be, $A_1 + B_0$ and, therefore, 8,000 + 1,000, that is 9,000 monetary units. The graph remained as follows:

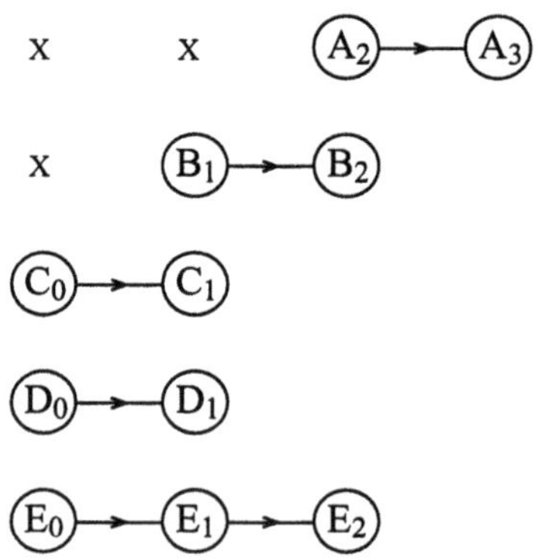

B_0 has been selected

Below we have shown all the successive decision taken:

These are as follows:

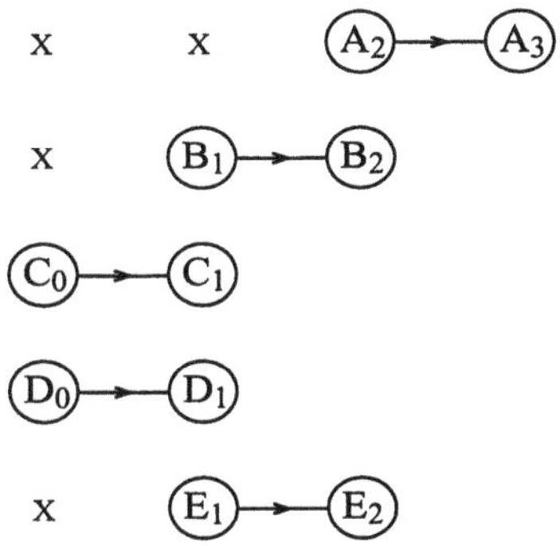

E_0 has been selected

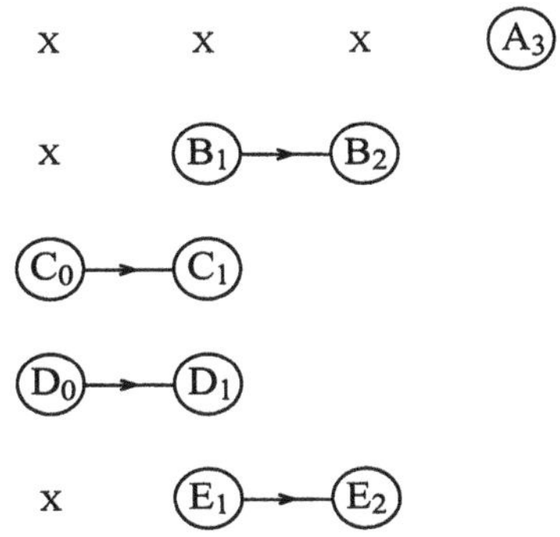

A_2 has been selected

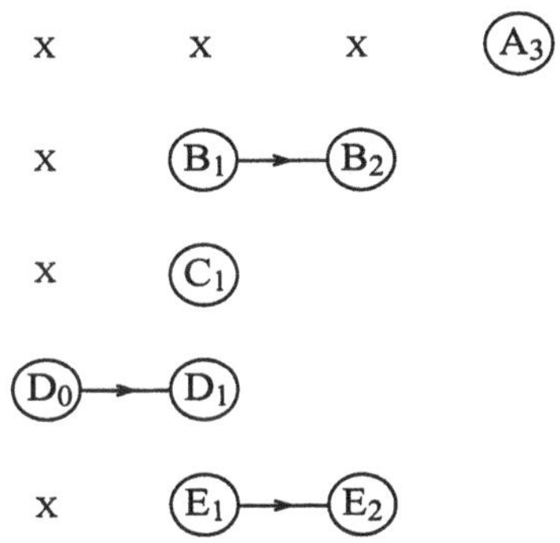

C_0 has been selected

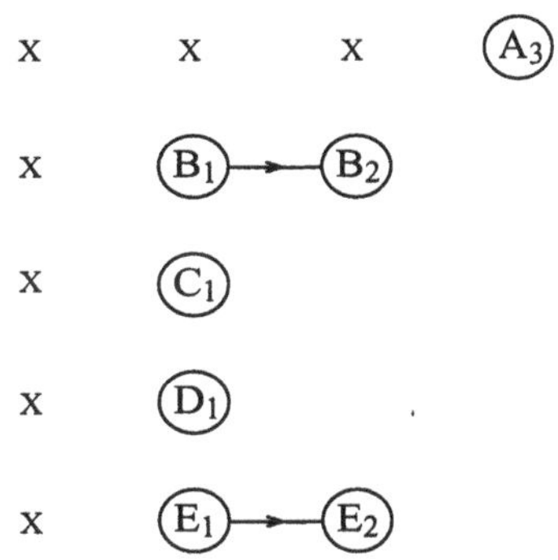

D_0 has been selected

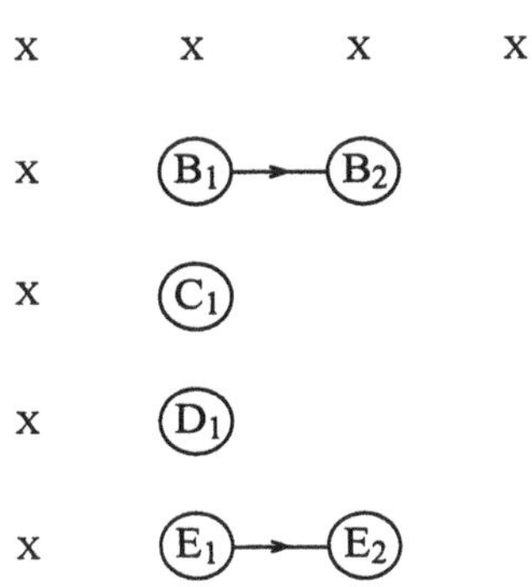

A_3 has been selected

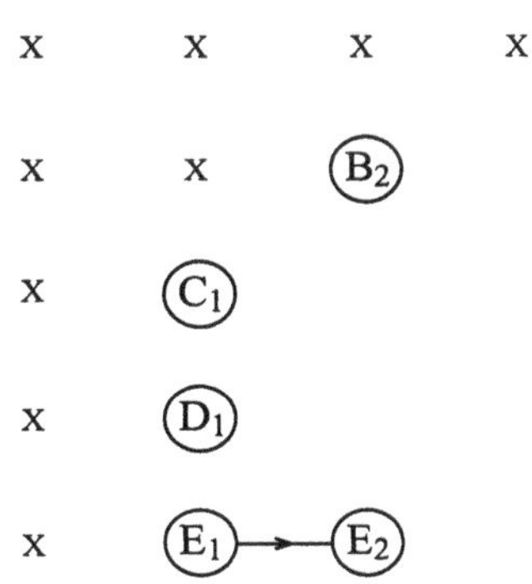

B_1 has beem selected

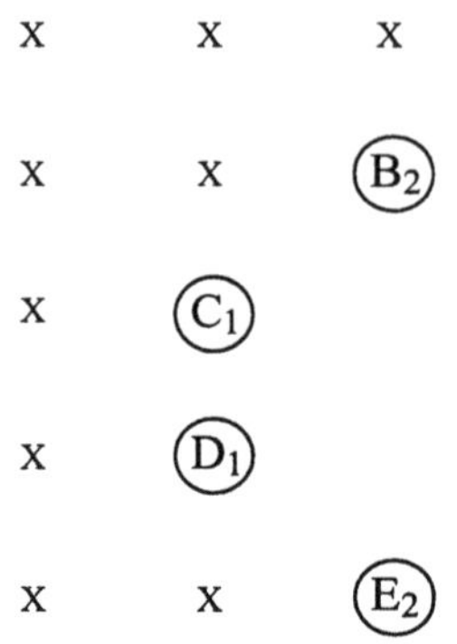

E_1 has beem selected

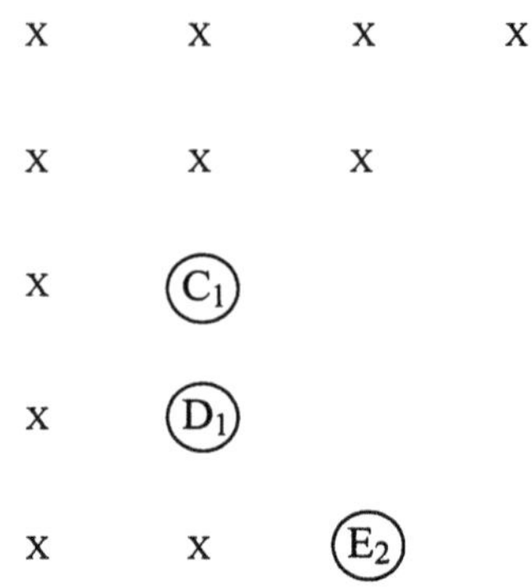

B_2 has been selected

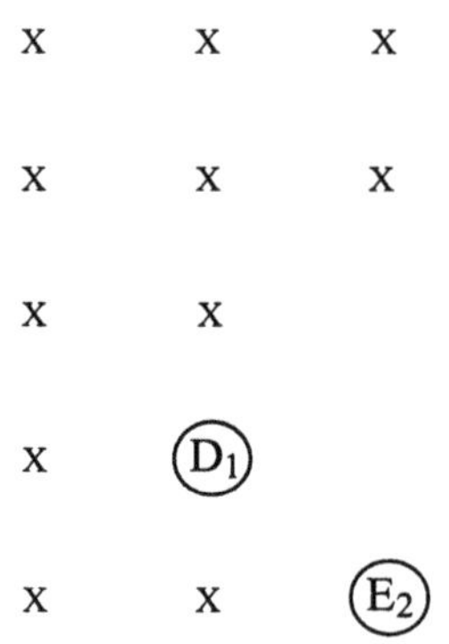

C_1 has been selected

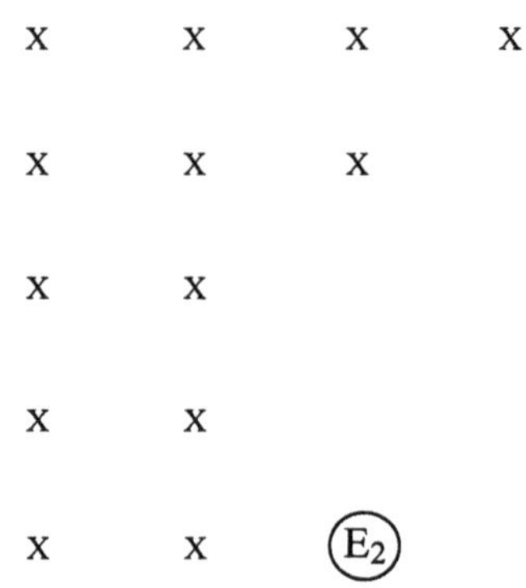

D_1 has been selected

Obviously the last selection consisted in starting up decision centre E, other Sections, at the maximum level E_2. For this a very large amount of resources would have to be available.

With these sequences we now possessed a basic element for the budgeting activity of the Club. In accordance with the decisions taken, the possible budget levels, from the lowest to the highest, were as follows:

Budget level I

$$A_0 = 6.000$$

Budget level II

$$A_1 = 8.000$$

Budget level III

$$A_1 + B_0 = 8.000 + 1.000 = 9.000$$

Budget level IV

$$A_1 + B_0 + E_0 = 8.000 + 1.000 + 800 = 9.800$$

Budget level V

$$A_2 + B_0 + E_0 = 9.000 + 1.000 + 800 = 10.800$$

NBudget level VI

$$A_2 + B_0 + C_0 + E_0 = 9.000 + 1.000 + 500 + 800 = 11.300$$

Budget level VII

$$A_2 + B_0 + C_0 + D_0 + E_0 = 9.000 + 1.000 + 500 + 400 + 800 = 11.700$$

Budget level VIII

$$A_3 + B_0 + C_0 + D_0 + E_0 = 12.000 +.1.000 + 500 + 400 + 800 = 14.700$$

Budget level IX

$$A_3 + B_1 + C_0 + D_0 + E_0 = 12.000 + 1.500 + 500 + 400 + 800 = 15.200$$

Budget level X

$$A_3 + B_1 + C_0 + D_0 + E_1 = 12.000 + 1.500 + 500 + 400 + 1.000 = 15.400$$

Budget level XI

$$A_3 + B_2 + C_0 + D_0 + E_1 = 12.000 + 3.000 + 500 + 400 + 1.000 = 16.900$$

Budget level XII

$$A_3 + B_2 + C_1 + D_0 + E_1 = 12.000 + 3.000 + 1.200 + 400 + 1.000 = 17.600$$

Budget level XIII

$$A_3 + B_2 + C_1 + D_1 + E_1 = 12.000 + 3.000 + 1.200 + 800 + 1.000 = 18.000$$

Budget level XIV

$$A_3 + B_2 + C_1 + D_1 + E_2 = 12.000 + 3.000 + 1.200 + 800 + 1.500 = 18.500$$

In this way what we have arrived at is a wide range of budgets, from the more modest, with only one section functioning, that is football at the lowest level, with an expenditure of 6,000 monetary units, right up to the highest level, which includes the maximum potential possible in all sections. This requires having available 18,500 monetary units. Now what must be done is to select, among all of these levels, the one that can be covered by the income obtained from the different sources of financing.

This last and definite phase no longer depended on those responsible for the Decision Centres, but from the work carried out by the Club Manager for obtaining monetary means. As is well known, these may be sourced from, in a sporting club, membership fees, season ticket holders, merchandising, granting of television rights, advertising in its different modes, and all kinds of sponsorships

James, the Club Manager was asked to put together all information available on the collection of resources, quantifying in a realistic manner those arising from each source of financing. A few days later, we received a thorough report in which, the possibilities of the Club were justified with a wealth of detail. These then were summarised into the following concepts and figures:

Ticket sales, non members	800
Merchandising	1.300
Television rights	5.300
Advertising	800
Sponsorships	2.300
Membership fees and season tickets	4.500
Direct product sales	500
Other income	300
I = Total estimated income	15.800

It was quite clear then that if the estimated income was reached, 15,800 monetary units, that we have designated by I, it would be possible to accept the level represented by:

$$A_3 + B_1 + C_0 + D_0 + E_1 = 15.400$$

with which there would be surplus of 400 monetary units.

On the other hand, the following level could not be covered, which means expenses for:

$$A_3 + B_2 + C_0 + D_0 + E_1 = 16.900$$

Mathematicians like to express this situation as follows:

$$A_3 + B_1 + C_0 + D_0 + E_1 < I < A_3 + B_2 + C_0 + D_0 + E_1$$

The distribution of the budget figures by Decision Centres was then established in the following numbers:

Football	A_3	=	12,000	monetary units
Basketball	B_1	=	1,500	" "
Handball	C_0	=	500	" "
Roller skate hockey .	D_0	=	400	" "
Other sections	E_1	=	1,000	" "
	Surplus		400	" "
Budget applications			15,800	monetary units

In this way we initiated a new way of structuring budget activity, which would see the most exceptional decade of the history of the Club.

But the self discipline of the President, who always worked to get more and do things better, detected a difficulty which would curb the efficiency of the scheme we had drawn up. It was as a result of the meetings with the Decision Centres and Club Manager when he became aware of the difficulties of arriving at exact numbers, both for expenses and income estimated for each level. There were frequent statements such as "more or less such a figure", "between this amount and that amount", "maybe it will be somewhat more than this amount"...., which under the necessary requirements were specified in determined amounts, not always to the total agreement of those responsible for estimating the same, and who on occasions felt most uncomfortable. Josep Lluis, intuitively felt that what was required was to give the scheme we had thought out a certain amount of *controlled flexibility.*

The Fuzzy Zero Based Budgeting

For some seasons, budget management of the Club was based on the scheme we outlined above which, with the suggestions and ideas of the President improved, year by year, but always left us with a nasty taste in our mouths on not having been able to solve the problem of the uncertainty that was always present in the data that went to make up the income and expense accounts. Therefore, the answer from some time back was in our own hands! In fact in the book which saw the start to a series of works in collaboration with Professor Kaufmann(3) we had included a section in one of the chapters on the "Fuzzy Zero Base Budgeting (F.Z.B.B.) What we were looking for was something to allow the estimates made by each Decision Centre and also by the Club Manager, in our case, to be expressed by *means of uncertain* values. The procedure to be followed would continue to be the same.

What had to be agreed upon first was the type of figure to be chosen for putting a value to each one of the expense levels of the different "sections". We talked about *confidence intervals* and *triplets*, but in the end we decided on the use of "fuzzy triangular numbers" (F.T.N). Let us recall that the F.T.N is an uncertain number that expresses the *lower limit* below which the expense will *not* take place, a *maximum presumption* representing what we feel with the greatest intensity will occur and an *upper* limit above which logically no expense will arise. This is why the fuzzy triangular number is normally shown. among other forms, in the so called ternary form, expressed by three numbers a_1, a_2, a_3 placed within brackets, as shown below:

$$(a_1, a_2, a_3)$$

where a_1 is the lower limit or extreme, a_2 the maximum presumption and a_3 lthe upper limit. We should also remember that the *presumptions* we will mark with values from *zero* to *one* in this way giving values to a letter μ which would represent them) within the segment [0, 1]. The greater the presumption, the closer to one will be the assigned number, and the less presumption they will be more removed from one and therefore closer to zero. This is why a fuzzy triangular number can be drawn in a graphical representation by means of a triangle such as we have shown below:

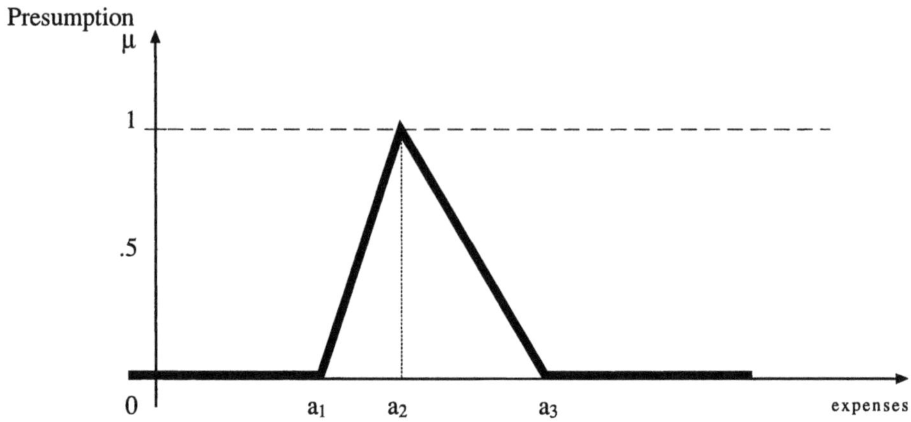

We knew all about this, but it was necessary to explain it to all the people responsible for the Decision Centres. We started with the president of the football section. That year we had just won the Spanish Championship, "The League" and everyone was highly elated and it seemed that everything was possible. The President, always cautious and prudent put a stop to these "attacks " of madness by contributing the necessary calmness. The *sequences* that had been provided in earlier years by the successive budgetary levels had not varied, on not positioning the Club within the social nor the sporting context. But what had varied, and significantly, were the costs of maintaining the team, which even at that time had generated an expansion spiral. When the new manner of estimating expenses was decided upon, the section president gave a sigh of relief which was soon cut short when Josep Lluis made it quite clear that "an increase in flexibility, meant an increase in responsibility". Under no circumstance or any pretext whatsoever would it be possible to exceed the *upper limit.* And every expense that was susceptible to increasing the overall figure above the *maximum* presumption would be looked at "under a microscope". The same scene was repeated with all the presidents of the basketball, handball, roller skate hockey and other sections. A few days later we received the corresponding data which we reproduce below:

A = Football

$$\underset{\sim}{A}_0 = (6.000, 6.500, 8.000)$$

$$\underset{\sim}{A}_1 = (8.500, 9.000, 9.500)$$

$$\underset{\sim}{A}_2 = (9.500, 10.000, 11.000)$$

$$\underset{\sim}{A}_3 = (12.000, 14.000, 15.000)$$

B = Basketball

$$\underset{\sim}{B}_0 = (1.000, 1.200, 1.500)$$

$$\underset{\sim}{B}_1 = (1.500, 2.000, 2.500)$$

$$\underset{\sim}{B}_2 = (3.000, 4.000, 5.000)$$

C = Handball

$$\underset{\sim}{C}_0 = (600, 650, 700)$$

$$\underset{\sim}{C}_1 = (1.200, 1.500, 2.000)$$

D = Roller Skate Hockey

$$\underset{\sim}{D}_0 = (500, 600, 650)$$

$$\underset{\sim}{D}_1 = (800, 900, 1.000)$$

E = Other sections

$$\underset{\sim}{E}_0 = (800, 850, 900)$$

$$\underset{\sim}{E}_1 = (900, 1.000, 1.100)$$

$$\underset{\sim}{E}_2 = (1.400, 1.600, 1.700)$$

George the president of the handball section saw that under each of the capital letters we placed a swung dash. He was right when he commented to his companions that this must mean that it referred to a fuzzy number. In this way it is distinguished from an exact number.

With this information we were now ready to calculate the successive budgets according to their increasing levels of requirement, as we had been doing over the last few years but now with the F.T.N. we had arrived at. The process was very simple, as the sum was the only operator to be used, at least for now. We knew that in order to add F.T.N. it was sufficient to add the lower limits together, the maximum presumptions together and then the upper extremes or limits also together. The results corresponding to the successive frequencies were as follows:

Budget level I

$$\underset{\sim}{A}_0 = (6.000, 6.500, 8.000)$$

Budget level II

$$\underset{\sim}{A}_1 = (8.500, 9.000, 9.500)$$

Budget level III

$$\underset{\sim}{A}_1 (+) \underset{\sim}{B}_0 = (8.500, 9.000, 9.500) (+) (1.000, 1.200, 1.500) =$$

$$= (9.500, 10.200, 11.000)$$

Budget level IV

$$\underset{\sim}{A}_1 (+) \underset{\sim}{B}_0 (+) \underset{\sim}{E}_0 = (8.500, 9.000, 9.500) (+) (1.000, 1.200, 1.500) (+)$$
$$(+) (800, 850, 900) = (10.300, 11.050, 11.900)$$

Budget level V

$$\underset{\sim}{A}_2 (+) \underset{\sim}{B}_0 (+) \underset{\sim}{E}_0 = (9.500, 10.000, 11.000) (+) (1.000, 1.200, 1.500) (+)$$
$$(+) (800, 850, 900) = (11.300, 12.050, 13.400)$$

Budget level VI

$$\tilde{A}_2 (+) \tilde{B}_0 (+) \tilde{C}_0 (+) \tilde{E}_0 = (9.500, 10.000, 11.000) (+) (1.000, 1.200, 1.500) (+) (600, 650, 700) (+) (800. 850, 900) = = (11.900, 12.700, 14.100)$$

Budget level VII

$$\tilde{A}_2 (+) \tilde{B}_0 (+) \tilde{C}_0 (+) \tilde{D}_0 (+) \tilde{E}_0 = (9.500, 10.000, 11.000) (+) (+) (1.000, 1.200, 1.500) (+) (+) (600, 650, 700) (+) (500, 600, 650) (+) (+) (800, 850, 900) = = (12.400, 13.300, 14.750)$$

Budget level VIII

$$\tilde{A}_3 (+) \tilde{B}_0 (+) \tilde{C}_0 (+) \tilde{D}_0 (+) \tilde{E}_0 = (12.000, 14.000, 15.000) (+) (+) (1.000, 1.200, 1.500) (+) (+) (600, 650, 700) (+) (500, 600, 650) (+) (+) (800, 850, 900) = = (14.900, 17.300, 18.750)$$

Budget level IX

$$\tilde{A}_3 (+) \tilde{B}_1 (+) \tilde{C}_0 (+) \tilde{D}_0 (+) \tilde{E}_0 = (12.000, 14.000, 15.000) (+) (+) (1.500, 2.000, 2.500) (+) (+) (600, 650, 700) (+) (500, 600, 650) (+) (800, 850, 900) = (15.400, 18.100, 19.750)$$

Budget level X

$$\tilde{A}_3 (+) \tilde{B}_1 (+) \tilde{C}_0 (+) \tilde{D}_0 (+) \tilde{E}_1 = (12.000, 14.000, 15.000) (+) (+) (1.500, 2.000, 2.500) (+) (+) (600, 650, 700) (+) (500, 600, 650) (+) (+) (900, 1.000, 1.100) = = (15.500, 18.250, 19.950)$$

Budget level XI

$$\underset{\sim}{A}_3 (+) \underset{\sim}{B}_2 (+) \underset{\sim}{C}_0 (+) \underset{\sim}{D}_0 (+) \underset{\sim}{E}_1 = (12.000, 14.000, 15.000) (+)$$
$$(+) (3.000, 4.000, 5.000) (+)$$
$$(+) (600, 650, 700) (+) (500, 600, 650) (+)$$
$$(+) (900, 1.000, 1.100) =$$
$$= (17.000, 20\ 250, 22.450)$$

Budget level XII

$$\underset{\sim}{A}_3 (+) \underset{\sim}{B}_2 (+) \underset{\sim}{C}_1 (+) \underset{\sim}{D}_0 (+) \underset{\sim}{E}_1 = (12.000, 14.000, 15.000) (+)$$
$$(+) (3.000, 4.000, 5.000) (+)$$
$$(+) (1.200, 1.500, 2.000) (+) (500, 600, 650)$$
$$(+) (900, 1.000, 1.100) =$$
$$= (17.600, 21.100, 23.750)$$

Budget level XIII

$$\underset{\sim}{A}_3 (+) \underset{\sim}{B}_2 (+) \underset{\sim}{C}_1 (+) \underset{\sim}{D}_1 (+) \underset{\sim}{E}_1 = (12.000, 14.000, 15.000) (+)$$
$$(+) (3.000, 4.000, 5.000) (+)$$
$$(+) (1.200, 1.500, 2.000) (+)$$
$$(+) (800, 900, 1.000) (+)$$
$$(+) (900, 1.000, 1.100) =$$
$$= (17.900, 21.400, 24.100)$$

Budget level XIV

$$\underset{\sim}{A}_3 (+) \underset{\sim}{B}_2 (+) \underset{\sim}{C}_1 (+) \underset{\sim}{D}_1 (+) \underset{\sim}{E}_2 = (12.000, 14.000, 15.000) (+)$$
$$(+) (3.000, 4.000, 5.000) (+)$$
$$(+) (1.200, 1.500, 2.000) (+)$$
$$(+) (800, 900, 1.000) (+)$$
$$(+) (1.400, 1.600, 1.700) =$$
$$= (18.400, 22.000, 24.700)$$

This part of the budgetary management, of extreme importance, was finished, but James, the excellent Club Manager asked how was it possible to resolve its uncertainty when estimating the total figure for resources for the season to be budgeted for. We resolved this question by stating formally: for each of the income components we will select the least that can be obtained and also the most. James saw this quite clearly and delivered the following estimate:

	Pessimistic position	Utopian position
Non member ticket sales	800	900
Merchandising	1,500	2,000
Television rights	5,500	6,000
Advertising	900	1,200
Sponsorships	2,700	2,800
Membership fees and season tickets	4,900	5,000
Direct product sales	1,100	1,200
Other income	600	900
Total	18,000	20,000

In the position we have called *pessimistic* we included what the Manager said he was *sure* that we were going to obtain, since for practically all of these items there were signed contracts. He also stated that it was totally *impossible* to arrive at the figures under the *utopian column*, and obviously any other figure. Speaking in technical terms, the "presumption" in *pessimism* was assigned a value of one and *utopia* a *zero*. James also reasoned that when passing from figures close to 18,000 to figures close to 20,000 the presumption of arriving at the same was increasingly less. We agreed on representing this manner of reasoning in a graphical representation:

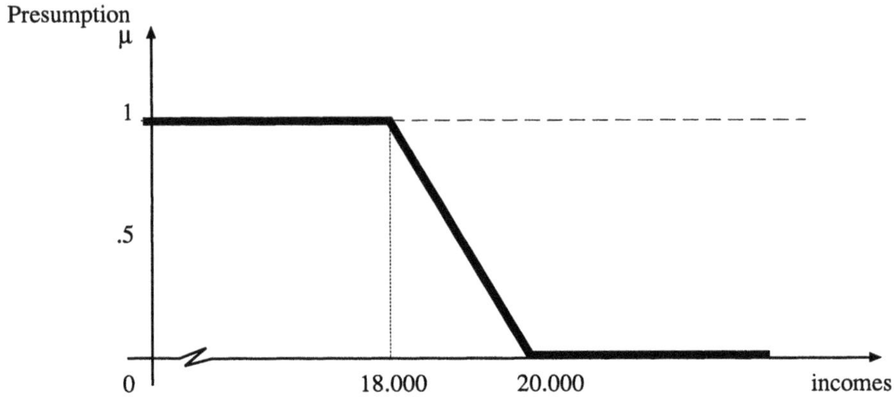

Now the moment of truth had arrived. Having gathered all the data from one and another part, a decision had to be made to accept one budget level over fourteen possibilities.

We have always felt that a good graphical presentation provides a perfect view of the problems and leads to adequate decisions. We then set out on the task of drawing a figure of the budget levels that could be selected, which we did with the following eight:

$$\underset{\sim}{A}_2 (+) \underset{\sim}{B}_0 (+) \underset{\sim}{C}_0 (+) \underset{\sim}{D}_0 (+) \underset{\sim}{E}_0 = (12.400,\ 13.300,\ 14.750)$$

$$\underset{\sim}{A}_3 (+) \underset{\sim}{B}_0 (+) \underset{\sim}{C}_0 (+) \underset{\sim}{D}_0 (+) \underset{\sim}{E}_0 = (14.900,\ 17.300,\ 18.750)$$

$$\underset{\sim}{A}_3 (+) \underset{\sim}{B}_1 (+) \underset{\sim}{C}_0 (+) \underset{\sim}{D}_0 (+) \underset{\sim}{E}_0 = (15.400,\ 18.100,\ 19.750)$$

$$\underset{\sim}{A}_3 (+) \underset{\sim}{B}_1 (+) \underset{\sim}{C}_0 (+) \underset{\sim}{D}_0 (+) \underset{\sim}{E}_1 = (15.500,\ 18.250,\ 19.950)$$

$$\underset{\sim}{A}_3 (+) \underset{\sim}{B}_2 (+) \underset{\sim}{C}_0 (+) \underset{\sim}{D}_0 (+) \underset{\sim}{E}_1 = (17.000,\ 20.250,\ 22.450)$$

$$\underset{\sim}{A}_3 (+) \underset{\sim}{B}_2 (+) \underset{\sim}{C}_1 (+) \underset{\sim}{D}_0 (+) \underset{\sim}{E}_1 = (17.600,\ 21.100,\ 23.750)$$

$$\underset{\sim}{A}_3 (+) \underset{\sim}{B}_2 (+) \underset{\sim}{C}_1 (+) \underset{\sim}{D}_1 (+) \underset{\sim}{E}_1 = (17.900,\ 21.400,\ 24.100)$$

$$\underset{\sim}{A}_3 (+) \underset{\sim}{B}_2 (+) \underset{\sim}{C}_1 (+) \underset{\sim}{D}_1 (+) \underset{\sim}{E}_2 = (18.400,\ 22.000,\ 24.700)$$

The triangles representing these are as follows:

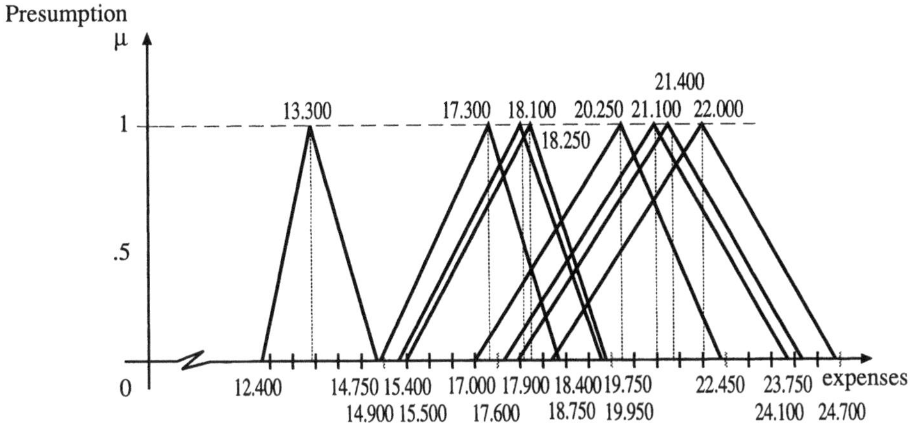

A quick glance at this "jumble of fuzzy triangular numbers" showed us that as the triangle represented higher budgetary levels it *moved towards* the *right*. This confirmed that the F.T.N. can be classified in increasing order, just as happens with ordinary numbers: *the further to the right the larger* and, at the same time, this opened up the door for us to be able to make a good decision. For this we included in one, the graphs representing the budgetary levels (the latter) and the one for "financial limitations" (the former). This gave us the following graphical representation:

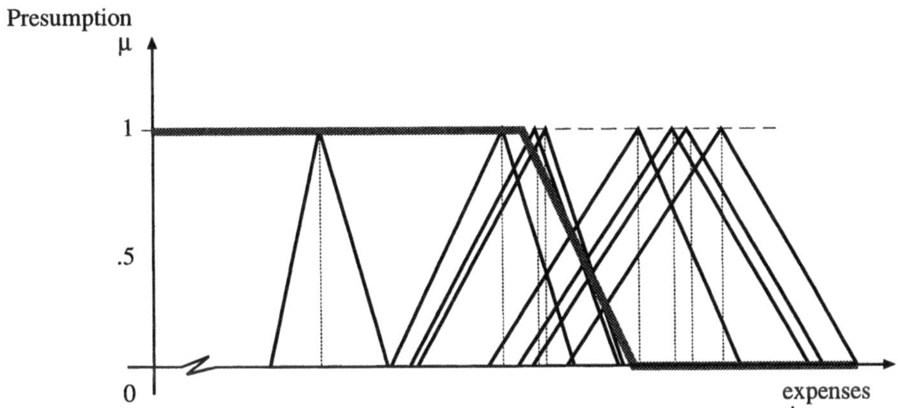

It was necessary to find a "criterion" on which to base ourselves for taking a decision. For this we did the following: the dotted line which we had called "the law of financial limitations" separated the field of the F.T.N. (triangles) into two areas. The left had side included triangles and parts of triangles that *fell within* the financial possibilities of the Club. Therefore they could be accepted without any kind of danger whatsoever. On the other hand the parts located to the right represented expenses that were more than likely not assumable. From this very simple observation a good criterion could be formed in order to facilitate an *index* capable of expressing in numbers the lack of budgetary coverage. For this reason we called it the *index of lack of budgetary coverage* and was defined as the "quotient between the area of the part of the triangle to the right of the dotted line representing the financial limitation and the total area of the triangle which expresses the studied budgetary level". Obviously the complement to the unit, will provide the index of "budgetary coverage". It all appeared to be so clear that we immediately commenced our calculations. And we did so with *budget level IX*, taking into account the fact that in the previous levels there was no doubt whatsoever as to the coverage provided. Therefore we represented the level studied in conjunction with the financial limitation by the following graphical representation:

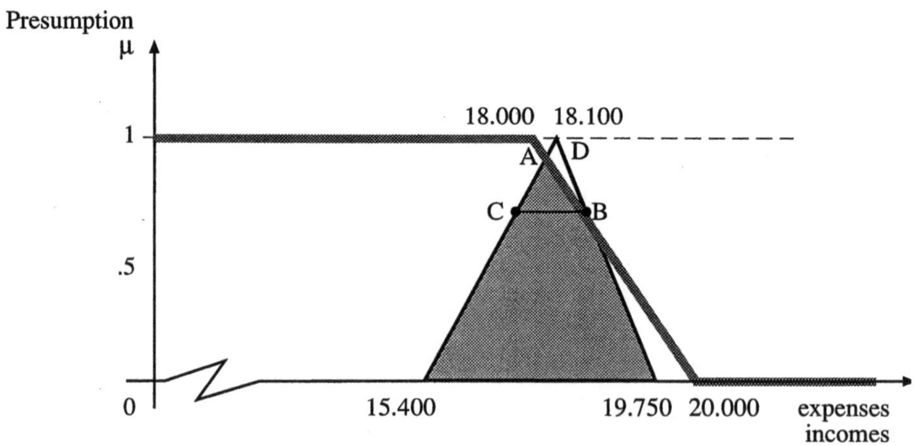

A mere glance was sufficient seeing that the *index of lack of coverage* was very small, therefore, the *index of coverage* would be very high. In spite of this we calculated it. And here we had to recall what we were taught in Secondary School, where we were told that between two points only one and only one straight line passes and that the equation of the straight line passing through two points is as follows:

$$\boxed{y - y_1 = \frac{y_2 - y_1}{x_2 - x_1}(x - x_1)}$$

In fact, in order to calculate the area of the "little triangle" outside the coverage it was necessary to arrive at the crossing over points of the sides of the triangle with the straight line representing the financial limitation, which we have called A and B.

- Equation of the left hand straight line of the triangle:

$$(x_1, y_1) = (15.400, 0)$$
$$(x_2, y_2) = (18.100, 1)$$

$$y - 0 = \frac{1 - 0}{18.100 - 15.400}(x - 15.400)$$

$$\boxed{y = \frac{x - 15.400}{2.700}}$$

- Equation of the right hand section of the triangle:

$$(x_1, y_1) = (18.100, 1)$$
$$(x_2, y_2) = (19.750, 0)$$

$$y - 1 = \frac{-1}{19.750 - 18.100}(x - 18.100)$$

$$\boxed{y = \frac{19.750 - x}{1.650}}$$

- Equation of the financial limitations straight line:

$$(x_1, y_1) = (18.000, 1)$$
$$(x_2, y_2) = (20.000, 0)$$

$$y - 1 = \frac{-1}{20.000 - 18.000}(x - 18.000)$$

$$\boxed{y = \frac{20.000 - x}{2.000}}$$

- Arriving at point A:

$$\frac{x-15.400}{2.700} = \frac{20.000-x}{2.000}$$

$$x = 18.043$$
$$y = 0.9785$$

- Arriving at point B:

$$\frac{19.750-x}{1.650} = \frac{20.000-x}{2.000}$$

$$x = 18.571$$
$$y = 0.7145$$

If from point B we draw a straight line parallel to the axis of the abscissa in the left hand side of the triangle we will arrive at *point C*, the coordinates for which are:

$$x = 17.329$$

$$y = 0.7145$$

We now move on to calculate the sought after area. For this we start by the area of the triangle BCD, which is as follows:

$$\widehat{BCD} = (18.571 - 17.329) \,.\, (1 - 0.7145)/2 = 177.3$$

and follow on with triangle ABC:

$$\widehat{ABC} = (18.571 - 17.329) \,.\, (0.9785 - 0.7145)/2 = 163.9$$

The difference between the two triangles gave us the sought after value of ABD:

$$\widehat{ABD} = \widehat{BCD} - \widehat{ABC} = 177.3 - 163.9 = 13.4$$

The area of the triangle representing budget level IX is:

$$\widehat{P_{IX}} = (19.750 - 15.400) \,.\, 1/2 = 2.175$$

Therefore the "Index of lack of coverage" I_{FC}, is:

$$I_{FC} = \frac{13{,}4}{2.175} = 0.006$$

and the "index of coverage":

$$I_C = 1 - 0.006 = 0.994$$

The very high value of I_C leaves no doubt whatsoever in accepting budget level IX, as the possibilities that income will not cover expenses is very remote. So we could now move on to budget level X.

We commenced the process once again along the same path we had just travelled. In this case the graph had the following form:

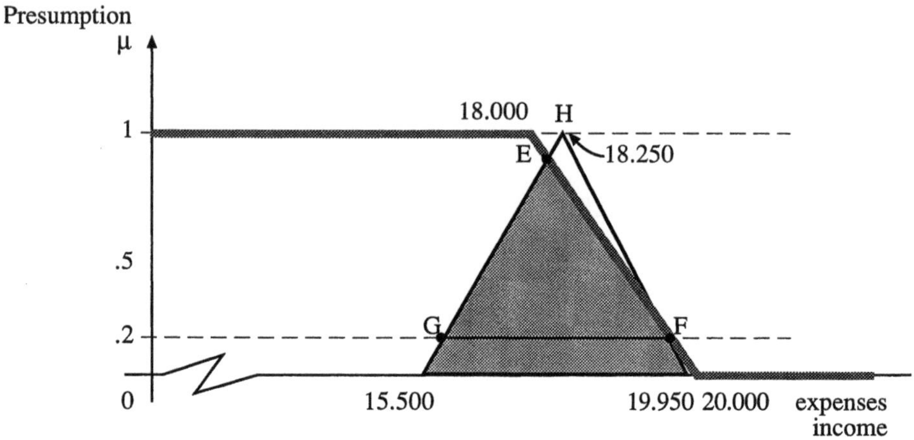

- Equation of the left hand side of the triangle:

$$y = \frac{x\text{-}15.500}{4.500}$$

- Equation of the financial limitations straight line:

$$y = \frac{20.000 - x}{2.000}$$

- The coordinates of point E are.

$$x = 18.105$$

$$y = 0.9474$$

- Equation of the right hand side of the triangle:

$$y = \frac{19.950 - x}{1.700}$$

- We repeat the equation of the financial limitations straight line

$$y = \frac{20.000 - x}{2.000}$$

- The coordinates for point F are:

$$x = 19.667$$

$$y = 0.1665$$

- The coordinates for point G are:

$$x = 16.249$$

$$y = 0.1665$$

In order to arrive at the area of triangle EFH, we must first find the area of triangle EFG:

$$\widehat{EFG} = (19.667 - 16.249) \,.\, (0.9474 - 0.1665) / 2 = 1.334,6$$

We then continue with triangle FGH:

$$\widehat{FGH} = (19.667 - 16.249) \,.\, (1 - 0.1665) / 2 = 1.424,4$$

The area of triangle EFH which we need to obtain, is:

$$\widehat{EFH} = \widehat{FGH} - \widehat{EFG} = 1.424,4 - 1.334,6 = 89,8$$

The area of the triangle expressing budget level X is:

$$\widehat{P_X} = (19.950 - 15.500) \,.\, 1/2 = 2.225$$

Therefore the "index of lack of coverage" I_{FC} is:

$$I_{FC} = \frac{89,8}{2.225} = 0.040$$

and the "index of coverage":

$$I_C = 1 - 0.04 = 0.96$$

The high level of the latter meant there were no problems for the immediate acceptance of budget level X. Therefore we moved on to the next level.

Before we continued we had a look at the graphical representation of budget level XI in conjunction with the law of financial limitation:

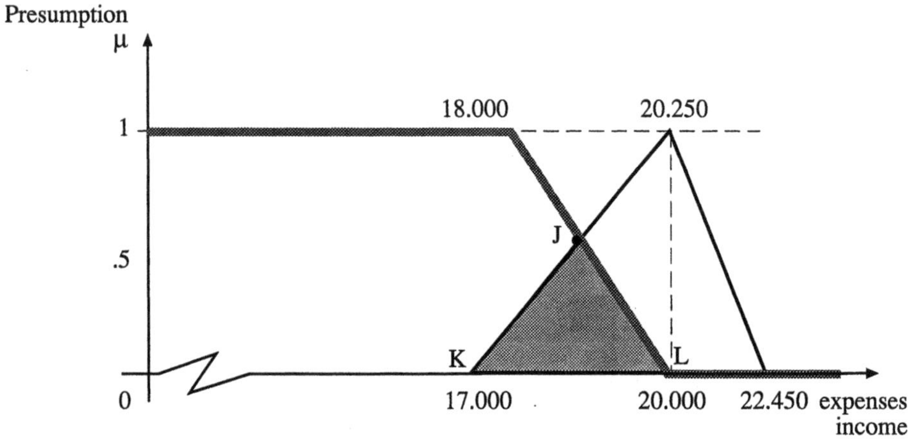

Having concluded this graphical representation, the President was very clear in his decision: he was not going to risk the stability of the Club, by setting out on an adventure with so few possibilities of success. Basically what we had was that there was a larger part of the triangle "outside" than "within" the financial *limits*. But to assure ourselves we did the corresponding calculations with the object of arriving at the necessary indices. Taking into account the fact that there is only one point of crossing with the straight line that expresses the law of financial limitation, here with the left hand side of the triangle, it will now be much simpler at the basic indices required for a decision.

- Equation of the left hand side of the triangle:

$$y = \frac{x\text{-}17.000}{3.250}$$

• Once again we reproduce the equation of the financial limitation straight line:

$$y = \frac{20.000 - x}{2.000}$$

• The coordinates for point J are:

$$x = 18.857$$

$$y = 0.5715$$

The area of the part of the triangle of the budgetary level covered by the financial limitation is:

$$\widehat{JKL} = (20.000 - 17.000) \,.\, 0.5715 / 2 = 857.3$$

The area of the triangle representing budget level XI is:

$$\widehat{P_{XI}} = (22.450 - 17.000) \,.\, 1/2 = 2.725$$

In this way we arrive at an "index of coverage" such as the following:

$$I_C = \frac{857.3}{2.725} = 0.31$$

and an "index of lack of coverage" as follows:

$$I_{FC} = 1 - 0.31 = 0.69$$

These results confirmed the initial idea arrived at by means of the graphical representation: The very elementary rules of prudence advised that we should desist from accepting a budget level, for which there was a very high possibility of it not being covered. It was in this way that the President made the correct decision, which would later allow for an increase in the Social Fund, due to a budget surplus. In this way then steps were taken to manage, in the best possible manner, budget level X which was approved and which allowed each section to operate and to aim at attain the following objectives:

Football: Win the "Champions League".

Basketball: Win the Spanish Championship.

Handball: Win the national Championship

Roller skate hockey: Win the national title.

Other sections: Maintain grass roots football and four non professional sports

For this the people responsible for each one of the *Decision Centres* would have the following financial means:

Decision Center	Not less than	Not more than	Warning number (maximum presumption)
Football	12.000	15.000	14.000
Basketball	1.500	2.500	2.000
Handball	600	700	650
Roller-skate hockey	500	650	600
Other sections	900	1.100	1.000
TOTALS	5.500	19.950	18.250

The majority of the "sections" achieved the desired objectives, but the greatest satisfaction felt by the members was winning the long sought after "European Cup" which was the achievement that the fans had for a very long time hoped for. And, thus for the first time in its history the Club became the finest entity in Europc and was considered as the best managed club in the world.

A test with fuzzy trapezoidal numbers

But enjoying the greatest historical triumph of the Club did not prevent our President from continuing to seek new ways for making it "even greater", as he was wont to say. One of these consisted in the continuous improvement of the economic-financial management, at the centre of which was the budget.

Throughout the continuous meetings we held with those responsible for the five decision centres, it was frequent stated by those present that it was difficult to assign only a single value for the "maximum presumption", that is to say the amount of money which it was assumed would be the most possible to spend at the corresponding budget level (the central number of the fuzzy triangular number). On many occasions we had to force a section president to represent thee different budget levels by means of F.T.N. Resolving this problem would allow our section presidents to feel more comfortable, and this very comfort would oblige them even more to keep to the figures, for the setting of which they had been provided with every facility. This was the reason for incorporating the "fuzzy trapezoidal numbers", in short F.Tp.N[4].

Accepting the fact that the people responsible for the different sections expressed their opinions by means of F.Tp.N. meant that they offered a value below which the expense would not be incurred, (lower extreme), a value above which there would be no expense (upper extreme), and two limits between which they felt with the greatest strength that the expense would be incurred (maximum presumption). This latter information means that the presumption is set down in a confidence interval.

In this way, if a fuzzy triangular number can be expressed by a triplet:

$$\underset{\sim}{a} = (a_*, a, a^*)$$

in which a_* is the lower extreme, a^* the upper extreme and a the maximum presumption: for a fuzzy trapezoidal number we would have:

$$\underset{\sim}{a} = (a_*, [a_1, a_2]\ a^*)$$

in which, on this occasion the maximum presumption is represented by a confidence interval $[a_1, a_2]$.

For greater clarity and comparison reasons below we show the two types as graphical representation:

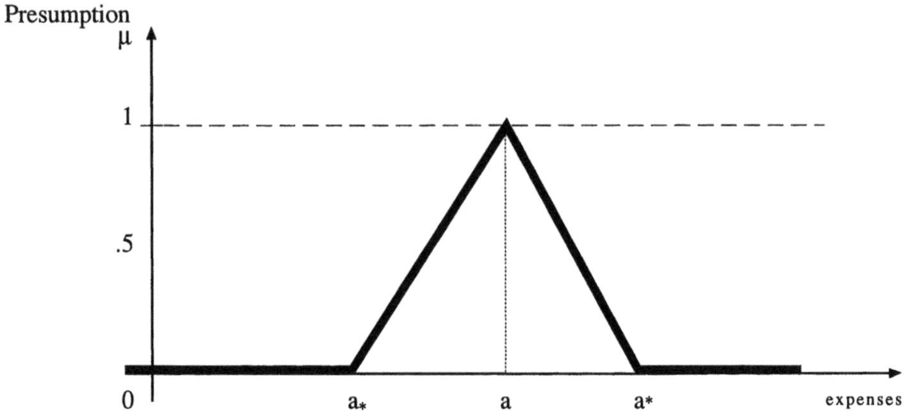

Fuzzy triangular number

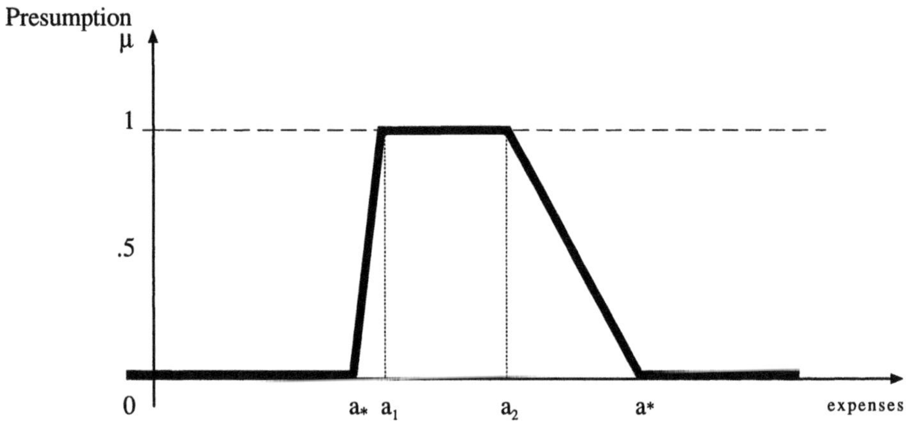

Fuzzy trapezoidal number

From a theoretical point of view, there was no problem in the passage from the use of a F.T.N to the use of a F.Tp.N. It was also clear that we should make a trial run with this new fuzzy number with the next Club budgets. And here also was an added advantage: both if the people responsible provided their information by means of F.T.N, or by F.Tp.N. both types could be expressed and later calculations made with them, as F.Tp.N. This we saw throughout the process.

Having accepted for a further year the same sequences with identical levels for each decision centre, after the succession of meetings the following estimates were accepted:

A = Football

$\underset{\sim}{A}_0 = (6.000, [6.200, 6.600], 8.000)$

$\underset{\sim}{A}_1 = (8.000, [8.500, 9.000], 10.000)$

$\underset{\sim}{A}_2 = (10.000, [10.500, 11.000], 12.000)$

$\underset{\sim}{A}_3 = (12.000, [14.000,15.000], 16.000)$

B = Basketball

$\underset{\sim}{B}_0 = (1.100, [1.200, 1.400], 1.600)$

$\underset{\sim}{B}_1 = (1.500, [1.800, 2.200], 2.600)$

$\underset{\sim}{B}_2 = (3.000, [3.500, 4.200], 5.500)$

C = Handball

$\underset{\sim}{C}_0 = (600, 650, 700) = (600,[650, 650], 700)$ (5)

$\underset{\sim}{C}_1 = (1.400, [1.500, 1.600], 2.000)$

D = Roller-skaye hockey

$\underset{\sim}{D}_0 = (500, [580, 620], 650)$

$\underset{\sim}{D}_1 = (750, [850, 920], 1.000)$

E = Other sections

$\underset{\sim}{E}_0 = (800, 850, 900) = (800, [850, 850], 900)$

$\underset{\sim}{E}_1 = (900, 1.000, 1.100) = (900, [1.000, 1.000], 1.100)$

$\underset{\sim}{E}_2 = (1.300, [1.500, 1.700], 1.800)$

For a further year we proceeded to make the calculations leading to obtaining the successive budget levels for the Club. With the experience we had accumulated we commenced the operation with budget level VII knowing full well that the previous six levels were not in line with the requirements of the Club nor with the resources that were going to be generated. We arrived at the following results:

Budget level VII

$$\underset{\sim}{A}_2 (+) \underset{\sim}{B}_0 (+) \underset{\sim}{C}_0 (+) \underset{\sim}{D}_0 (+) \underset{\sim}{E}_0 =$$

= (10.000, [10.500, 11.000], 12.000 (+)

(+) (1.100, [1.200, 1.400], 1.600) (+) (600,]650, 650], 700) (+)

(+) (500, [580, 620], 650) (+) (800, [850, 850], 900) =

= (13.000, [13.780, 14.520], 15.850)

Budget level VIII

$$\underset{\sim}{A}_3 (+) \underset{\sim}{B}_0 (+) \underset{\sim}{C}_0 (+) \underset{\sim}{D}_0 (+) \underset{\sim}{E}_0 =$$

= (12.000, [14.000, 15.000], 16.000) (+)

(+) (1.100, [1.200, 1.400], 1.600) (+) (600,[650, 650], 700) (+)

(+) (500, [580, 620], 650) (+) (800, [850, 850], 900) =

= (15.000, [17.280, 18.520], 19.850)

Budget level IX

$$\underset{\sim}{A}_3 (+) \underset{\sim}{B}_1 (+) \underset{\sim}{C}_0 (+) \underset{\sim}{D}_0 (+) \underset{\sim}{E}_0 =$$

= (12.000, [14.000, 15.000], 16.000) (+)

(+) (1.500, [1.800. 2.200], 2.600) (+) (600, [650, 650], 700) (+)

(+) (500, [580, 620], 650) (+) (800, [850, 850], 900) =

= (15.400, [17.880, 19.320], 20.850)

Budget level X

$\underset{\sim}{A}_3 (+) \underset{\sim}{B}_1 (+) \underset{\sim}{C}_0 (+) \underset{\sim}{D}_0 (+) \underset{\sim}{E}_1 =$

= (12.000, [14.000, 15.000], 16.000) (+)

(+) (1.500, [1.800, 2.200], 2.600) (+) (600, [650, 650], 700) (+)

(+) (500, [580, 620], 650) (+) (900, [1.000, 1.000], 1.100) =

= (15.500, [18.030, 19.470], 21.050)

Budget level XI

$\underset{\sim}{A}_3 (+) \underset{\sim}{B}_2 (+) \underset{\sim}{C}_0 (+) \underset{\sim}{D}_0 (+) \underset{\sim}{E}_1 =$

= (12.000, [14.000,15.000], 16.000) (+)

(+) (3.000, [3.500, 4.200], 5.500) (+) (600, [650, 650], 700) (+)

(+) (500, [580, 620], 650) (+) (900, [1.000, 1.000], 1.100) =

= (17.000, [19.730, 21.470], 23.950)

Budget level XII

$\underset{\sim}{A}_3 (+) \underset{\sim}{B}_2 (+) \underset{\sim}{C}_1 (+) \underset{\sim}{D}_0 (+) \underset{\sim}{E}_1 =$

= (12.000, [14.000, 15.000], 16.000) (+)

(+) (3.000, [3.500, 4.200], 5.500) (+) (1.400, [1.500, 1.600], 2.000)

(+) (500, [580, 620], 650) (+) (900, [1.000, 1.000], 1.100) =

= (17.800, [20.580, 22.420], 25.250)

Budget level XIII

$\underset{\sim}{A}_3 (+) \underset{\sim}{B}_2 (+) \underset{\sim}{C}_1 (+) \underset{\sim}{D}_1 (+) \underset{\sim}{E}_1 =$

= (12.000, [14.000, 15.000], 16.000) (+)

(+) (3.000, [3.500, 4.200], 5.500) (+) (1.400, [1.500, 1.600], 2.000)

(+) (750, [850, 920], 1.000) (+) (900, [1.000, 1.000], 1.100) =

= (18.050, [20.850, 22.720], 25.600)

Budget level XIV

$\underset{\sim}{A}_3 (+) \underset{\sim}{B}_2 (+) \underset{\sim}{C}_1 (+) \underset{\sim}{D}_1 (+) \underset{\sim}{E}_2 =$

= (12.000, [14.000, 15.000], 16.000) (+)

(+) (3.000, [3.500, 4.200], 5.500) (+) (1.400, [1.500, 1.600], 2.000)

(+) (750, [850, 920], 1.000) (+) (1.300, [1.500, 1.700], 1.800) =

= (18.450, [21.350, 23.320], 26.500)

Over the last few sporting seasons, under the direction of the Club President, and thanks to his imagination and fierce tenacity, the Club had managed to significantly increase resources obtained from financing sources other than members and season ticket holders. This led him to a certain amount of unpopularity which he assumed bravely, placing the institutional interests before his own personal interests. Also what was achieved was to reduce uncertainty by securing and closing agreements in the medium term, which were also open to future increases. The Club manager presented the following estimates:

	Pessimistic position	*Utopian position*
Non member ticket sales	900	1,000
Merchandising	2,000	2,500
Television rights	6,200	6,500
Advertising	1,300	1,500
Sponsoships	3,000	3,100
Members fees and season tickets	5,100	5,200
Direct product sales	1,500	1,600
Other income	2,000	2,100
Totals	22,000	23,500

As we did in previous years, we showed this information in the form of a graphical representation, as follows:

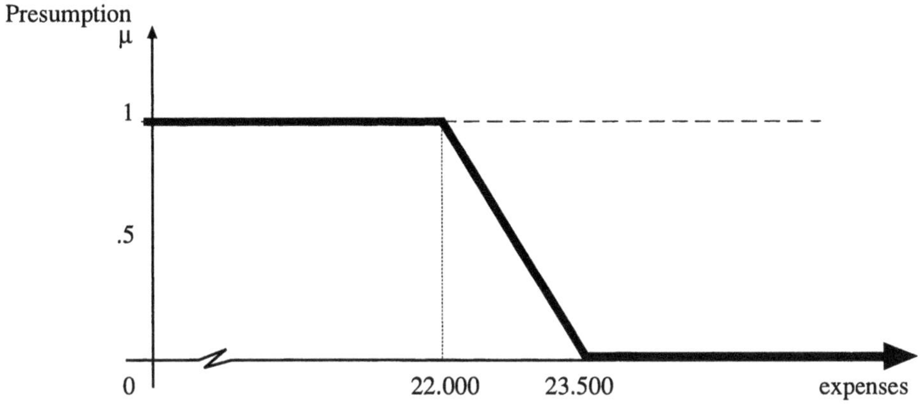

In this way we arrived at the last part of the process, not before showing, also in the form of a graphical representation, the budget levels that were susceptible to being chosen. In that season they were the last five.

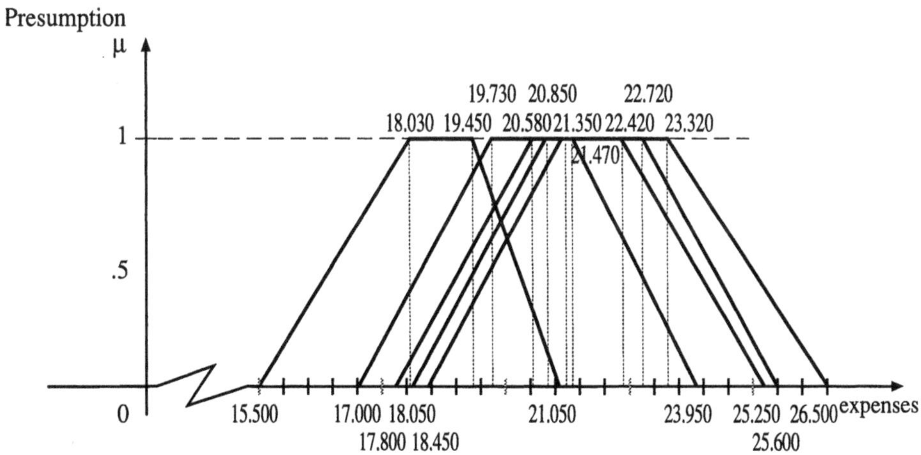

We then moved on to unite in a single graphical representation this whole range of fuzzy trapezoidal numbers and the law of financial limitation with the following result:

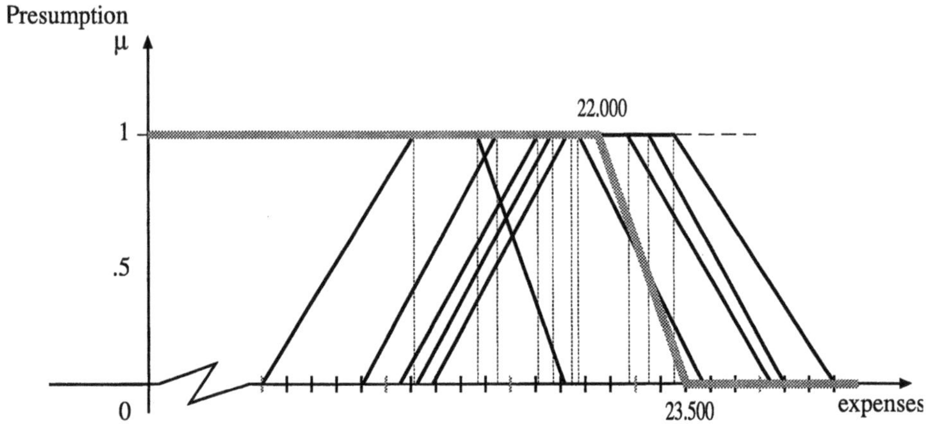

We were now in the position of using a decision criterion. We decided to continue with the "index of lack of coverage" and its complement the "index of coverage".

It was quite obvious that we should accept budget level X, without reservation, with level XI appearing to be assumable. In order to verify this we commenced calculation with this budget level. We then drew a graph which included the budget level and the financial limitation.

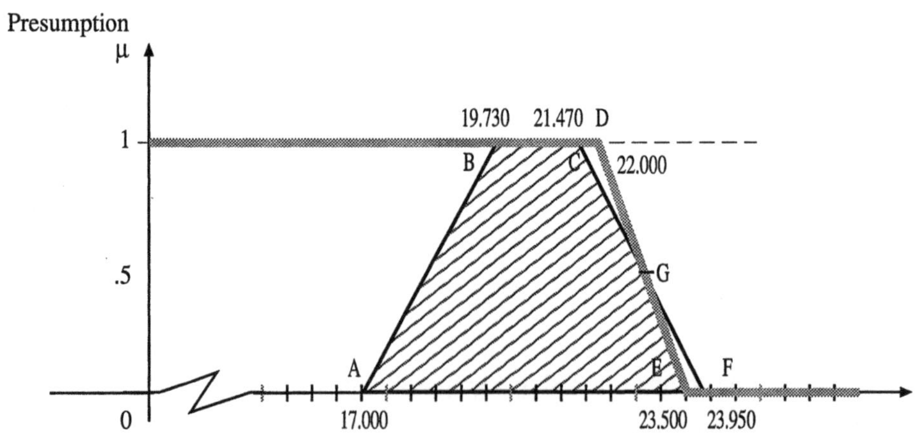

In order to arrive at the corresponding indices required, on this occasion, the prior calculation of the coordinates for point G. For this we arrived at the equations of the straight lines CF from the trapezoid relative to the budget level and DE corresponding to the law of financial limitation. These were as follows:

$$y = \frac{2.480}{23.950}x$$

$$y = \frac{1.500}{23.500}x$$

and the sought after coordinates:

$$x = 22.812$$

$$y = 0.46$$

It was then easy to calculate the desired indices. Therefore the *"index for lack of coverage"* was found as the quotient between the area of triangle EFG and that for the trapezoid ABCF.

$$\widehat{EFG} = 1/2\ (23.950 - 23.500)\ .\ (0.46) = 103.5$$

$$\widehat{ABCF} = 1/2\ [(23.950 - 17.000) + (21.470 - 19.730)]\ .\ (1) = 4.345$$

In this way:

$$I_{FC} = \frac{\widehat{EFG}}{\widehat{ABCF}} = \frac{103{,}5}{4345} = 0{,}024$$

$$I_C = 1 - 0.024 = 0.976$$

The high value of the "coverage index" made acceptance of budget level XI almost automatic. We then looked at the possibility of accepting the following budget level with the graphical representation below:

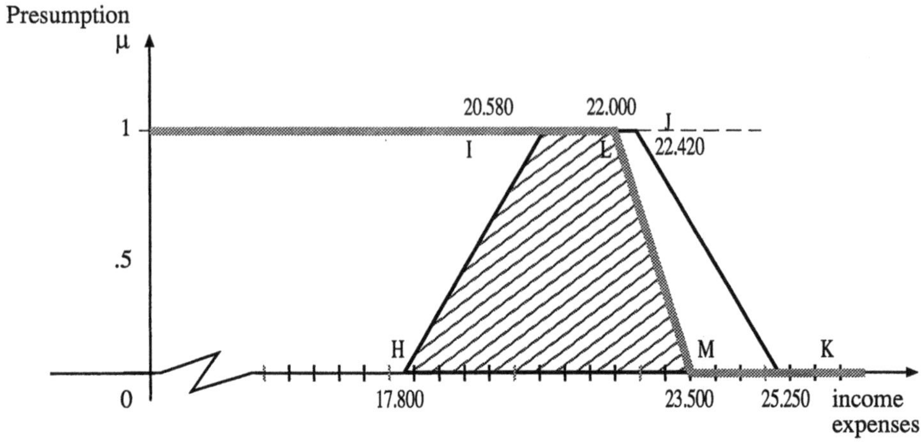

In view of the clarity of the figure, we immediately moved on to calculate the "coverage index" which, in this case, did not require any further operations other than arriving at the quotient between the shaded trapezoid and the trapezoid representing the F.Tp.N. as follows:

$$I_C = \frac{\overline{HILM}}{\overline{HIJK}} = \frac{1/2[(23.500\text{-}17.800) + (22.000\text{-}20.580)]\cdot(1)}{1/2[(25.250\text{-}17.800) + (22.420\text{-}20.580)]\cdot(1)} = \frac{7.120}{9.290} = 0{,}77$$

then arriving at the *"index for lack of coverage"* is immediate:

$$I_{FC} = 1 - 0.77 = 0.23$$

It was not easy to take a decision when faced with indices which were located at the limits of what prudence advised. But the President did not disappoint us and personally assumed the responsibilities of his office and encouraged everyone to work along the line of austerity which was so characteristic of his long and fruitful years as head of the destinies of the Club. Therefore, budget level XII was accepted, which meant aiming for the following objectives:

Football:	Win the "Champions League"
Basketball:	Win the European Championship

Handball: Win the European competition.

Roller- skate hockey: Win the Spanish league.

Other sections: Maintain grass roots football and a further four sports.

Each one of the decision centres would have to manage, and they did so, the following monetary resources:

Decision Centre	Not less than	Not more than	Warning number (maximum presumption)
Football	12.000	16.000	[14.000, 15.000]
Baasketball	3.000	5.500	[3.500, 4.200]
Handball	1.400	2.000	[1.500, 1.600]
Roller-skate hockey	500	650	[580, 620]
Othe sections	900	1.100	[1.000, 1.000]
TOTALS	17.800	25.250	[20.580, 22.420]

Looking back, we saw how the world of sport had changed, particularly football. The financial needs of the clubs had multiplied over a few years, and in order to maintain their position in the elite international competitions required professionalism not only in the purely sporting aspects but also in the organisational structures of those entities dedicated to sport. We were convinced, and continue to be so, that in order to survive successfully in the sporting-spectacle jungle it was necessary to base the sporting institutions on very solid foundations, the basic pillars of which were the sporting, economic-financial, social and organisational spheres. Josep Lluis was very much aware what it meant to maintain a company, in constant growth, solvent. And he applied all his wisdom to converting his Club into the best and most admired club in the world. No sporting entity had ever presented such a well balanced and solvent balance sheet and a more brilliant profit and loss account than those provided by the President at his farewell as President of the Club. He left his office not because of tiredness or because his task had finished, but as a last act of service, resigning from his office so that the jealousies, hate and envy of those who are incapable of recognising the merits of others and the cravings of the media and political groups would not damage even further the legitimate prestige that the Club had attained on the five continents.

REFERENCES

[1] KAUFMANN, A. and GIL ALUJA, J.: *Introducción de la teoría de los subconjuntos borrosos a la gestión de las empresas*. publ. Milladoiro. Santiago de Compostela 1986, page.54.

[2] KAUFMANN, A. and GIL ALUJA, J.: *Introducción de la teoría de los subconjuntos borrosos a la gestión de las empresas*. publ. Milladoino. Santiago de Compostela 1986, page 56

[3] KAUFMANN, A. and GIL ALUJA, J.: *Introducción de la teoría de los subconjuntos borrosos a la gestión de las empresas*. publ. Milladoiro. Santiago de Compostela 1986 pages 59 - 66.

[4] KAUFMANN, A. and GIL ALUJA, J.: *Técnicas de gestión de empresas. Previsiones, decisiones y estrategias*. publ. Pirámide Madrid 1992. pages 200-205.

[5] In this case the zero budget level was estimated by means of a F.T.N. which, for later operations, we transferred to the trapezoidal form.

V ECONOMIC TREATMENT OF FIXED ASSETS

Fixed assets included within production activity

Increasingly, businesses, institutions and also individual professionals, are obliged to resort to the purchase of more expensive and sophisticated equipment, be this for the offices, or the factory, with the object of competing, in quality and price, with a presence that invades the available markets with a high dosage of commercial aggressiveness. But technological advances, that are increasingly faster and more profound add to the traditional problems of investment, the always important problem of knowing the *optimum moment* for renewal of operating equipment.

For the study of this problem, writers on the subject, start out from the premise of the existence of the equipment (machine, production installations, computers, copiers, ...) and they bring up the subject of their renewal or substitution at a certain time, acting on economic decision criteria. This could mean the acquisition of a machine and then the problem arises at what future moment should the old equipment be retired, or of an already existing piece of equipment and then the problem is if it should continue or a new one be acquired.

There is no need to point out that what we have just stated is a very short or partial summary of the vast content of the so-called theory of the renewal of equipment. It is said that an existing fixed asset looses its *economic usefulness* when the newer and more modern one that can be acquired allows for reducing costs. One question should be heavily underlined at this point: That the criteria on which the renewal studies are based should be actual and *accounting criteria.* For this reason there is no significance in the state of the depreciation of the existing equipment[1].

The study of equipment renewal is a decisional field that permits acting in a adequately economic way on the realities of businesses, institutions and professional activities. And this is so because, "although it is possible that surrounding the optimum moment for renewal there were to be no important differences between current operational costs with the new or existing asset, as the decision moves away from the optimum moment, increases in the same can become increasingly accelerated"[2]. Let us not forget that in operating costs, depreciation occupies a prominent position, as a consequence of the gradual but insistent substitution of human work by that of machines.

The determination of the optimum moment for equipment renewal is a constantly present problem in economic activity. Just a few months ago I had the opportunity of cooperating with my friend Bob in finding the solution to a question which, according to him, did not let him sleep. Robert, Bob to his friends, was a companion in those happy years of secondary school and with him we shared adventures that were typical of the youth and times in which we lived. Son of a German father and Spanish mother, who I feel inherited the best of the central European character and the purest essences of the Latin temperament. Thoughtful and a dreamer, he commenced his working life in Germany during the seventies, he became successful as an expert for musically implementing images for the film and advertising industries. He became an authority in this field with the use of computers. He had an eventful sentimental life with two marriages and two divorces, a rare occurrence in such a good, straightforward and quiet person. His friends always felt that the Latin sensitivity must not have fit in too well in those freezing nights alongside the Baltic. Now in his maturity he has returned to our country. Here he appears to have settled down in his personal life without closing his German company which he directs from the distance with the aid of one of his sons.

Bob is obsessed with his complex collection of computers and electronic complements. The investment required for a change or renewal of all this equipment is very high, even though later the "expenses" (always using his terminology) would be very much decreased. But it also happens, he says, that in stead of waiting a year in order to acquire the more modern electronic equipment I buy it now, in a very short time it will become antiquated again. And if I continue with this reasoning I am going to remain very soon with machines that will not allow me to be minimally competitive. He complains that he finds himself "up against the wall".

The approach taken by my friend was correct and his confusion quite normal, if we take into account the fact that the general structure of the problem of renewal is very complex. I think that his mind was laid to rest by reading a few pages of a recently published book[3], in which the fact is brought to light that in order for there to be a better understanding of the subject, it is necessary to consider two aspects:

1º Cause for renewal.

2º Techniques for treating renewal

If we consider the first of these, it can be said[4] that, apart from the physical *destruction* of the equipment, the main reasons that justify renewal, which in practice are mixed up, are:

a) *Physical wear and tear*, hat justifies the withdrawal of equipment when operating cost exceed a certain level.

b) *Obsolescence,* which appears as a result of technical progress and which makes useable equipment give way to more modern equipment.

These two aspects of renewal have given rise to expressions such as "functional ageing" and "economic ageing".

Terborgh, proposes an interesting definition, which is very useful relative to the causes for renewal: *inferiority of service*. This term includes both the physical wear and tear and obsolescence and allows for economically comparing a worn out and antiquated piece of equipment with a newer and more modern one.

"Bob" began to see things clearly in that, in order to resolve "his" problem it was necessary to take things step by step and set his ideas on a solid base. Now then, something was worrying him and this did not allow these concepts to come together in a precise manner. I remembered then that during my university days, I had used a graphical representation which was owed to Desrousseaux[5]. With a few slight modifications this would do the task of showing a German-Latin the heart of the problem of renewal.

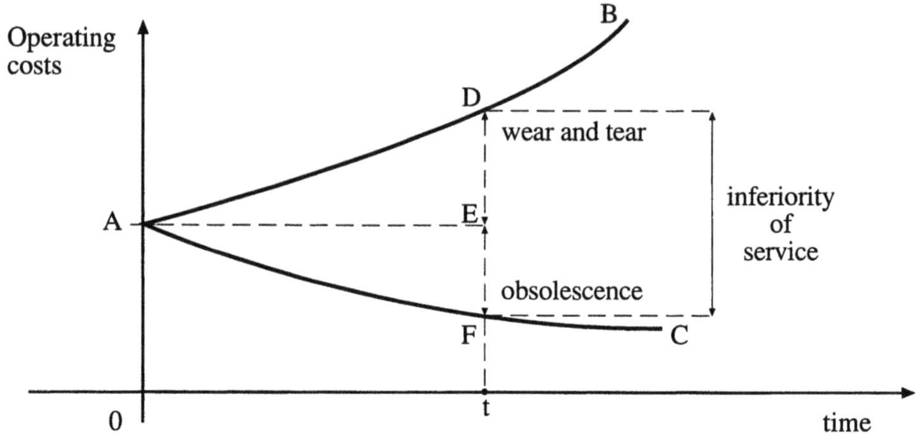

We began by explaining the significance of curves AB and AC. The first of these brings to light the *growing* development of expenses I the operating account (excepting depreciations) , as a consequence of the fact that, as the machine is more used, breakdowns and other necessary costs for maintaining it in service, are more frequent and mean higher payments. All the points along this line refer to a single piece of equipment, purchased at moment 0 with the technology existing at that time. Obviously what was chosen then was considered the best.

The second curve, AC, is intended to represent an infinite set of points from which the infinite *growing curves* start out and which would show the growing development of expenses in the operating account (excepting depreciations), if at each instant in time the best new equipment were to have been purchased. Therefore, if every year the was purchased, we would have for the first year 0 the already stated curve AB; for year 1, GH; for year 2, IJ, for year 3, KL; ... We have shown this in the following graphical representation:

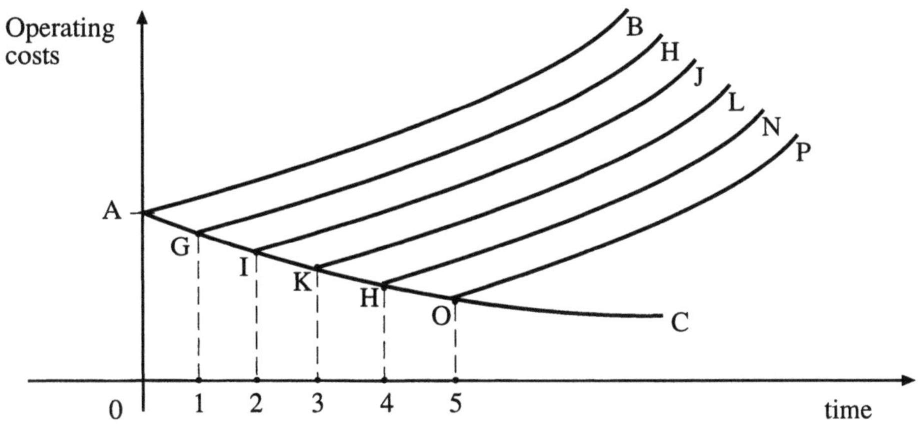

With these explanations the significance and role played by the three principal notions on which the theory of renewal is based appeared quite clear. In fact, we returned to the first of these two graphical representations and in it we saw that, in any moment during the life of a piece of equipment, for instance the moment designated by t, a situation of *service inferiority* occurs for the "old and antiquated" machine relative to the better "new and more modern" one. This is quantified by the dimension FD. And this is so because the "expenses" with a modern machine which *commences to operate* at t are equal to tF, while if we continue to operate with the old machine (the age of which is 0t) requires expenses such as tD. The resulting difference, obviously is, tD – tF = FD. The first of these concepts then was correctly assumed. The rest was easy.

When the machine in service was new (but with moment 0 technology) operating expenses were 0A, or what amounts to the same, to tE. Therefore comparing the two new machines (just recently put into service) the one with the technology of year 0 and the other with the technology of year t, the difference in expenses is tE – tF = FE. This figure then shows us the economic consequences of *obsolescence*.

Finally if we compare the operating expenses there were, generated by using the machine acquired at moment 0, when it was *new* and when it had been in operation during period 0t, that is to say *aged*, we arrive at the economic consequences of wear and tear. In our case these are shown by the figure tD – tE = ED. If we add the consequences of obsolescence FE and those of wear and tear ED, we will arrive at the *inferiority of service* FD.

"Bob" had patiently followed this story with his characteristic good humour. Nor a single objection or question. I asked myself if this was due to his intelligence or to my teaching qualities. I opted for the first of these, which appeared to be more prudent, taking into account what was waiting for us. It did not take long for Bob's German character to come to the fore, and as a steam roller, slowly but inexorably, he inquired about the manner of using, "applying" in his language, these ideas to his problem. To put this another way, which were the *techniques* to be used in order to get to know the *optimum moment* for the renewal of his complex array of computers and complementary equipment.

On looking through texts on the subject we became aware of the fact that, in a summarised manner, the studies on machinery renewal had led to certain *models* in which the necessary information for their use was expressed numerically by means of values considered as certain, through probabilities, and latterly, by using *uncertain numbers*. Bob was fully aware of my activities in the field of research into uncertain systems and he feared that I would immediately lead him into the use of complicated and even esoteric methods, only destined to lofty souls chosen by the Gods of Olympus. Nothing was further from the truth. We have always believed in things simple, although it may be very difficult to explain things in a simple manner and on the other hand, very easy to make difficult. And in this sense it could be of some use for my friend if he knew about, even superficially, the techniques up to then for solving the problems of renewal. For this we resorted to reading just a few paragraphs of a recently published book[6].

"In classical works on the subject, for example in Schneider's [7] book, the problem of the optimum moment for renewal is treated, considering a static economy. In these models, *wear and tear takes* on a principal role, (the hypothesis of a seasonal nature eliminates the appearance of technical progress), and on not considering the phenomenon of obsolescence, their capacity for use in evolving economic systems is diminished. In this respect, it is significant that, in these studies, renewal should be carried out with an identical machine to the one in service"... Currently, given the context in which economic activities are taking place, "there can be no doubt of the interest of t*echnical progress* in the decision to renew. In fact, even considering continued use of the machines, which causes or could cause, important wear and tear, technical progress will normally modify the time in which the old machines should be retired from service. If technical progress is slow, the machine will not show a particularly important obsolescence and, in that case there will be a certain interest in acquiring equipment that is resistant to wear and tear. When technical progress is accelerated, the use of the machine is decreased in time. In this case, it would appear to be more convenient to purchase machines that are suitable for frequent renewal, and under certain circumstances not considering *long lasting* machines". Having said all this we feel obliged to make special mention of the "theory of the adverse minimum", development and divulging of which is owed to a group headed up by Terborgh. "Its basis can be found in the determination, for a certain machine, of the costs arising from its *inferiority of service,* which (as we already know) includes wear and tear, if the economic system is static or else wear and tear and obsolescence, if the economic system is evolving, as a consequence of the appearance of technical progress".

Moving on from the conceptual point of view to the numerical, it will be seen that traditionally it was assumed that the economic values intervening in the problem of renewal were *certain.* Therefore one started out on the precise knowledge of the realities. "But the lack of historical data on all the cloudiness that covers the future, they increasingly confide new applications of the theory of fuzzy sub-sets and its multiple variations.. The opening up of these new horizons allowed for an updated orientation in works on renewal. And new models have appeared which based within the framework of an evolving economy, allow for treatment of the problems arising as a consequence of the use of modern equipment. For this there are available, apart from classical techniques, those arising from the development of *multivalent logic* which, also in other fields, represents important advances for management studies".

"Bob" has seen throughout this reading, that a concept appeared constantly: That was *inferiority of service*. And he had reached the conclusion of the importance of considering the fact that certain increased costs occur by using the old and *antiquated equipment* relative to the *new and modern equipment*. He then added if I calculate *the average cost arising from the inferiority of service* for each year, I will have a valuable element to help me determine the precise moment for renewal. In fact, he had discovered one of the aspects to be found in the origin of the criterion used for the formulation of the "adverse minimum method". According to this method the adverse minimum of a machine is known as the lowest value of a cost formed by the average adversity plus certain capital charges relative to time. The date that makes this cost minimum will determine the optimum moment for renewal.

Let us consider that this prior phase, meant to clarify the problem of renewal and focus Bob's ideas, has come to an end. Obvious proof of this is that he himself had reached the correct conclusion of working, "in a certain fashion", with the minimisation of a component of the average adversity. Now what was missing was to determine how we could introduce into the calculations the counterpart of the inferiority of service, that is the cost of acquiring the new and modern equipment which would allow him to eliminate the undesired *inferiority*.

The adverse minimum method

"Bob" was anxious and wanting to move on and "see" specific results, even if only approximate. We had to do something so as not to discourage him. We were therefore left no other alternative than to begin to estimate the values representing the inferiority of service and the estimated cost of the new and more modern set of computers and complementary equipment. He did not take too long to call me so that we could look at the information he had gathered. He was very surprised at the volume acquired year after year by repair and operating costs in general, which grew in marked progression. But what was most surprising was the "loss" incurred due to maintaining equipment with operating speeds and possibilities of carrying out tasks that were increasingly reduced relative to the newer and more modern computers, the dynamic memory chips of which were foreseen would attain spectacular speeds. Meticulous and ordered he showed me a table, which we discussed and on which we made certain observations. After the corresponding modifications it remained as follows:

Year	Wear and tear expenses	Losses due to obsolescence	Annual inferiority of	Average inferiority
1	50	-	50	50
2	100	180	280	165
3	200	400	600	310
4	450	950	1.400	583
5	1.000	3.000	4.000	1.266

We considered that, throughout the first year, technical progress would not be an issue, as we assumed that the equipment he had acquired was the best and most modern existing in the market. Its acquisition cost, paid in cash, was 2,000 monetary units. Likewise, Bob felt, that over the next few years, interest rates would remain around 5%. Based on this very precise data, we drew up another chart in which for each year we included annual depreciation, assuming that the equipment would be retired from service at the end of that period, and the average financial expenses. We arrived at the following chart:

Year	Average annual machine cost	Average financial expenses	Average capital charges
1	2.000	2.000 x 0,05 = 100	2.100
2	1.000	1/2 3.000 x 0.05 = 75	1.075
3	667	1/3 3.667 x 0.05 = 61	728
4	500	1/4 4.167 x 0.05 = 52	552
5	400	1/5 4.567 x 0.05 = 46	446

On adding the two components of the "average adversity" we arrived at the following chart:

Year	Average inferiority	Average capital charges	Average adversity	
1	50	2.100	2.150	
2	165	1.075	1.240	
3	310	728	1.038	←
4	583	552	1.135	
5	1.266	446	1.712	

By this route we had arrived at an initial solution according to which the set of computers and ancillary equipment to be purchase at moment 0 should be retired from service three years after their start up. Bob had done a good job in gathering the data relative to the previous years, but what was far more meritorious was his search for sufficiently valid information for *estimating future values.*

But in spite of the satisfaction of having in his hand a good aid for a correct decision on renewal, his central European spirit encouraged hi to look for generalisations for so many numbers and the following question was asked: based on this scheme would it not be possible to find a formula that could be used on other occasions? It would appear that this task was not too difficult. For this we were going to use the following terminology:

M: payment for the computers and ancillary equipment

d_1, d_2, d_3, d_4, d_5:
amounts for the service inferiority for years 1, 2, 3, 4, 5.

i = interest rate

x^*_t = average adversity in year t, t=1, 2, 3, 4, 5.

The average inferiority could be expressed as follows:

$d_1^* = d_1$

$d_2^* = 1/2\ (d_1 + d_2)$

$d_3^* = 1/3\ (d_1 + d_2 + d_3)$

$d_4^* = 1/4\ (d_1 + d_2 + d_3 + d_4)$

$d_5^* = 1/5\ (d_1 + d_2 + d_3 + d_4 + d_5)$

and in general:

$$d_t^* = \frac{1}{t} \sum_{j=1}^{t} d_j$$

Average capital costs are:

$M_1^* = M + M \cdot i$

$M_2^* = 1/2\ M + 1/2\ (M \cdot i + 1/2\ M \cdot i)$

$M_3^* = 1/3\ M + 1/3\ (M \cdot i + 1/2\ M \cdot i + 1/3\ M \cdot i)$

$M_4^* = 1/4\ M + 1/4\ (M \cdot i + 1/2\ M \cdot i + 1/3\ M \cdot i + 1/4\ M \cdot i)$

$M_5^* = 1/5\ M + 1/5\ (M \cdot i + 1/2\ M \cdot i + 1/3\ M \cdot i + 1/4\ M \cdot i + 1/5\ M \cdot i)$

and in general:

$$M_t^* = \frac{M}{t} + \frac{1}{t}\sum_{j=1}^{t}\frac{1}{j}M\cdot i$$

What remains now to be done is the addition of the average inferiority and the average capital costs:

$$x^* = d_t^* + M_t^* = \underbrace{\frac{1}{t}\sum_{j=1}^{t} d_j}_{\text{average service inferiority}} + \underbrace{\frac{1}{t}M + \frac{1}{t}\sum_{j=1}^{t}\frac{1}{j}\cdot M\cdot i}_{\text{average capital cost}}$$

This formula, although interesting from a practical point of view, is prone to certain defects and vagueness, above all relative to the service inferiority for each year, which are added even though corresponding to different periods and, therefore, are heterogeneous figures. In other words, the interest rate plays a certain role in capital costs but is not included in service inferiority. In this way then, the results arrived at could be admitted as a good *approximation*. It was obvious then that we should delve deeper into the selected path in order to attempt to get a little closer to a certain perfection. This led us directly to the adverse minimum method.

The first contribution by George Terborgh took place in 1946 with a book [8] the development and application of which was carried out at the research service of the Machinery and Allied Products Institute (MAPI), directed by Terborgh himself. The work carried out at MAPI [9] concentrated on the study of the behaviour of machines throughout their useful life. The object of these works was perfectly suited to the objectives sought by Bob, since they consisted in determining, from an economic point of view, the *date of retirement of a machine which commenced working at moment zero*. It appeared clear that to continue with the machine as from a certain moment, less profits would be obtained in the operating account (without considering depreciation) compared to those foreseen with a new and more modern machine. Nevertheless, by acting in this way, the outlay necessary for the purchase of the new machine would be avoided. These considerations encouraged us to find a solution to our problem by means of one of the schemes drawn up by Terborgh. The information assumed by Bob was sufficient for advancing along this road, which should end at the determination of the renewal date marked by the *average minimum adversity.*

Terborgh's proposal consisted in finding the average of the purchasing values of the machine and the inferiorities, updated by the interest rates, and later minimised.

For greater clarity we have put all the calculations together in tables. The first of these includes the total updated values of the adversity, that is the accumulated and updated adversity:

Year	Equipment costs	Annual inferiority	Annual updating rate	Updated inferiority	Accumulation of updated inferiority	Sum of equipment and accumulated updated inferiority
1	2.000	50	0.95238	47,62	47,62	2.047,62
2	2.000	280	0.90703	253,97	301,59	2 301,59
3	2.000	600	0.86384	518,30	819,89	2 819,89
4	2.000	1.400	0.82270	1 151,78	1.971,67	3 971,67
5	2.000	4.000	0.78353	3 134,12	5.105,79	7 105,79

The last column of the above table shows the total adversity for each year. Now we had to find the average adversity. For this we chose a weighted average, in which the weights would be the annual updating rates, the values for which we already had as we had used them in the fourth column of the above table:

Year	Annual updating rate	Sum of updating rates	Sum of equipment and accumulated updated inferiority	Average adversity
1	0.95238	0.95238	2 047,62	2 150
2	0.90703	1.85941	2 301,59	1 238
3	0.86384	2.72325	2 819,89	1 035 ←
4	0.82270	3.54595	3 971,67	1 120
5	0.78353	4.32948	7 105,79	1 641

Once again these calculations, by strictly following the most orthodox guidelines, took us to the same result which consisted in setting the third year for the date of renewal. But also we had verified that all the values we had arrived at were not too far removed from the values arrived at by the preceding simplified path. We felt satisfied on seeing that, intuitively, it was also possible to find behaviour rules of a great value. Before "Bob" could say; and the formula? I told him that there already existed a valid expression of a general nature.

For the accumulation of total adversity:

$$x_1 = M + d_1 (1+i)^{-1}$$

$$x_2 = M + d_1 (1+i)^{-1} + d_2 (1+i)^{-2}$$

$$x_3 = M + d_1 (1+i)^{-1} + d_2 (1+i)^{-2} + d_3 (1+i)^{-3}$$

$$x_4 = M + d_1 (1+i)^{-1} + d_2 (1+i)^{-2} + d_3 (1+i)^{-3} + d_4 (1+i)^{-4}$$

$$x_5 = M + d_1 (1+i)^{-1} + d_2 (1+i)^{-2} + d_3 (1+i)^{-3} + d_4 (1+i)^{-4} + d_5 (1+i)^{-5}$$

and in general

$$x_t = M + \sum_{j=1}^{t} d_j (1+i)^{-j} \qquad t = 1, 2, 3, \ldots$$

For the successive average adversities:

$$x_1^* = (M + d_1(1+i)^{-1}) : (1+i)^{-1}$$

$$x_2^* = (M + d_1(1+i)^{-1} + d_2(1+i)^{-2}) :((1+i)^{-1} + (1+i)^{-2})$$

$$x_3^* = (M + d_1(1+i)^{-1} + d_2(1+i)^{-2} + d_3(1+i)^{-3}) :((1+i)^{-1} + (1+i)^{-2} + (1+i)^{-3})$$

$$x_4^* = (M + d_1(1+i)^{-1} + d_2(1+i)^{-2} + d_3(1+i)^{-3} + d_4(1+i)^{-4}) : \\ :((1+i)^{-1} + (1+i)^{-2} + (1+i)^{-3} + (1+i)^{-4})$$

$$x_5^* = (M + d_1(1+i)^{-1} + d_2(1+i)^{-2} + d_3(1+i)^{-3} + d_4(1+i)^{-4} + d_5(1+i)^{-5}) : \\ :((1+i)^{-1} + (1+i)^{-2} + (1+i)^{-3} + (1+i)^{-4} + (1+i)^{-5})$$

and in general

$$x_t^* = \left(M + \sum_{j=1}^{t} d_j (1+i)^{-j}\right) : \sum_{j=1}^{t} (1+i)^{-j} \qquad t = 1, 2, 3, \ldots$$

This expression is well known and can be found in many of the texts that treat the subject of renewal [10], which gave us a certain sense of well being. In a certain way, we could say that the problem had been solved. But… was it possible that the data assumed by Bob could not be contested? Was it possible to state with total reliability that the estimated figures would occur in reality and were we sure of this fact?

Obviously the questions that had been approached touched on one of the most crucial aspects of today's society: uncertainty. Neither Bob nor anybody else was capable of estimating, with total certainty, the economic consequences of future events. In this way, both the figures assigned for wear and tear, and those for obsolescence and interest rates could only be accepted as a *presumption* by my friend, the "maximum" presumption if you like, but, when all is said and done, only that. Bob then said that at last my researcher *vein* had finally emerged. And with resignation, he prepared himself for following the path of uncertainty.

The fuzzy average adversity method

The truth is that all we had done up to this point would make it very easy to transfer from the sphere of certainty to that of uncertainty. We ourselves drew up a model[11] which would be of great use. What it entailed, in a first approximation, was to estimate the service inferiorities by means of confidence triplets, in stead of doing so by means of precise numbers, as we had done up to now. In this way the d_j would be substituted by:

$$(d_j^{(1)}, d_j^{(2)}, d_j^{(3)})$$

It would be necessary, although briefly, to explain to Bob the significance or representation if you like, of confidence triplets. All triplets are formed by an *lower extreme or limit* beneath which it is felt that the real value will not appear, a *maximum of presumption*, which value represents with the greatest strength that this will in reality occur, and an *upper extreme or limit* above which it is also felt that the real phenomena will not occur. With the confidence triplet the uncertainty between the extremes is limited and a value is established between the two limits, or coinciding with one of them, which is the closest estimate. The two extremes $d_j^{(1)}$, $d_j^{(3)}$ can be very separated (a lot of uncertainty) or very close (little uncertainty). The worth of the expert providing the triplet can be seen when providing extremes that are close and, obviously, does not make the error of reality *bouncing* outside the limits. Therefore, it is better to have more uncertainty and to be sure of not making an error. This is a fundamental principle of uncertainty.

As we saw it, the path to be followed was clear. The values which Bob had provided before could be considered as the maximum presumptions of the triplets. The only thing that remained to be done was to place a value to the left (lower limit) and to the right (upper limit), always considering that reality would not pass these limits either below or above. The rest was simple numerical calculation, that is, confidence triplet arithmetic. The only exception to the rule was to be found in the set of new and modern equipment for which Bob had a *closed offer* and, therefore, the information provided a precise number which, as we have seen, was 2,000 monetary units. But this caused no problem whatsoever, because a precise number can also be expressed by means of a confidence triplet. In our case it would be (2,000, 2,000, 2,000). With this simple aside we could now commence operating, starting out with the new information:

Year	Equipament cost	Anual inferiority	Updating rate from each year
1	(2.000, 2.000, 2.000)	(36, 50,.55)	(0.94787, 0.95238, 0.95420)
2	(2.000, 2.000, 2.000)	(250, 280, 300)	(0.89760, 0.90703, 0.91136)
3	(2.000, 2.000, 2.000)	(550, 600, 620)	(0.84920, 0.86384, 0.87129)
4	(2.000, 2.000, 2.000)	(1.300, 1.400, 1.500)	(0.80264, 0.82270, 0.83377)
5	(2.000, 2.000, 2.000)	(3.800, 4.000, 4.200)	(0.75793, 0.78353, 0.79863)

The estimated interest rates which gave rise to the annual updating rates were:

year 1: (4.8%, 5%, 5.5%)

year 2: (4.7%, 5%, 5.6%)

year 3: (4.6%, 5%, 5.7%)

year 4: (4.5%, 5%, 5.8%)

year 5: (4.4%, 5%, 5.9%)

We then arrived at the accumulated total adversity:

Year	Updated inferiority	Accumulation of the updated inferiority	Accumulated total adversity
1	(34.12, 47.62, 52.48)	(34.12, 47.62, 52.48)	(2034.12, 2047.62, 2052.48)
2	(224.40, 253.97, 273.41)	(258.52, 301.59, 325.89)	(2258.52, 2301.59, 2325.89)
3	(467.06, 518.30, 540.20)	(725.58, 819.89, 866.09)	(2725.58, 2819.89, 2866.09)
4	(1043.43, 1151.78, 1250.66)	(1769.01, 1971.67, 2116.75)	(3769.01, 3971.67, 4116.76)
5	(2880.13, 3134.12, 3354.25)	(4649.14, 5105.79, 5471.00)	(6649.14, 7105.79, 7471.00)

And from here we found the average adversity:

Year	Accumulated total adversity	Sum of updating rates	Average adversity
1	(2034.12, 2047.62, 2052.48)	(0.94787, 0.95238, 0.95420)	(2146, 2150, 2151)
2	(2258.52, 2301.59, 2325.89)	(1.84547, 1.85941, 1.86556)	(1224, 1238, 1247)
3	(2725.58, 2819.89, 2866.09)	(2.69467, 2.72325, 2.73685)	(1011, 1035, 1047) ←
4	(3769.01, 3971.67, 4116.76)	(3.49731, 3.54595, 3.57062)	(1078, 1120, 1153)
5	(6649.14, 7105.79, 7471.00)	(4.25524, 4.32948, 4.36925)	(1563, 1641, 1710)

The result we arrived at was highly satisfactory. We could see that the average adversity for each period, expressed by means of confidence triplets had as maximum presumption the same values we had found when the calculations were done in the sphere of certainty, as was only to be expected. Also the *minimum adversity* took place during the third year, just as had happened when we used precise numbers. But now we had available some complementary but very valuable information. In fact what the result showed was that although retirement from service of the equipment would mean an average adversity of 1035, under the most favourable circumstances it could be reduced even to 1011, and in the worst of events it might even reach 1047, very nearly the same figure we would come up with in the most optimistic case if renewal were to take place in year four. One thing appeared to be obvious: inferiority would be located within the interval [1011, 1047].

The generalised description of this scheme was not too difficult, although operating within the sphere of uncertainty means certain nuances, in this case, relative to the quotient used for finding the average adversity from total adversity. We have pointed this out below.

For each year we arrived at the total accumulated adversity as follows:

$$(x_1^{(1)}, x_1^{(2)}, x_1^{(3)}) = M\ (+)\ (d_1^{(1)}, d_1^{(2)}, d_1^{(3)})\ (\cdot)\ \left(\frac{1}{1+i_1^{(3)}}, \frac{1}{1+i_1^{(2)}}, \frac{1}{1+i_1^{(1)}}\right)$$

$$\begin{aligned}(x_2^{(1)}, x_2^{(2)}, x_2^{(3)}) = {} & M\ (+)\ (d_1^{(1)}, d_1^{(2)}, d_1^{(3)})\ (\cdot)\ \left(\frac{1}{1+i_1^{(3)}}, \frac{1}{1+i_1^{(2)}}, \frac{1}{1+i_1^{(1)}}\right)\ (+) \\ & (+)\ (d_2^{(1)}, d_2^{(2)}, d_2^{(3)})\ (\cdot)\ \left(\frac{1}{(1+i_1^{(3)})(1+i_2^{(3)})}, \frac{1}{(1+i_1^{(2)})(1+i_2^{(2)})}, \frac{1}{(1+i_1^{(1)})(1+i_2^{(1)})}\right)\end{aligned}$$

$$\begin{aligned}(x_3^{(1)}, x_3^{(2)}, x_3^{(3)}) = {} & M\ (+)\ (d_1^{(1)}, d_1^{(2)}, d_1^{(3)})\ (\cdot)\ \left(\frac{1}{1+i_1^{(3)}}, \frac{1}{1+i_1^{(2)}}, \frac{1}{1+i_1^{(1)}}\right)\ (+) \\ & (+)\ (d_2^{(1)}, d_2^{(2)}, d_2^{(3)})\ (\cdot)\ \left(\frac{1}{(1+i_1^{(3)})(1+i_2^{(3)})}, \frac{1}{(1+i_1^{(2)})(1+i_2^{(2)})}, \frac{1}{(1+i_1^{(1)})(1+i_2^{(1)})}\right)(+) \\ & (+)\ (d_3^{(1)}, d_3^{(2)}, d_3^{(3)})\ (\cdot)\ \left(\frac{1}{(1+i_1^{(3)})(1+i_2^{(3)})(1+i_3^{(3)})}, \frac{1}{(1+i_1^{(2)})(1+i_2^{(2)})(1+i_3^{(2)})}, \frac{1}{(1+i_1^{(1)})(1+i_2^{(1)})(1+i_3^{(1)})}\right)\end{aligned}$$

By following the same process, which requires adding the corresponding products, we will find $(x_4^{(1)}, x_4^{(2)}, x_4^{(3)})$ and $(x_5^{(1)}, x_5^{(2)}, x_5^{(3)})$. Then it is easy to arrive at the general expression:

$$(x_t^{(1)}, x_t^{(2)}, x_t^{(3)}) = M\ (+)\sum_{j=1}^{t} (d_j^{(1)}, d_j^{(2)}, d_j^{(3)})\ (\cdot)\ \left(\frac{1}{\prod_{j=1}^{t}(1+i_j^{(3)})}, \frac{1}{\prod_{j=1}^{t}(1+i_j^{(2)})}, \frac{1}{\prod_{j=1}^{t}(1+i_j^{(1)})}\right)$$

I made "Bob" see that in order to pass from the interest rate $(i_j^{(1)}, i_j^{(2)}, i_j^{(3)})$ to the updating rate $\left(\frac{1}{1+i_j^{(3)}}, \frac{1}{1+i_j^{(2)}}, \frac{1}{1+i_j^{(1)}}\right)$, we had resorted to the "normal" division of triplets, which consists of *crossed over* division of the extremes. On the other hand, in order to determine the average adversity (which we have expressed below) the operator used was the *Minkowski* division, which consists in finding the quotient between the lower extremes divided by themselves, the maximum presumption by themselves and also the upper extremes by themselves. As we have stated on numerous occasions, the *nature of the problem* itself will determine the type of division to be used.

We now move on to the general formulation of the average adversity:

$$(x_1^{(1)}, x_1^{(2)}, x_1^{(3)})^* = (x_1^{(1)}, x_1^{(2)}, x_1^{(3)})\ (\dot{m})\left(\frac{1}{1+i_1^{(3)}}, \frac{1}{1+i_1^{(2)}}, \frac{1}{1+i_1^{(1)}}\right)$$

$$(x_2^{(1)}, x_2^{(2)}, x_2^{(3)})^* = (x_2^{(1)}, x_2^{(2)}, x_2^{(3)}) \; (\dot{\cdot}_{m}) \Big[\Big(\frac{1}{1+i_1^{(3)}}, \frac{1}{1+i_1^{(2)}}, \frac{1}{1+i_1^{(1)}} \Big) (+)$$

$$(+) \Big(\frac{1}{(1+i_1^{(3)})(1+i_2^{(3)})}, \frac{1}{(1+i_1^{(2)})(1+i_2^{(2)})}, \frac{1}{(1+i_1^{(1)})(1+i_2^{(1)})} \Big) \Big]$$

$$(x_3^{(1)}, x_3^{(2)}, x_3^{(3)})^* = (x_3^{(1)}, x_3^{(2)}, x_3^{(3)}) \; (\dot{\cdot}_{m}) \Big[\Big(\frac{1}{1+i_1^{(3)}}, \frac{1}{1+i_1^{(2)}}, \frac{1}{1+i_1^{(1)}} \Big) (+)$$

$$(+) \Big(\frac{1}{(1+i_1^{(3)})(1+i_2^{(3)})}, \frac{1}{(1+i_1^{(2)})(1+i_2^{(2)})}, \frac{1}{(1+i_1^{(1)})(1+i_2^{(1)})} \Big) (+)$$

$$(+) \Big(\frac{1}{(1+i_1^{(3)})(1+i_2^{(3)})(1+i_3^{(3)})}, \frac{1}{(1+i_1^{(2)})(1+i_2^{(2)})(1+i_3^{(2)})}, \frac{1}{(1+i_1^{(1)})(1+i_2^{(1)})(1+i_3^{(1)})} \Big) \Big]$$

... ...

The general expression would be as follows:

$$(x_t^{(1)}, x_t^{(2)}, x_t^{(3)})^* = (x_t^{(1)}, x_t^{(2)}, x_t^{(3)}) \; (\dot{\cdot}_{m}) \sum_{j=1}^{t} \Big(\frac{1}{\prod_{j=1}^{t}(1+i_j^{(3)})}, \frac{1}{\prod_{j=1}^{t}(1+i_j^{(2)})}, \frac{1}{\prod_{j=1}^{t}(1+i_j^{(1)})} \Big)$$

Having reached this point it appeared that the problem was resolved. And we had even found an expression that was susceptible to being used in a wider range of real situations. We were aware of the fact that the use of confidence triplets as uncertain numbers was one of the many possibilities provided by management operating techniques. Selection was supported on the simplicity and ease of calculation and also in their efficiency of having available the maximum presumption which has played the role of a certain number in determinist schemes. Apart from that, the model is extensive to confidence intervals, fuzzy triangular numbers[12], fuzzy trapezoidal numbers, and any of their generalisations. We had reached our sought after destination .. . or that is what it seemed.

Management of fixed asset maintenance

For "Bob" among the positive effects of having made the studies for determining the optimum time for renewal of his computer installations, was to become aware of the very high cost of maintenance. In fact, repairs, contracts for materials and spare parts supply, "preventive" substitution of the more fragile components, ..., constituted a high percentage of operating costs. His question then was on the possibility a carrying out a *correct* management of these expenses. He knew perfectly well that the operation of his computers and complementary equipment causes considerable wear and tear that meant periodic revisions. According to him, that without a doubt the implementation of *preventive maintenance* reduced the expenses as it delayed physical ageing, avoided breakdowns, and decreased operating difficulties in future periods. Therefore in professional circles it is admitted that adequate revisions will avoid stoppages to the activity and always more costly future repairs, which increase operating costs.

According to Bob's expert opinion, revisions can take place under two different situations:

a) in a "preventive" manner,

b) when the equipment has been repaired as a consequence of a breakdown.

This approach reminded me of a subject I had studied in my youth, obviously with the means available at that time [13], but above all, a section of a much more recently published book [14]. We had there certain precedents that we would take advantage of, particularly the latter.

We were aware of something very important: the useful life of the equipment was 3 years. We agreed that we would establish a particular watch over the activities of the work means throughout these periods and that, prior the end of each year, we would have available good information relative to the operation. According to Bob we could expect answers under four sections:

E_1 = There has been a serious breakdown

E_2 = There has been failures in the principal components

E_3 = Operating difficulties have been detected

E_4 = The equipment has operated normally

It is normal when an incident occurs, such as those described under E_1, E_2, E_3 *(abnormal operation)*, that urgent immediate repair takes place and that a *certain revision* should take place at the end of the accounting period. What in fact takes place as a normal rule is that a repair automatically included a revision. On the other hand, if the equipment has functioned *normally* (section E_4), then it is up to the responsible person to *carry out or not* the preventive revision. This type of "philosophy" was carried out during all and each one of the years of useful life of the equipment. Revisions tended to be of two types, either partial or total.

This led us to express that at the end of each year two decisions had to be taken leading to one of the following situations:

E_1'=Equipment totally revised

E_2'= Equipment partially revised

E_3'= Equipment not revised

We had defined the problem, but a quick graphic look at it would better fix our ideas. For this we resorted to the following graph:

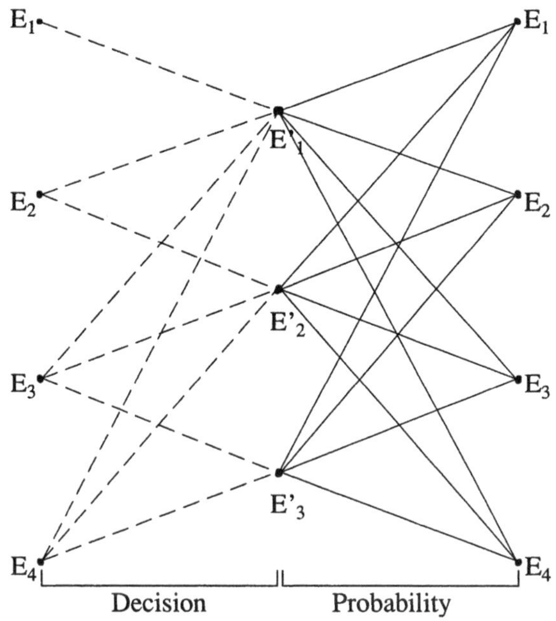

The following is a list of the operating terms of the process.

- With a very serious breakdown, there would be no choice but for total revision.

- If there had been failures in main components, there would be a choice of two decisions: total revision or partial revision.

- If operating difficulties had been detected there could be a choice of three decisions: Total revision, partial revision or no revision.

- Normal operation would also allow for a three way choice: total revision, partial revision or no revision.

This led us to express the transition between conditions E_j, j = 1, 2, 3, 4, to conditions E_k'= 1, 2, 3 would depend on Bob's *decision.* And it was precisely these decisions which would determine one or other maintenance costs. This then in fact, would be the aspect where our cooperation was required. Because passing from the E_k', k = 1, 2, 3 to the E_j, j = 1, 2, 3, 4, were independent of the will of my friend. In the very best of cases they could be expressed, just as we did, by means of *probabilities.* In fact, *always to a greater or lesser degree*, anything could happen (an important breakdown, failures in the main components, operating difficulties or normal activity).

Therefore what remained was to resolve the most delicate part which consisted in assigning expenses and probabilities to the possible transitions from one state to another. We started with the expenses arising from revisions, the change in status of which took place by decisions, from E_j, j = 1, 2, 3, 4 to E_k', k =1, 2, 3. These expenses were expressed by the following confidence triplets:

↗	E_1'	E_2'	E_3'
E_1	(180, 200, 210)		
E_2	(60, 80, 90)	(30, 40, 50)	
E_3	(50, 70, 80)	(10, 20, 25)	(0, 0, 0)
E_4	(25, 30, 40)	(5, 10, 15)	(0, 0, 0)

$E_j \rightarrow E_k$'

Even though we were aware of the arbitrariness we committed, confident in the efficiency and professionalism of Bob, but above all else in his wide experience, I let myself be convinced in order to determine the changes in status from commencement to the end of the year by means of probabilities. In this way the following were established for the transitions from E_k' to E_j,:

↗	E_1	E_2	E_3	E_4
E_1'	0	0	.1	.9

First period
$E_1' \rightarrow E_j$

↗	E_1	E_2	E_3	E_4
E_1'	.1	.1	.2	.6
E_2'	.1	.2	.3	.4
E_3'	.3	.4	.2	.1

Second period
$E_k' \rightarrow E_j$

↗	E_1	E_2	E_3	E_4
E_1'	.1	.1	.3	.5
E_2'	.2	.3	.3	.2
E_3'	.6	.3	.1	0

Third period
$E_k' \rightarrow E_j$

It was obvious that in the first period no decision whatsoever could be taken since the equipment was new and therefore revised at source. The transition that commenced at E_1' (equipment totally revised).

Likewise, we also estimated by confidence triplets the repair costs and production stoppages, expressed by means of the transition from $E_k' = 1, 2, 3$ to E_j, $j = 1, 2, 3, 4$, considered equal for each period:

$$E_k' \rightarrow E_1 : (650, 700, 800) = (b_{k1}^{(1)}, b_{k1}^{(2)}, b_{k1}^{(3)})$$

$$E_k' \rightarrow E_2 : (200, 250, 280) = (b_{k2}^{(1)}, b_{k2}^{(2)}, b_{k2}^{(3)})$$

$$E_k' \rightarrow E_3 : (60, 100, 120) = (b_{k3}^{(1)}, b_{k3}^{(2)}, b_{k3}^{(3)})$$

$$E_k' \rightarrow E_4 : (0, 0, 0) = (b_{k4}^{(1)}, b_{k4}^{(2)}, b_{k4}^{(3)})$$

Finally, we had already estimated for the studies we had done previously, the interest rates which gave rise to the updating rates which let us recall were as follows:

Year	Types of interest	Types of annual update
1	(4.8%, 5%, 5.5%)	(0.94787. 0.95238, 0.95420)
2	(4.7%, 5%, 5.6%)	(0.94697, 0.95238, 0.95511)
3	(4.6%, 5%, 5.7%)	(0.94607, 0.95238, 0.95602)

The process we had come up with was represented by means of a sequential graph, in which *all* the conditions through which the equipment could pass during the whole of its useful life, established according to our calculation as three years.

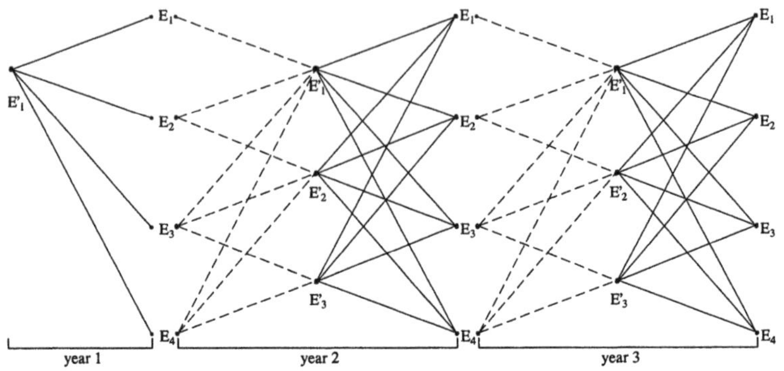

Having stated the problem in these terms, we commenced the calculation process, from the beginning to the end, just as is done in Dynamic Programming. For this we started at the commencement of the third year. The possible conditions were:

E_1: the equipment has suffered a severe breakdown during the previous period and, in accordance with established criteria, a total revision must be made. Therefore there is a transition to E_1'. Once this task has been carried out, what happens in the following period (and last year) is dependent upon probabilities. Therefore an addition must be made of the cost of the total revision plus the mathematical expectation of the cost of repair and stoppage during the third period [15] .Below we have shown the operations that were done:

$$\underset{\sim}{E}_{11}^{(3)} = (180, 200, 210)\ (+)\ \begin{bmatrix} .1 & .1 & .3 & .5 \end{bmatrix}\ (\cdot)\ \begin{bmatrix} (650, 700, 800) \\ (200, 250, 280) \\ (60, 100, 120) \\ (0, 0, 0) \end{bmatrix}$$

$$(\cdot)\ (0.94607, 0.95238, 0.95602) =$$

$$= (180, 200, 210)\ (+)\ [(65, 70, 80)\ (+)\ (20, 25, 28)\ (+)\ (18, 30, 36)\ (+)$$
$$(+)\ (000)]\ (.)\ (0.94607, 0.95238, 0.95602)$$

$$= (180, 200, 210)\ (+)\ (103, 125, 144)\ (.)\ (0.94607, 0.95238, 0.95602)$$

$$= (277, 319, 348) \leftarrow$$

E_2: the equipment has suffered breakdowns in the main components during the previous period. The alternative is limited, in this case, to either *total revision* or *partial revision*. The decision will end up in either E_1' or a E_2'.

$$\underset{\sim}{E}_{21}^{(3)} = (60, 80, 90) \ (+) \begin{bmatrix} .1 & .1 & .3 & .5 \end{bmatrix} (\cdot) \begin{bmatrix} (650, 700, 800) \\ (200, 250, 280) \\ (60, 100, 120) \\ (0, 0, 0) \end{bmatrix} (\cdot)$$

$$(\cdot) \ (0.94607, 0.95238, 0.95602) =$$

$$= (60, 80, 90) \ (+) \ (103, 125, 144) \ (.) \ (0.94607, 0.95238, 0.95602) =$$

$$= (157, 199, 228) \leftarrow$$

$$\underset{\sim}{E}_{22}^{(3)} = (30, 40, 50) \ (+) \begin{bmatrix} .2 & .3 & .3 & .2 \end{bmatrix} (\cdot) \begin{bmatrix} (650, 700, 800) \\ (200, 250, 280) \\ (60, 100, 120) \\ (0, 0, 0) \end{bmatrix} (\cdot)$$

$$(\cdot) \ (0.94607, 0.95238, 0.95602) =$$

$$= (30, 40, 50) \ (+) \ [(130, 140, 160) \ (+) \ (60, 75, 84) \ (+) \ (18, 30, 36) \ (+)$$
$$(+) \ (0, 0, 0)] \ (.) \ (0.94607, 0.95238, 0.95602) =$$

$$= (30, 40, 50) \ (+) \ (208, 245, 280) \ (.) \ (0{,}94607, 0.95238, 0.95602) =$$

$$= (227, 273, 318)$$

Should we find ourselves at the end of the second period (commencement of the third) in E_2, it is obvious that Bob should select the transition from E_2 to E_1', as there is a cost expectation at that time which is lower than with the other alternative. Therefore the optimum sub-strategy would be:

$$\boxed{E_2 \rightarrow E_1'}$$

E_3 : the equipment had operating difficulties during the last year. The decision to be taken then faced with the three alternatives is, *total revision, partial revision, and no revision.* Therefore the transition could be up either, E_1', E_2'and E_3'

$$\underset{\sim}{E}_{31}^{(3)} = (50, 70, 80)\ (+)\ \begin{array}{|c|c|c|c|}\hline .1 & .1 & .3 & .5 \\ \hline \end{array}\ (\cdot)\ \begin{array}{|c|}\hline (650, 700, 800) \\ \hline (200, 250, 280) \\ \hline (60, 100, 120) \\ \hline (0, 0, 0) \\ \hline \end{array}\ (\cdot)$$

(·) (0.94607, 0.95238, 0.95602) =

= (50, 70, 80) (+) (103, 125, 144) (.) (0.94607, 0.95238, 0.95602) =

= (147, 189, 218) ←

$$\underset{\sim}{E}_{32}^{(3)} = (10, 20, 25)\ (+)\ \begin{array}{|c|c|c|c|}\hline .2 & .3 & .3 & .2 \\ \hline \end{array}\ (\cdot)\ \begin{array}{|c|}\hline (650, 700, 800) \\ \hline (200, 250, 280) \\ \hline (60, 100, 120) \\ \hline (0, 0, 0) \\ \hline \end{array}\ (\cdot)$$

(·) (0.94607, 0.95238, 0.95602) =

= (10, 20, 25) (+) (208, 245,280) (.) (0.94607, 0.95238, 0.95602) =

= (207, 253, 293)

$$\underset{\sim}{E}_{33}^{(3)} = (0, 0, 0)\ (+)\ \begin{array}{|c|c|c|c|}\hline .6 & .3 & .1 & 0 \\ \hline \end{array}\ (\cdot)\ \begin{array}{|c|}\hline (650, 700, 800) \\ \hline (200, 250, 280) \\ \hline (60, 100, 120) \\ \hline (0, 0, 0) \\ \hline \end{array}\ (\cdot)$$

(·) (0.94607, 0.95238, 0.95602) =

= (0, 0, 0) (+) [(390, 420, 480) (+) (60, 75, 84) (+) (6, 10, 12) (+)
(+) (0, 0, 0)] (.) (0.94607, 0.95238, 0.95602)

= (0, 0, 0) (+) (456, 505, 576) (.) (0.94607, 0.95238, 0.95602)

= (431, 481, 551)

If at this commencement of the third year the status of the equipment were such that it was located at E_3, the decision would leave no doubt whatsoever and the transition should be made to E_1', which would mean total revision. Below we have written the optimum sub-strategy:

$$\boxed{E_3 \rightarrow E_1'}$$

E_4 : the equipment has functioned normally throughout the previous period. Here also the decision can be taken to either make a *total revision*, a *partial revision* or *no revision*, which means the transition E_1', E_2' and E_3', respectively:

$$\underset{\sim}{E}_{41}^{(3)} = (25, 30, 40)\ (+)\ \begin{bmatrix} .1 & .1 & .3 & .5 \end{bmatrix}\ (\cdot)\ \begin{bmatrix} (650, 700, 800) \\ (200, 250, 280) \\ (60, 100, 120) \\ (0, 0, 0) \end{bmatrix}\ (\cdot)$$

$$(\cdot)\ (0.94607, 0.95238, 0.95602) =$$

$$= (25, 30, 40)\ (+)\ (103, 125, 144)\ (.)\ (0.94607, 0.95238, 0.95602) =$$

$$= (122, 149, 178) \leftarrow$$

$$\underset{\sim}{E}_{42}^{(3)} = (5, 10, 15)\ (+)\ \begin{bmatrix} .2 & .3 & .3 & .2 \end{bmatrix}\ (\cdot)\ \begin{bmatrix} (650, 700, 800) \\ (200, 250, 280) \\ (60, 100, 120) \\ (0, 0, 0) \end{bmatrix}\ (\cdot)$$

$$(\cdot)\ (0.94607, 0.95238, 0.95602) =$$

$$= (5, 10, 15)\ (+)\ (208, 245, 280)\ (.)\ (0.94607, 0.95238, 0.95602) =$$

$$= (202, 243, 283)$$

$$\underset{\sim}{E}_{43}^{(3)} = (0, 0, 0)\ (+) \begin{array}{|c|c|c|c|}\hline .6 & .3 & .1 & 0 \\ \hline \end{array}\ (\cdot) \begin{array}{|c|}\hline (650, 700, 800) \\ \hline (200, 250, 280) \\ \hline (60, 100, 120) \\ \hline (0, 0, 0) \\ \hline \end{array}\ (\cdot)$$

$$(\cdot)\ (0.94607, 0.95238, 0.95602) =$$

$$= (0, 0, 0)\ (+)\ (456, 505, 576)\ (.)\ (0.94607, 0.95238, 0.95602) =$$

$$= (431, 481, 551)$$

Once again we find that if the equipment were to be found at position E_4, it was necessary to pass through E_1', that is to carry out the revision on the transition from E_4 to E_1' as this incurred less costs than the other alternatives. Therefore the solution set forth below was convenient:

$$\boxed{E_4 \rightarrow E_1'}$$

We feel we should do a summary that will give us an overall view of the third year.

At the end of the second year, or, if you like, at the commencement of the third year we had:

$$\underset{\sim}{E}_{11}^{(3)} = (277,319, 348)$$

$$\underset{\sim}{E}_{21}^{(3)} = (157, 199, 228)$$

$$\underset{\sim}{E}_{31}^{(3)} = (147, 189, 218)$$

$$\underset{\sim}{E}_{41}^{(3)} = (122, 149, 178)$$

and therefore, the optimum sub-strategy consisted in making the following transitions:

. If we were to be at E_1, we would decide on the transition to E_1': $E_1 \rightarrow E_1'$

. If we were to be at E_2 we would decide on the transition to E_1': $E_2 \rightarrow E_1'$

. If we were to be at E_3 we would decide on the transition to E_1': $E_3 \rightarrow E_1'$

. If we were to be at E_4 we would decide on the transition to E_1': $E_4 \rightarrow E_1'$

At the beginning of the third year, *whatever the situation found at the end of the second year* the decision should be to do a total revision. We have represented this sub-strategy in the following figure:

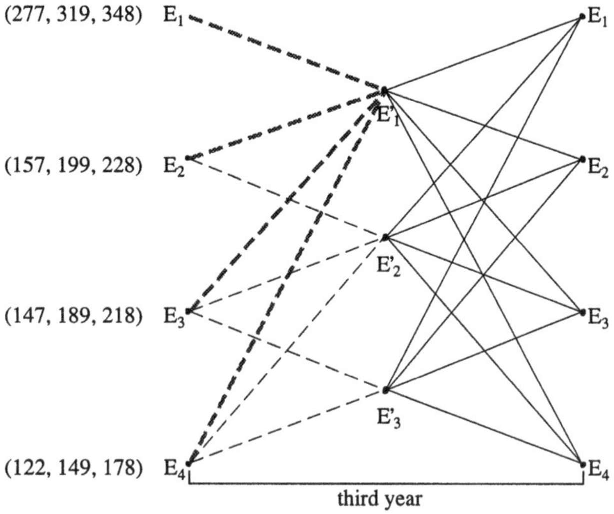

The process we had proposed meant at that time, passing over to the second year, obviously adding the confidence triplets located at the beginning of the third year (or what amounted to the same thing, at the end of the second year), which were designated by $\underset{\sim}{E}_{11}^{(3)}$, $\underset{\sim}{E}_{21}^{(3)}$, $\underset{\sim}{E}_{31}^{(3)}$, y $\underset{\sim}{E}_{41}^{(3)}$. Calculations would be similar to those for the third period, but now adding those uncertain numbers (triplets). And that is what we did:

E_1: We began assuming a serious breakdown during the first year. The decision did not allow for any alternatives: a total revision had to be carried out. The cost, accumulating the second and third years, obviously discounted at the commencement of the second year, would be as follows:

$$\underset{\sim}{E}_{11}^{(2)} = (180, 200, 210)\ (+)\ \begin{bmatrix} .1 & .1 & .2 & .6 \end{bmatrix}\ (\cdot)\ \begin{bmatrix} (650, 700, 800)\ (+)\ (277, 319, 348) \\ (200, 250, 280)\ (+)\ (157, 199, 228) \\ (60, 100, 120)\ (+)\ (147, 189, 218) \\ (0, 0, 0)\ (+)\ (128, 149, 178) \end{bmatrix}\ (\cdot$$

$(\cdot)\ (0.94697, 0.95238, 0.95511) =$

$$= (180, 200, 210)\ (+) \begin{bmatrix} .1 & .1 & .2 & .6 \end{bmatrix} (\cdot) \begin{bmatrix} (927, 1019, 1148) \\ (357, 449, 508) \\ (207, 289, 338) \\ (128, 149, 178) \end{bmatrix} (\cdot)$$

$(\cdot)\ (0.94697, 0.95238, 0.95511) =$

$= (180, 200, 210)\ (+)\ [(92.7, 101.9, 114.8)\ (+)\ (35.7, 44.9, 50.8)\ (+)$
$(+)\ (41.4,.57.8,.67.6)\ (+)\ (76.8, 89.4, 106.8)]\ (.)$
$(.)\ (0.94697, 0.95238, 0.95511) =$

$= (180, 200, 210)\ (+)\ (246.6, 294, 339.2)\ (.)$
$(.)\ (0.94697, 0.95238, 0.95511) =$

$= (414, 480, 534) \leftarrow$

E_2 : This situation was characterised by the existence of breakdowns in the main components during the first year. The accumulated cost at the beginning of the second year depended from the decision of *total or partial* revision:

$$\underset{\sim}{E}_{21}^{(2)} = (60, 80,90)\ (+) \begin{bmatrix} .1 & .1 & .2 & .6 \end{bmatrix} (\cdot) \begin{bmatrix} (927, 1019, 1148) \\ (357, 449, 508) \\ (207, 289, 338) \\ (128, 149, 178) \end{bmatrix} (\cdot)$$

$(0.94697, 0.95238, 0.95511) =$

$= (294, 360, 414) \leftarrow$

$$\underset{\sim}{E}_{22}^{(2)} = (30, 40, 50)\ (+) \begin{bmatrix} .1 & .2 & .3 & .4 \end{bmatrix} (\cdot) \begin{bmatrix} (927, 1019, 1148) \\ (357, 449, 508) \\ (207, 289, 338) \\ (128, 149, 178) \end{bmatrix} (\cdot)$$

$(\cdot)\ (0.94697, 0.95238, 0.95511) =$

$= (30, 40, 50)\ (+)\ [(92.7, 101.9, 114.8)\ (+)\ (71.4, 89.8, 101.6)\ (+)$

$(+)\ (62.1, 86.7, 101.4)\ (+)\ (51.2, 59.6, 71.2)]\ (.)$

$(.)\ (0.94697, 0.95238, 0.95511) =$

$= (30, 40, 50)\ (+)\ (277.4, 338, 389)\ (.)\ (0.94697, 0.95238, 095511) =$

$= (293, 362, 422)$

The comparison between the two triplets represented by $\tilde{E}_{21}^{(2)}$ and $\tilde{E}_{22}^{(2)}$ brought to light the fact that, in spite that the lower extreme $\tilde{E}_{22}^{(2)}$ was lower than the lower extreme of $\tilde{E}_{21}^{(2)}$, the maximum presumption, and the upper extreme of the latter was lower than the extremes of $\tilde{E}_{22}^{(2)}$. Although this case was clear maybe other were not quite so clear. For this reason I explained to Bob that a very simple criterion existed for ordering triplets. It consisted in reducing each triplet to an exact number by multiplying the extremes by the unit and the maximum presumption by two, add these products and divide the result by four. The values found (exact numbers) automatically supplied the sought after order. In this case we had:

$$\overline{\tilde{E}_{21}^{(2)}} = \frac{294 + (2)\,.\,(360) + 414}{4} = 357 \qquad \overline{\tilde{E}_{22}^{(2)}} = \frac{293 + (2)\,.\,(362) + 422}{4} = 360$$

$\tilde{E}_{21}^{(2)}$was selected as the optimum as it showed, overall, the lowest costs:

$$\boxed{E_2 \rightarrow E_1'}$$

E_3: This state brings to light the fact that during the first year operating difficulties were encountered. The decision could be any of the three alternatives: *total revision, partial revision and no revision:*

$$\underset{\sim}{F}_{31}^{(2)} = (50, 70, 80) \ (+) \begin{bmatrix} .1 & .1 & .2 & .6 \end{bmatrix} (\cdot) \begin{bmatrix} (927, 1019, 1148) \\ (357, 449, 508) \\ (207, 289, 338) \\ (128, 149, 178) \end{bmatrix} (\cdot)$$

$$(\cdot)\ (0.94697, 0.95238, 0.95511) =$$

$$= (284, 350, 404)$$

$$\underset{\sim}{F}_{32}^{(2)} = (10, 20, 25) \ (+) \begin{bmatrix} .1 & .2 & .3 & .4 \end{bmatrix} (\cdot) \begin{bmatrix} (927, 1019, 1148) \\ (357, 449, 508) \\ (207, 289, 338) \\ (128, 149, 178) \end{bmatrix} (\cdot)$$

$$(\cdot)\ (0.94697, 0.95238, 0.95511) =$$

$$= (273, 342, 397) \leftarrow$$

$$\underset{\sim}{F}_{33}^{(2)} = (0, 0, 0) \ (+) \begin{bmatrix} .3 & .4 & .2 & .1 \end{bmatrix} (\cdot) \begin{bmatrix} (927, 1019, 1148) \\ (357, 449, 508) \\ (207, 289, 338) \\ (128, 149, 178) \end{bmatrix} (\cdot)$$

$$(\cdot)\ (0.94697, 0.95238, 0.95511) =$$

$$= (0, 0, 0)\ (+)[(278.1, 305.7, 344.4)\ (+)\ (142.8, 179.6, 203.2)\ (+)$$

$$(+)\ (41.4, 57.8, 67.8)\ (+)\ (12.8, 14.9, 17.8)]\ (.)$$

$$(.)\ (0.94697, 0.95238, 0.95511) =$$

$$= (0, 0, 0)\ (+)\ (475.1, 558, 633.2)\ (.)\ (0.94697, 0.95238, 0.95511) =$$

$$= (450, 531, 605)$$

The lowest triplet is $\underset{\sim}{E}_{32}^{(2)}$ which represents the following transition

$$\boxed{E_3 \rightarrow E_2'}$$

E_4: Normal operation during the first year allows for any of the three possible decisions: *total revision, partial revision and no revision.* The cost for this could be:

$$\underset{\sim}{E}_{41}^{(2)} = (25, 30, 40)\ (+) \begin{array}{|c|c|c|c|}\hline .1 & .1 & .2 & .6 \\ \hline\end{array}\ (\cdot) \begin{array}{|c|}\hline (927, 1019, 1148) \\ \hline (357, 449, 508) \\ \hline (207, 289, 338) \\ \hline (128, 149, 178) \\ \hline\end{array}\ (\cdot)$$

$$(\cdot)\ (0.94697, 0.95238, 0.95511) =$$

$$= (259, 310, 364) \leftarrow$$

$$\underset{\sim}{E}_{42}^{(2)} = (5, 10, 15)\ (+) \begin{array}{|c|c|c|c|}\hline .1 & .2 & .3 & .4 \\ \hline\end{array}\ (\cdot) \begin{array}{|c|}\hline (927, 1019, 1148) \\ \hline (357, 449, 508) \\ \hline (207, 289, 338) \\ \hline (128, 149, 178) \\ \hline\end{array}\ (\cdot)$$

$$(\cdot)\ (0.94697, 0.95238, 0.95511) =$$

$$= (268, 332, 387)$$

$$\underset{\sim}{E}_{43}^{(2)} = (0, 0, 0)\ (+) \begin{array}{|c|c|c|c|}\hline .3 & .4 & .2 & .1 \\ \hline\end{array}\ (\cdot) \begin{array}{|c|}\hline (927, 1019, 1148) \\ \hline (357, 449, 508) \\ \hline (207, 289, 338) \\ \hline (128, 149, 178) \\ \hline\end{array}\ (\cdot)$$

$$(\cdot)\ (0.94697, 0.95238, 0.95511) =$$

$$= (450, 531, 605)$$

The triplet that gives the minimum cost is $\underset{\sim}{E}_{41}^{(2)}$, which indicates and optimum transition from E_4 hasta E_1', which we represent as follows:

$E_4 \rightarrow E_1$'

In this way we arrived at the conclusion that at the end of the first year we had the following as optimum accumulated costs from the first period:

$\underset{\sim}{E}_{11}^{(2)}$= (414, 480, 534)

$\underset{\sim}{E}_{21}^{(2)}$= (294, 360, 414)

$\underset{\sim}{E}_{32}^{(2)}$= (273, 342, 397)

$\underset{\sim}{E}_{41}^{(2)}$= (259, 310, 364)

the optimum sub-strategy being:

If we found ourselves at E_1 we should decide to pas over to E_1.

"	"	E_2	"	"	E_1.
"	"	E_3	"	"	E_2.
"	"	E_4	"	"	E_1.

At the beginning of year two the solution in the majority consisted also in carrying out a *total revision*, except when the condition were to be E_3, when a *partial revision* would be carried out. A graphical representation of this would help us to see the situation more clearly:

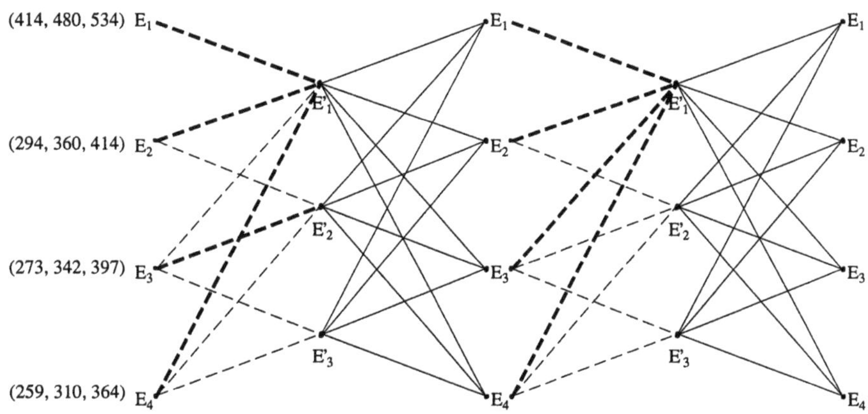

Having reached this point, what only remained was to pass from the end of the first year to the start up time of the equipment. We had to assume that all the equipment which had just been purchased was *totally revised*, which meant nothing else than reflecting everyday reality. For this reason there existed a single start out state which was E_1' equipment *new and revised commence their to operate*. What occurred throughout this period could be formalised by means of probabilities. With the following calculations the process ended:

$$\underset{\sim}{A} = \begin{array}{|c|c|c|c|}\hline 0 & 0 & .1 & .9 \\ \hline \end{array} (\cdot) \begin{array}{|c|}\hline (650, 700, 800)\ (+)\ (414, 480, 534) \\ \hline (200, 250, 280)\ (+)\ (294, 360, 414) \\ \hline (60, 100, 120)\ (+)\ (273, 342, 397) \\ \hline (0, 0, 0)\ (+)\ (259, 310, 364) \\ \hline \end{array} (\cdot)$$

$$(\cdot)\ (0.94787, 0.95238, 0.95420) =$$

$$= \begin{array}{|c|c|c|c|}\hline 0 & 0 & .1 & .9 \\ \hline \end{array} (\cdot) \begin{array}{|c|}\hline (1064, 1180, 1334) \\ \hline (494, 610, 694) \\ \hline (333, 442, 517) \\ \hline (259, 310, 364) \\ \hline \end{array} (\cdot)$$

$$(\cdot)\ (0.94787, 0.95238, 0.95420) =$$

$$= (0, 0, 0)\ (+)\ (0, 0, 0)\ (+)\ (32, 42, 49)\ (+)\ (234, 266, 313) =$$

$$= (266, 308, 362)$$

This result warranted a brief comment, but before this it would be useful to repeat the complete sequential scheme marking the sub-strategies and, therefore, establishing the *maintenance strategy:*

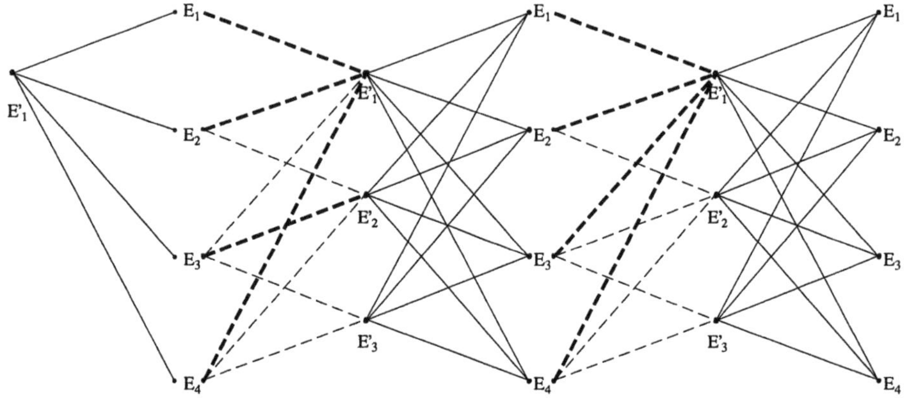

"Bob" now had the necessary elements for managing his fixed assets correctly, in the first place determining the date for renewal (three years) and from this important data establishing an optimum strategy for maintenance.

A complementary use could be arrived at from the valued result of the optimum strategy: the composition of the estimates that Bob had made for "wear and tear expenses", which were 50, 100 and 200 monetary units for the first three years, that is a total of 350 monetary units. These values, let us recall, had been taken as the maximum presumption on doing the calculations within the sphere of uncertainty. This brief commentary made for an interesting comparison between the *optimum strategy* and the *initial idea* that Bob had relative to the maintenance of the equipment. According to him, during the three years he would have a cost *discounting* moment zero of:

$$B = (50) \,.\, (0.95238) + (100) \,.\, (0.90703) + (200) \,.\, (0.86384) = 311$$

I was agreeably surprised on seeing that there was very nearly a coincidence between the maximum presumption of the triplet representing maintenance costs following the optimum strategy (308 monetary units), and the updated value provided at the beginning of our calculations (311 monetary units). Bob smiled maliciously: his long experience, his technical knowledge and, added by me, his intelligence put to the service of the task, made him foresee frequent total revisions for his equipment. But now he had the conviction that his previous actions were correct and, above all else, he arrived at something that was very important for him: to feel much happier in his early retirement in the city where he lived his youth, since the person responsible for his company, with far less experience, now had certain rules of behaviour that guaranteed its future with total guarantees of following on in a position of privilege that it enjoyed in the market, that is characterised by very fierce competition.

REFERENCIAS

[1] SCHNEIDER, E: *Economía política y económica de la empresa.* publ. Sagitario Barcelona, 1968, page 397.

[2] KAUFMANN, A. and GIL ALUJA, J.: *Introducción de la teoría de los subconjuntos borrosos a la gestión de la empresas.* publ. Milladoiro Santiago Compostela 1986, page 88.

[3] GIL ALUJA, J.: *Invertir en la incertidumbre.* publ. Pirámide. Madrid 1997, pages 378-383.

[4] ABRAHAM, C. and THOMAS, A.: *Microéconomie: décisions optimales dans l'entreprise et dans la nation.* publ. Dunod. Paris 1966, pages 325-327.

[5] DESROUSSEAUX, J.: L'evolution économique et le comportement industriel. Ed. Dunod. Paris 1965. page 176.

[6] GIL ALUJA, J.: *Invertir en la incertidumbre.* publ. Pirámide. Madrid 1997, pages 381-383.

[7] SCHNEIDER, E.: *Teoría de la inversión.* publ. Ateneo. Buenos Aires 1956, page. 87 y ss.

[8] TERBORGH, G.: *Dynamic Equipment Policy.* publ Mc. Graw-Hill. New York, 1946.

[9] TERBORGH, G.: *Business Investment Policy. Machinery and Allied Products Institute.* Washington 1958.

TERBORGH, G.: Business Investment Management. Machinery and Allied Products Institute. Washington 1967.

[10] GIL ALUJA, J.: *Invertir en la incertidumbre.* publ. Pirámide. Madrid 1997, page.386.

[11] GIL ALUJA, J.: *FAMM (Fuzzy Adverse Minimum Method). Proceedings of the AEDEM European Conference.* Minsk 20-23 Septiembre 1993.

[12] GIL ALUJA, J.: *Invertir en la incertidumbre.* publ. Pirámide. Madrid 1997, page 394-399.

[13] GIL ALUJA, J.: *El estudio dinámico de la elección de las inversiones.* Revista Técnica Contable. Febrero 1967.

[14] KAUFMANN, A. and GIL ALUJA, J.: *Nuevas técnicas para la dirección estratégica.* publ. Universidad de Barcelona. Barcelona 1991, page 251-263

[15] It can be presumed that although the revision costs take place at the beginning of year 3, payment for the repair will take place at the end of the period, therefore it should be discounted at the beginning in order to be able to do the addition.

VI MANAGEMENT OF HUMAN RESOURCES

Incorporation of man into labour activity

Since time immemorial, the biblical curse, "you will earn your bread by the sweat of your brow", has hung like a millstone around man's neck. And some with more sweat and other with less, no one, or nearly no-one, has freed themselves from rendering their efforts in exchange for a handful of rice, or a pound of caviar. This contribution has undergone changes throughout time and does not have the same configuration in one geographical area or another, but nobody can have any doubt whatsoever on the importance of the labour factor in the organisation of business and institutions. What perhaps may have been forgotten too much is its incidence on the emotional stability of each individual and as a consequence of this, in social coexistence. It is not possible to ignore the idea that the number of hours given over to production activity represents the greater part of our time. To arrive at correct harmony between character, the manner of being of the worker and the tasks that must be carried out, will have an inexorable repercussion on the results of the business or institution, as well as in the family surroundings.

This is an element of capital importance in the life of thinking persons, which has motivated the fact that "its study has been done from very diverse angles and in it have intervened psychologists, sociologists, politicians, jurists, economists, engineers, ... It has been repeated ad nauseam that a good human team is the most appreciated asset of any community. In businesses and institutions a conglomerate of human beings intervene who carry out the most diverse activities, from foundation and promotion, through development, and right up to liquidation. It is men and women, with their work and decisions who lead to success and also failure. It should not come as a surprise then that one of the objectives in every business or institution should be the hiring, training and condition of a coordinated, efficient, dynamic, polyvalent and at times stable human group"[1].

But the scientific and technical revolution, already present in the last century and leading player in the new century, is causing new ways of looking at the adaptation of people to the work place with the object of providing satisfaction both for the men and women as for the objectives of the businesses or institutions. "Current times are very far removed from those times when the same activity could be carried out during ones whole life: Mobility and adaptability are absolutely necessary. Attaining quality, the obligatory will for innovation at all levels, flexibility and variety in employment as well as the human will for promotion constitute basic behaviour. Further than the necessary social morality, efficiency is also imperative. The problems to be resolved are not simple and become increasingly complicated as time goes by"[2].

The first contact between *man and business or institution* takes place when a candidate requests a job, or when the business or institution calls that worker it considers suitable for carrying out a certain task. The operations leading to contracting human resources are known under the generic name of "personnel selection" and consist of "selecting a person for a job with a determined profile, that can be defined by means of certain measurements or valuations that allow comparison with the qualities of the candidate"[3].

Tasks leading to the incorporation to specific jobs of adequate candidates are normal during the life of a business and institution. All of us who in some way walk the paths of economic activity within production units have at one time or another found ourselves involved in "selection" tasks, which were resolved with different levels of success depending on the instruments available at each time. Fortunately, today we possess a whole arsenal of elements that are capable of providing a magnificent answer to the important challenges that appear in the filed of human resources management. We are not going to refer to any of the "solutions" that, in times gone by, we gave to problems of selection requested during our professional activity, because just to recall them makes us blush. It is clear that current means are not even remotely what we had available then. On the other hand it could serve as an illustration to mention an experience I went through some years ago, as a result of the promotion of my friend Pepe in Public Administration.

I coincided with Pepe for the first time in the classrooms of the Institute where I studied in my city. He was born in Madrid, son and grandson of public servants. From his very earliest youth he had laid down his roots in this Catalan area where he felt, and feels perfectly at home. His distinguished bearing, from his very youth, showed that he was brought up in highly cultural surroundings. The admiration he aroused in us, in no way prevented his incorporation into the "band of ruffians" formed by the group of *poor students*, very wont to pranks, forgetful about attending classes and in this way loosing the *interesting* explanations of our teachers. Pepe followed in the footsteps of his forebears, and made an early entry into public service and embarking on an ascending path that ended with his appointment as Chief Telegrapher in a large Catalan city, with the express objective of totally reorganising the personnel. This meant taking on a whole set of actions that went from selecting people from other offices and temporary workers, to the training of groups destined for other towns of less importance but responsible to him. The rules of action that were officially available for these cases, dated from "Napoleonic times", according to him. Therefore, in the meeting organised for celebrating his promotion, between drink and drink, we undertook to do "something" to remedy this unpardonable abandonment. We were able to count on the always efficient cooperation of Consol, Pepe's wife.

Personnel selection

The first and most acute problem was the selection of the most adequate candidates, from among those made available to him by the Administration. This was a very frequent approach, for which we had a good solution, checked out by past experience. The phases to be followed were as follows:

1) *Analysis of the tasks* which would be assigned to each public servant in the work place, listing the qualities, characteristics or peculiarities considered as significant for carry out a good job.

2) Establishing an *ideal profile* for each job, consisting of setting, for each quality, characteristic or peculiarity, the level, which "should" be possessed by the *perfect public servant.*

3) Draw an actual profile of each candidate, that is a "descriptor" of the public servants or persons that could be contracted, from screening, tests, interviews, references, etc., setting the levels possessed for each quality, characteristic or peculiarity.

4) *Comparision* between the ideal profile of each job and that of the candidates, in order to find the greater or lesser *"closeness"* between the *ideal requirements and actual possibilities.*

5) Taking the decisions leading to the formation of the payroll for the telegraph office.

The first task to be carried out by Pepe and Consol consisted in specifying the jobs in a strict manner, independently one from the others. In this sense it was clear to them that the one fact that relative to the new incorporations they had five new employees to consider. One of these was special in that the requirement was for a higher responsibility. The other four, although with different characteristics, were susceptible to being covered by less qualified personnel. After the necessary consultations, they decided on, for the position of head of section the following characteristics and peculiarities:

C_1 = Personnel leadership qualities.

C_2 = Flexibility in relations with the public.

C_3 = Dedication to the work.

C_4 = Knowledge of the tasks of his employees.

C_5 = Technical instrument training.

C_6 = Capacity for computer learning.

Obviously this was not an exhaustive list, but both sustained that these six points could be included in the essential "properties" and also in the most significant.

Having covered this first phase the next was to move on to establishing the *ideal profile*, setting for each one of the characteristics, qualities or peculiarities the *desired level* or level considered as optimum. In this specific point it was necessary to point out, very clearly, what was the significance of the *desired level*. In fact, one of the first meanings would tend to indicate that the specified level or degree is ideal, and it is as bad not to attain it as to exceed it (this required penalising the lowest levels as well as the highest). A second meaning would bring to light that it is only bad not to attain the required level, but indifferent if the level is attained or exceeded. (This means penalising the lowest levels, but not penalising nor strengthening the highest levels). To make this distinction was important because the treatment to be given to the qualities, characteristics or peculiarities, under one meaning or the other, would be different.

From these premises, and after an interchange of opinions, the conclusion was reached to include under the first meaning (penalise both below the level and above the level) characteristics, qualities and peculiarities C_2 and C_3, arguing that for the first of these an excess of flexibility would hinder the fluidity of the service, and for the second, that an excess of dedication would create tension among the personnel. This did not exclude giving a high level as the ideal to be attained. In this way included under the second meaning (only penalise not attaining the ideal level) qualities, characteristics or peculiarities ideal level) qualities, characteristics or peculiarities C_1, C_4, C_5, C_6.

We had now come to the time to start on the *second phase*. For this, it was necessary to *put values* on each of the desired levels. As is well known, we have acquired the habit of using a semantic scale with eleven values (hendecagonal system) from zero to one (in tenths) in such a way that when greater the need to possess a "property" the *nearer* to one will be the assigned number, while when *less* possessed the less near to one it will be, or in other words, the closer to zero. With the object of making the assignment of valuations easier in [0, 1] we have drawn up the following semantic scale:

1 : Excellent, extraordinary
0.9 : Very, very good
0.8 : Very good
0.7 : Good
0.6 : Quite good
0.5 : Not so good
0.4 : Quite bad
0.3 : Bad
0.2 : Very bad
0.1 : Very, very bad
0 : Terrible, nil

From this criterion Pepe and Consol provided the following information:

For	$C_1 = 0.8$	For	$C_2 = 0.7$
"	$C_4 = 0.7$	"	$C_3 = 0.8$
"	$C_5 = 1$		
"	$C_6 = 0.6$		

In this way, the words normally used by the experts entrusted with establishing the level would be converted, from this table into values from the segment [0, 1]. In our case, the semantic expression "very good qualities for handling personnel" had been numerically transformed into 0.8; "good flexibility in relations with the public" into 0.7. And so on for each and every one of the qualities, characteristics and peculiarities.

The description of the particular work place then offered no difficulty. The notion of the fuzzy sub-set was particularly suitable for this task. Therefore the following fuzzy sub-set acted as the descriptor for the job profile:

	C_1	C_2	C_3	C_4	C_5	C_6
$\underset{\sim}{T}$	.8	.7	.8	.7	1	.6

Mathematicians have acquired the habit of representing a fuzzy sub-set like the one we have shown as follows:

	C_1	C_2	C_3	C_4	C_5	C_6
$\underset{\sim}{T}$	μ_1	υ_2	υ_3	μ_4	μ_5	μ_6

In a practical manner and theoretically we knew the ideal profile. What was missing then was to find the *"actual"* profile of each of the candidates that the Administration had made available to the new Head Telegrapher, for occupying the position of Head of Section. On this occasion we had six names available. Pepe entrusted a specialised company to carry out the screening and test, and personally reserved for himself the final interview. Form the report he received and his own opinion sufficient data was extracted in order to describe each one of the six candidates by means of the corresponding fuzzy sub-sets, just as was done for the ideal profile. The result was as follows:

	C_1	C_2	C_3	C_4	C_5	C_6
$\underset{\sim}{P}_1$	.4	1	.6	.8	.7	.5

	C_1	C_2	C_3	C_4	C_5	C_6
$\underset{\sim}{P}_2$	.6	.3	.7	.6	.8	.7

	C_1	C_2	C_3	C_4	C_5	C_6
$\underset{\sim}{P}_3$	.8	.9	.7	.9	.8	.7

	C_1	C_2	C_3	C_4	C_5	C_6
$\underset{\sim}{P}_4$	.7	.4	.6	.7	.5	.6

	C_1	C_2	C_3	C_4	C_5	C_6
$\underset{\sim}{P}_5$	.9	.8	.8	.7	.9	.7

	C_1	C_2	C_3	C_4	C_5	C_6
$\underset{\sim}{P}_6$	.5	.6	.4	.6	.7	.8

Having finished this third stage, what remained to be done was to resolve the last phase the characteristics of which were full of a marked technical aspect, since the matter brought up had as its objective finding one or more means that would be capable of numerically expressing the "closeness" of the profile of each candidate relative to the ideal profile. And in our problem this matter had become complicated due to the existence of two different groups of qualities, characteristics or peculiarities acting along differentiated criteria. What technical elements could be used for one or the other case? Did we have any experience in order to make a translation of the problem we were trying to resolve? Fortunately the answers were optimistic and we would be able to break them down with a certain ease.

a) Relative to the first meaning, which stated that it was as bad to not attain as it was to exceed the required level, we could use the mathematical concept of distance, and among the possibilities, we were inclined to use, because of its simplicity and operability, the "Hamming distance". This notion adapted to "perfection" for the treatment of qualities, characteristics or peculiarities

C_2 and C_3. Arriving at it did not bring up any difficulties. It was sufficient to arrive at the differences existing, in absolute value, between the level of the required quality in the ideal profile and the level possessed by the candidate under study. What should be taken into account is the *absolute value,* since what occurs is a "positive" distancing both if the difference is positive and negative. Let us remember that this is only valid in the qualities, characteristics or peculiarities that were governed by this criterion, so that the differences calculated in this way, were only applicable to characteristics C_2 and C_3 as follows:

For P_1 : $|.7 - 1| + |.8 - .6| = 0.5$

" P_2 : $|.7 - .3| + |.8 - .7| = 0.5$

" P_3 : $|.7 - .9| + |.8 - .7| = 0.3$

" P_4 : $|.7 - .4| + |.8 - .6| = 0.5$

" P_5 : $|.7 - .8| + |.8 - .8| = 0.1$

" P_6 : $|.7 - .6| + |.8 - .4| = 0.5$

In the theoretical sphere it is customary to represent these levels by means of the symbols listed below:

$\mu_T(C_u)$: level ideal for C_u.

$\mu_{Pj}(C_u)$: level possessed of C_u, by P_j.

u : properties for which both lacking and excess are penalised.

This allowed for formulating the following general expression:

$$\sum_u |\mu_T(C_u) - \mu_{Pj}(C_u)|$$

In this way we arrived at the distancing of each employee relative to the ideal profile of the job, for qualities, characteristics or peculiarities C_2 and C_3. Now we had to move on and resolve the problems for the remainder.

b) With regard to the second meaning, according to which it was bad not to attain the required level, but exceeding the level was not penalised, what was adequate was a simple calculation consisting in arriving at the differences of level, for each quality, characteristic or peculiarity, between the required level and reality (without the absolute value) and substitute the negative values with zeros. This would be valid for C_1, C_4, C_5 and C_6, and the results were:

For $P_1 : [0 \vee (.8 - .4)] + [0 \vee (.7 - .8)] + [0 \vee (1 - .7)] + [0 \vee (.6 - .5)] = 0.8$

" $P_2 : [0 \vee (.8 - .6)] + [0 \vee (.7 - .6)] + [0 \vee (1 - .8)] + [0 \vee (.6 - .7)] = 0.5$

" $P_3 : [0 \vee (.8 - .8)] + [0 \vee (.7 - .9)] + [0 \vee (1 - .8)] + [0 \vee (.6 - .7)] = 0.2$

" $P_4 : [0 \vee (.8 - .7)] + [0 \vee (.7 - .7)] + [0 \vee (1 - .5)] + [0 \vee (.6 - .6)] = 0.6$

" $P_5 : [0 \vee (.8 - .9)] + [0 \vee (.7 - .7)] + [0 \vee (1 - .9)] + [0 \vee (.6 - .7)] = 0.1$

" $P_6 : [0 \vee (.8 - .5)] + [0 \vee (.7 - .6)] + [0 \vee (1 - .7)] + [0 \vee (.6 - .8)] = 0.7$

With these simple calculations we determined the distance between each employee and the ideal profile, in qualities, characteristics or peculiarities C_1, C_4, C_5 and C_6.

We could also generalise these operations for theoretical effects by calling:

$\mu_T (C_v)$: ideal level for C_v.

$\mu_{Pj} (C_v)$: level possessed of C_v, by P_j.

v: properties for which only lacking is penalised

In this way we can write:

$$\sum_{V} \left| 0 \vee [\mu_T (C_v) - \mu_{Pj} (C_v)] \right|$$

It was patently obvious then that in order to arrive at the total distance, taking into account all the qualities, characteristics or peculiarities it would be sufficient to add the values of the two criteria found for each employee. The result was:

$$\text{For } P_1 : 0.5 + 0.8 = 1.3$$
$$\text{“ } P_2 : 0.5 + 0.5 = 1.0$$
$$\text{“ } P_3 : 0.3 + 0.2 = 0.5$$
$$\text{“ } P_4 : 0.5 + 0.6 = 1.1$$
$$\text{“ } P_5 : 0.1 + 0.1 = 0.2$$
$$\text{“ } P_6 : 0.5 + 0.7 = 1.2$$

These values now allowed us to establish an order of preference between the employee–candidates. To show the generalisation it was sufficient to do the sum of the previous expressions, that is:

$$\sum_u |\mu_T(C_u) - \mu_{Pj}(C_u)| + \sum_v |0 \vee [\mu_T(C_v) - \mu_{Pj}(C_v)]|$$

Given that the values arrived at brought to light a “distancing” of each one of the candidates compared to the sought after profile, the *preferences* fell to the l*owest values*. In this way the order of preference was, from high to low:

$$P_5 \rightarrow P_3 \rightarrow P_2 \rightarrow P_4 \rightarrow P_6 \rightarrow P_1$$

The solution to this first problem appeared to be quite clear and preference for public servant P_5 left room for no discussion. Even in the event that this candidate should fail, candidate P_3 was readily available as a good substitute. The remainder were very far removed.

Now then, even though what I was about to say to my friends would affect the conclusions arrived at, we felt that it was a positive move to explain to them that, at least from a technical point of view, the *absolute values* did not constitute the end of the calculations. The custom was to convert them into *relative values*. In this way the result would be converted into an index, always located between zero and one. The conversion was simple: it was sufficient to divide the absolute values by the number of qualities, characteristics or peculiarities, which, in our case were six. We then arrived at:

For $P_1 : 1.3 / 6 = 0.217$
" $P_2 : 1.0 / 6 = 0.167$
" $P_3 : 0.5 / 6 = 0.083$
" $P_4 : 1.1 / 6 = 0.183$
" $P_5 : 0.2 / 6 = 0.033$
" $P_6 : 1.2 / 6 = 0.200$

As it could be in no other way, the order was exactly the same. Nevertheless with these *distancing indices* it was immediately possible to find the *closeness indices*. For this all we had to do was to calculate their "complement to the unit". Therefore:

For $P_1 : 1- 0.217 = 0.783$
" $P_2 : 1- 0.167 = 0.833$
" $P_3 : 1- 0.083 = 0.917$
" $P_4 : 1- 0.183 = 0.817$
" $P_5 : 1- 0.033 = 0.967$
" $P_6 : 1- 0.200 = 0.800$

The selection based on the *closeness indices* obviously means finding an order of priority from the *highest values* to the *lowest values*. On doing this the same list of preferences would appear that we had found in the previous calculations.

These indices are called "indices of maximum and minimum level" and are the result of an interesting work presented at an AEDEM Congress that took place in Reggio di Calabria[4]. The author finds a general expression, which for the "distancing index" of each candidate is:

$$\eta(\underset{\sim}{T}, \underset{\sim}{P}_j) = \frac{1}{u+v}\left(\sum_u |\mu_T(C_u) - \mu_{Pj}(C_u)| + \sum_v |0 \vee [\mu_T(C_v) - \mu_{Pj}(C_v)]|\right)$$
$$j = 1, 2, 3\ldots$$

As a consequence of this the "closeness index" is formed as follows:

$$\upsilon(\underset{\sim}{T}, \underset{\sim}{P}_j) = 1 - \eta(\underset{\sim}{T}, \underset{\sim}{P}_j)$$

We had hardly finished this brief technical exposition, justly attributed to its author, when the first very interesting question arose. Is this always valid, even when the importance of each quality, characteristic or peculiarity were to be different? The answer could be nothing else but negative. Even on many occasions when introducing this new element the order of preference undergoes a significant change. This caused a certain amount of consternation in Pepe and Consol, which could only be clarified by introducing into the study "degree of importance" for the different qualities, characteristics or peculiarities under consideration.

In order to do a good job in this sense a very careful consideration had to be given to which were the most significant tasks to be carried out by the Head of Section. In short, what was expected of him. From this we would arrive at the importance of each property relative to the remainder, which should be numbered by means of values, that is, by assigning subjective numbering. It was agreed, as was our custom, to use the hendecagonal system in the segment [0, 1] and to establish the following semantic scale:

1 : Vital, absolutely important

0.9 : Very, very important

0.8 : Very important

0.7 : Important

0.6 : Quite important

0.5 : Moderate importance

0.4 : Low importance

0.3 : Hardly important

0.2 : Very little importance

0.1 : Almost dispensable

0 : Nil importance, totally dispensable

It was agreed that the phrases we had selected showed the vocabulary that was normally used by the experts who would be asked for an opinion. In fact, my friends did not make a mistake. The replies offered were not substantially removed from the words used in the scale. It was possible then to propose the following as *degrees of importance* for each of the qualities, characteristics or peculiarities:

$$\omega_1 = 1; \quad \omega_2 = 0.9; \quad \omega_3 = 0.5; \quad \omega_4 = 0.7; \quad \omega_5 = 0.6; \quad \omega_6 = 0.8$$

The sum of these valuations was:

$$\sum_{i=1}^{6} \omega_i = 1 + 0.9 + 0.5 + 0.7 + 0.6 + 0.8 = 4.5$$

In this way, the relative importance of each of the qualities, characteristics or peculiarities was calculated as follows:

$$z_1 = \frac{\omega_1}{\sum_{i=1}^{6} \omega_i} = 0.22; \qquad z_2 = \frac{\omega_2}{\sum_{i=1}^{6} \omega_i} = 0.20; \qquad z_3 = \frac{\omega_3}{\sum_{i=1}^{6} \omega_i} = 0.11;$$

$$z_4 = \frac{\omega_4}{\sum_{i=1}^{6} \omega_i} = 0.16; \qquad z_5 = \frac{\omega_5}{\sum_{i=1}^{6} \omega_i} = 0.13; \qquad z_6 = \frac{\omega_6}{\sum_{i=1}^{6} \omega_i} = 0.18$$

As by construction the sum of the ω_i, $i = 1, 2, \ldots, 6$ was equal to the unit, these valuations allowed us to arrive at a *convex weighting*. For this, all that was need was to multiply each difference (o maximisation of the difference) by the "weights" we had arrived at and do the addition. The convexity of the weighting avoids having to do, later on, a division as happened in the previous case.

The greater part of the structure of the calculations was already done, so therefore we took advantage, to a greater degree, of what we had. We then did:

For P_1:

$(0.22) . [0 \vee (.8 - .4)] + (0.20) . |.7 - 1| + (0.11) . |.8 -.6| + (0.16) . |0 \vee (.7 - .8)] +$

$+ (0.13) . [0 \vee (1 -.7)] + (0.18) . [0 \vee (.6 - .5)] =$

$= 0.088 + 0.060 + 0.022 + 0 + 0.039 + 0.018 =$

$= 0.227$

For P_2:

$(0.22) . [0 \vee (.8 - .6)] + (0.20) . |.7 - .3| + (0.11) . |.8 -.7| + (0.16) . [0 \vee (.7 - .6)] +$

$+ (0.13) . [0 \vee (1 -.8)] + (0.18) . [0 \vee (.6 - .7)] =$

$= 0.044 + 0.080 + 0.011 + 0.016 + 0.026 + 0 =$

$= 0.177$

For P_3:

$(0.22) . [0 \vee (.8 - .8)] + (0.20) . |.7 - .9| + (0.11) . |.8 -.7| + (0.16) . [0 \vee (.7 - .9)] +$

$+ (0.13) . [0 \vee (1 -.8)] + (0.18) . [0 \vee (.6 - .7)] =$

$= 0 + 0.040 + 0.011 + 0 + 0.026 + 0 =$

$= 0.077$

For P_4:

$(0.22) . [0 \vee (.8 - .7)] + (0.20) . |.7 - .4| + (0.11) . |.8 -.6| + (0.16) . [0 \vee (.7 - .7)] +$

$+ (0.13) . [0 \vee (1 -.5)] + (0.18) . [0 \vee (.6 - .6)] =$

$= 0.022 + 0.060 + 0.022 + 0 + 0.065 + 0 =$

$= 0.169$

For P_5:

$(0.22) . [0 \vee (.8 - .9)] + (0.20) . |.7 - .8| + (0.11) . |.8 -.8| + (0.16) . [0 \vee (.7 - .7)] +$

$+ (0.13) . [0 \vee (1 -.9)] + (0.18) . [0 \vee (.6 - .7)] =$

$= 0 + 0.020 + 0 + 0 + 0.013 + 0 =$

$= 0.033$

For P_6:

$(0.22) . [0 \vee (.8 - .5)] + (0.20) . |.7 - .6| + (0.11) . |.8 -.4| + (0.16) . [0 \vee (.7 - .6)] +$

$+ (0.13) . [0 \vee (1 -.7)] + (0.18) . [0 \vee (.6 - .8)] =$

$= 0.066 + 0.020 + 0.044 + 0.016 + 0.039 + 0 =$

$= 0.185$

In our case, the valuations we arrived at do not differ essentially from those we arrived at, implicitly, when we considered the same importance for all the qualities, characteristics or peculiarities. Nevertheless, in the order of preference there had been a change between public servants P_2 and P_4. The new order was now, from greater to lower:

$$P_5 \rightarrow P_3 \rightarrow P_4 \rightarrow P_2 \rightarrow P_6 \rightarrow P_1$$

The first problem approached by the Head Telegrapher was now closed and resolved, and in due course public servant-candidate P_5 would be incorporated into the position of Head of Section to everyone's satisfaction. But the curiosity of Consol made us re-do the previous general formula for the "distance index", in such a way that it should include the phrase of the different importance of the qualities, characteristics and peculiarities that were taken into account. The transformation was really very simple, and we arrived at the following expression:

$$\upsilon(\underset{\sim}{T}, \underset{\sim}{P}_j) = \sum_{u} (z_u) \left| \mu_T (C_u) - \mu_{Pj} (C_u) \right| + \sum_{v} (z_v) \left[0 \vee [\mu_T (C_V) - \mu_{Pj} (C_V)] \right]$$
$$j = 1, 2, 3 \ldots$$

Although the formula at first glance appeared to very complicated by the symbols that are normally used, in reality, it was very easily applied as my good friends had seen.

The problem of assigning human resources

A few days after having solved the problem of the incorporation of a new Head of Section into the Telegraph Administration, we became occupied in the need to take on public servants and contracted personnel for the jobs that we wanted to cover. Initially we followed the same process describing both the four positions to be covered and the candidates (public servants or contracted personnel) *by means of fuzzy sub-sets* from the referential of the qualities, characteristics or peculiarities considered as important, from a higher to a lower degree.

After the usual steps and with the cooperation of the general personnel services it was agreed, that in order to represent the ideal profiles for each one of the jobs T_1, T_2, T_3, T_4, we should consider the following qualities, characteristics and peculiarities

C_1: Integration into group work.
C_2: Capacity for adaptation to new tasks.
C_3: Speed in carrying out work.
C_4: Experience in the functions entrusted.
C_5: Training in telecommunications.
C_6: Capacity for relating to the public.
C_7: Computer knowledge.
C_8: Spirit of cooperation

The determination of the levels considered ideal for each of the C_j, j = 1, 2, ..., 8 allowed us to express the profiles for the four jobs:

	C_1	C_2	C_3	C_4	C_5	C_6	C_7	C_8
T_1	.8	.6	.7	.4	.9	1	.3	.9

	C_1	C_2	C_3	C_4	C_5	C_6	C_7	C_8
T_2	.9	.8	.8	.6	.7	.6	1	1

	C_1	C_2	C_3	C_4	C_5	C_6	C_7	C_8
T_3	.8	.7	.8	.3	.9	.8	.8	.7

	C_1	C_2	C_3	C_4	C_5	C_6	C_7	C_8
T_4	.5	.9	.7	.8	.6	1	.5	.6

The Administration proposed to the Head Telegrapher that he choose *five* employees for occupying these *four* jobs from the group formed by public servants and contracted personnel. One of these therefore, would remain without a job.

We then had to do the screening, tests, etc,... in order to value the level possessed by each candidate of each of the qualities, characteristics and peculiarities. The result made available to us a *description* of *each* candidate by means of the following fuzzy sub-sets:

	C_1	C_2	C_3	C_4	C_5	C_6	C_7	C_8
P_1	.4	.8	.5	.8	.9	.8	.7	.5

	C_1	C_2	C_3	C_4	C_5	C_6	C_7	C_8
P_2	.7	.5	.9	.6	.9	.6	.9	.8

	C_1	C_2	C_3	C_4	C_5	C_6	C_7	C_8
P_3	.8	.7	.6	.3	.8	.9	.5	.8

	C_1	C_2	C_3	C_4	C_5	C_6	C_7	C_8
P_4	.6	.9	.7	.6	.8	.7	.9	.6

	C_1	C_2	C_3	C_4	C_5	C_6	C_7	C_8
P_5	.3	.8	.9	.8	.6	.8	.7	.5

Just as we had done for the selection of the Head of Section we now had to separate the qualities, characteristics and peculiarities into two groups. The first of these included those for which not attaining the level as well as exceeding the level there would be a penalty. In this group we included C_3, C_4 y C_8. This meaning was justified because excessive speed inexorably leads to shoddy work. Likewise it was argued that too much experience was a breeding ground for routine. It was also insisted on that cooperation should not lead to servility, a source of lack of solidarity. In the group of qualities, characteristics and peculiarities for which only the lack of the level would be penalised C_1, C_2, C_5, C_6, C_7 were included.

With the object of having all the information available for commencing calculations, we requested the importance of each "property". The answers were put into the hendecagonal semantic scale, we had used before, which gave the following valuations in [0, 1]:

$$\omega_1 = 0.6;\ \omega_2 = 0.9;\ \omega_3 = 0.5;\ \omega_4 = 0.6;\ \omega_5 = 0.8;\ \omega_6 = 0.6;\ \omega_7 = 0.8;\ \omega_8 = 0.7$$

For greater ease in the calculations, just as we had done for resolving another problem, we made a convex weighting. Therefore, since the sum of the ω_i, i = 1. 2, ..., 8 was:

$$\sum_{i=1}^{8} \omega_i = 5.5$$

We arrived at the following as "weights" for each one of the qualities, characteristics and peculiarities:

$$z_1 = \frac{\omega_1}{\sum_{i=1}^{8}\omega_i} = 0.109\,; \quad z_2 = \frac{\omega_2}{\sum_{i=1}^{8}\omega_i} = 0.164\,; \quad z_3 = \frac{\omega_3}{\sum_{i=1}^{8}\omega_i} = 0.091\,; \quad z_4 = \frac{\omega_4}{\sum_{i=1}^{8}\omega_i} = 0.109\,;$$

$$z_5 = \frac{\omega_5}{\sum_{i=1}^{8}\omega_i} = 0.145\,; \quad z_6 = \frac{\omega_6}{\sum_{i=1}^{8}\omega_i} = 0.109; \quad z_7 = \frac{\omega_7}{\sum_{i=1}^{8}\omega_i} = 0.145; \quad z_8 = \frac{\omega_8}{\sum_{i=1}^{8}\omega_i} = 0.128^{(5)}$$

We then were able to find, from all the information we had, the "separation index" [6] *for each* profile of the four jobs with *each* candidate, from the formula shown at the end of the previous section. We have shown below the operations for the first indices, and only the results for the remainder.

Distance of all the candidates from job T_1:

$$\begin{aligned} \upsilon(\underset{\sim}{T}_1, \underset{\sim}{P}_1) = {} & (0.109) \,.\, [0 \vee (.8 - .4)] + (0.164) \,.\, [0 \vee (.6 - .8)] + (0.091) \,.\, |.7 - .5| + \\ & + (0.109) \,.\, |.4 - .8| + (0.145) \,.\, [0 \vee (.9 - .9)] + (0.109) \,.\, [0 \vee (1 - .8)] + \\ & + (0.145) \,.\, [0 \vee (.3 - .7)\,] + (0.128) \,.\, |.9 - .5| = \\ & = 0.0436 + 0 + 0.0182 + 0.0436 + 0 + 0.0218 + 0 + 0.0512 = \\ & = 0.1784 \end{aligned}$$

$$\begin{aligned} \upsilon(\underset{\sim}{T}_1, \underset{\sim}{P}_2) = {} & (0.109) \,.\, [0 \vee (.8 - .7)] + (0.164) \,.\, [0 \vee (.6 - .5)] + (0.091) \,.\, |.7 - .9| + \\ & + (0.109) \,.\, |.4 - .6| + (0.145) \,.\, [0 \vee (.9 - .9)] + (0.109) \,.\, [0 \vee (1 - .6)] + \\ & + (0.145) \,.\, [0 \vee (.3 - .9)\,] + (0.128) \,.\, |.9 - .8| = \\ & = 0.0109 + 0.0164 + 0 + 0.0182 + 0.0218 + 0 + 0.0436 + 0 + 0.0128 = \\ & = 0.1237 \end{aligned}$$

$$\upsilon(\underset{\sim}{T}_1, \underset{\sim}{P}_3) = (0.109) . [0 \vee (.8 - .8)] + (0.164) . [0 \vee (.6 - .7)] + (0.091) . |.7 - .6| +$$
$$+ (0.109) . |.4 -.3| + (0.145) . [0 \vee (.9 - .8)] + (0.109) . [0 \vee (1 - .9)] +$$
$$+ (0.145) . [0 \vee (.3 - .5)] + (0.128) . |.9 - .8| =$$
$$= 0 + 0 + 0.0091 + 0.0109 + 0.0145 + 0.0109 + 0 + 0.0128=$$
$$= 0.0582$$

$$\upsilon(\underset{\sim}{T}_1, \underset{\sim}{P}_4) = 0.1292$$
$$\upsilon(\underset{\sim}{T}_1, \underset{\sim}{P}_5) = 0.2328$$

Distance of all candidates from job T_2:

$$\upsilon(\underset{\sim}{T}_2, \underset{\sim}{P}_1) = 0.2111 \ , \ \upsilon(\underset{\sim}{T}_2, \underset{\sim}{P}_2) = 0.1202 \ , \ \upsilon(\underset{\sim}{T}_2, \underset{\sim}{P}_3) = 0.1763 \ ,$$
$$\upsilon(\underset{\sim}{T}_2, \underset{\sim}{P}_4) = 0.1075 \ , \ \upsilon(\underset{\sim}{T}_2, \underset{\sim}{P}_5) = 0.2183$$

Distance of all candidates from job T_3:

$$\upsilon(\underset{\sim}{T}_3, \underset{\sim}{P}_1) = 0.1655 \ , \ \upsilon(\underset{\sim}{T}_3, \underset{\sim}{P}_2) = 0.1201 \ , \ \upsilon(\underset{\sim}{T}_3, \underset{\sim}{P}_3) = 0.0890 \ ,$$
$$\upsilon(\underset{\sim}{T}_3, \underset{\sim}{P}_4) = 0.1018 \ , \ \upsilon(\underset{\sim}{T}_3, \underset{\sim}{P}_5) = 0.2017$$

Distance of all candidates from job T_4:

$$\upsilon(\underset{\sim}{T}_4, \underset{\sim}{P}_1) = 0.0801 \ , \ \upsilon(\underset{\sim}{T}_4, \underset{\sim}{P}_2) = 0.1748 \ , \ \upsilon(\underset{\sim}{T}_4, \underset{\sim}{P}_3) = 0.1329 \ ,$$
$$\upsilon(\underset{\sim}{T}_4, \underset{\sim}{P}_4) = 0.0545 \ , \ \upsilon(\underset{\sim}{T}_4, \underset{\sim}{P}_5) = 0.0910$$

In this way we had all the distances for each candidate from each job. What we now had to do was to "organise" as best we could all this information. For this we resorted to the matrix form, much loved by mathematicians. What this is, is simply a table formed by columns and rows, in such a way that the jobs are placed in the rows and the candidates in the columns. What we arrive at here is to identify the "distance" by row and column of which it is a member. Thus, for example, the intersection of row T_3 with column P_4 brings to light the distance of candidate P_4 from job T_3. Acting just as we proposed we drew the following matrix:

$[\underset{\sim}{R}] =$

	P_1	P_2	P_3	P_4	P_5
T_1	0.1784	0.1237	0.0582	0.1292	0.2328
T_2	0.2111	0.1202	0.1763	0.1075	0.2183
T_3	0.1655	0.1201	0.0890	0.1018	0.2017
T_4	0.0801	0.1748	0.1329	0.0545	0.0910

By its very construction, this matrix is a "matrix of distance". It is very easy then to convert it into a "matrix of closeness". For this all we have to do is to substitute the values in each box by their respective complements to the unit. We then arrive at:

$[\underset{\sim}{\bar{R}}] =$

	P_1	P_2	P_3	P_4	P_5
T_1	0.8216	0.8763	0.9418	0.8708	0.7672
T_2	0.7889	0.8798	0.8237	0.8925	0.7817
T_3	0.8345	0.8799	0.9110	0.8982	0.7983
T_4	0.9199	0.8252	0.8671	0.9455	0.9090

Either of these two matrices was valid as a basis for determining the optimum assignment of the candidates to the jobs. In order to attain this objective it is customary to resort to an *algorithm*[7], that is, to a standardised procedure for doing the calculations. We had several algorithms available that were capable of leading us towards a solution of the problem. The better known were the algorithm of "assignment by eliminating rows and columns", the "Hungarian algorithm" and the "branch and bound algorithm"[8]. The latter two provided an optimum solution, while the first does *not always* arrive at the optimum solution, although, in any event, it provides a good result. The advantage was its great simplicity and ease of operation. Appreciating these properties I proposed to Pepe and Consol that we use the *algorithm for assignment by eliminating rows and columns*. This algorithm can be used either starting out from a distance matrix as well as from a closeness matrix. In an arbitrary fashion we decided to use the latter. The steps to be taken were as follows[9].

1) Find the element of the fuzzy relation $[\underset{\sim}{R}]$ with the highest value.

2) This element determines, by the row and column of which he is a member, the candidate to be assigned to the job.

3) Eliminate this element from the matrix, both in the corresponding row and column. We now have a matrix of a lower order.

4) We commence the process once again with this lower order matrix, finding the element with the highest value.

5) We do points 2) , 3) and 4) until we vacate the whole matrix.

6) There will be one candidate without a job.

Without further preamble we reproduce the matrix or fuzzy relation $[\overline{R}]$, surrounding the highest value.

$[\underset{\sim}{R}] =$

	P_1	P_2	P_3	P_4	P_5
T_1	0.8216	0.8763	0.9418	0.8708	0.7672
T_2	0.7889	0.8798	0.8237	0.8925	0.7817
T_3	0.8345	0.8799	0.9110	0.8982	0.7983
T_4	0.9199	0.8252	0.8671	(0.9455)	0.9090

The highest value is 0.9455 and corresponds to element (T_4, P_4). This leads us to doing the first assignment.

Candidate P_4 will be assigned to job T_4.

We then eliminated row T_4 and column P_4, thus remaining with a matrix of a lower order, in which we have surrounded the element, of those remaining, with the highest value.

	P_1	P_2	P_3	P_5
T_1	0.8216	0.8763	0.9418	0.7672
T_2	0.7889	0.8798	0.8237	0.7817
T_3	0.8345	0.8799	0.9110	0.7983

The highest value is 0.9418, corresponding to element (T_1, P_3), ptherefore we can make the second assignment:

Candidate P_3 will be assigned to job T_1.

We then eliminated row T_1 and column P_3 and the following matrix remained in which we had to surround the highest of the remaining values:

	P_1	P_2	P_5
T_2	0.7889	0.8798	0.7817
T_3	0.8345	0.8799	0.7983

Having arrived at this point, a problem arose which normally escapes from the mechanisms of the algorithm. Candidate P_2 was "practically" as good for being assigned to job T_2 as to job T_3, since the difference between *the two closeness* was only of a ten-thousandth, an insignificant number, above all in the sphere of uncertainty. What to do when faced with a situation such as this?

We feel the most prudent thing to do was to continue the process to the end in its two versions: first selecting element (T_2, P_2) and then element (T_3, P_2). Finally we would decide those assignments the *total sum of the closeness, which was the highest*. This we did:

a) First alternative:

	P_1	P_2	P_5
T_2	0.7889	0.8798	0.7817
T_3	0.8345	0.8799	0.7983

In this case the assignment was:

Candidate P_2 will be assigned to job T_2.

On eliminating row T_2 and column P_2 the following fuzzy relation remained, the highest value of which has been surrounded with a circle:

	P_1	P_5
T_3	(0.8345)	0.7983

The last assignment would be:

Candidate P_1 will be assigned to job T_3.

If we followed this alternative the algorithm would have ended here.

b) Second alternative:

	P_1	P_2	P_5
T_2	0.7889	0.8798	0.7817
T_3	0.8345	(0.8799)	0.7983

Now the assignment would be:

Candidate P_2 will be assigned to job T_3.

Eliminating row T_3 and column P_2, we arrived at the following fuzzy relation, the highest value of which has likewise been surrounded by a circle:

	P_1	P_5
T_2	(0.7889)	0.7817

Here we had the last assignment as:

Candidate P_1 will be assigned to job T_2.

With this assignment the process ended

Now the only thing left was to select between the first and the second alternative. For this all we had to do was add the "closeness" of the assignments arrived at in each case. These were as follows:

a) First alternative:

$$S_1 = 0.9455 + 0.9418 + 0.8798 + 0.8345 = 3.6016$$

b) Second alternative:

$$S_2 = 0.9455 + 0.9418 + 0.8799 + 0.7889 = 3.5561$$

It now became clear that the decision should fall to the first of these alternatives, with which final assignment would be as follows:

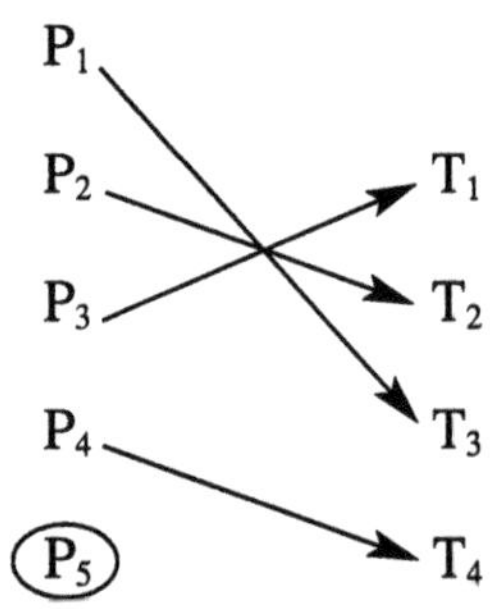

The candidate remaining without a job was P_5. And this was so whichever of the two alternatives was decided on.

The algorithm we had used had given us a "good" answer to the problem of assignment that Pepe need to resolve. The question that then arose was to know if this *good solution* was the *optimum solution*. In other words, could another better solution exist?. The answer to this question would be given if we compared the results with the use of one of the other algorithms. Among the possibilities we selected the so-called *Hungarian algorithm*.

The Hungarian algorithm for assignment

The immediate question asked by Pepe and Consol was only to be expected. Why is this algorithm known by the name of *Hungarian algorithm*? For some reason that is not very frequent among mathematicians, but in this case it was designated by the nationality of the person who established the reasoning behind its elaboration, D. König, in a very well known work[10].

The support on which this algorithm is applied is also a matrix of separation or matrix of closeness. We have adopted the habit of using the matrix of separation, with which the process of optimisation must follow criteria of *minimisation*, taking into account the fact that, under these circumstances, what is desired is the *least possible separation* between the descriptor of the job to be covered and the descriptor of the candidate to occupy the job. Therefore, following on with our custom, we take as a basis the fuzzy relation $[\underset{\sim}{\bar{R}}]$. Therefore as this matrix is not square (it has a different numbers of rows and columns) we resort to fiction by creating a *fictitious job*, F, which will consist of a "non job". The description of this "non job" should be as far removed from the remaining jobs as possible. As we are dealing with a matrix of *distance*, the furthest removed would be to give each quality, characteristic or peculiarity "level one". Having done this operation we now had the following square matrix:

$[\underset{\sim}{\bar{R}}'] =$

	P_1	P_2	P_3	P_4	P_5
T_1	0.1784	0.1237	0.0582	0.1292	0.2328
T_2	0.2111	0.1202	0.1763	0.1075	0.2183
T_3	0.1655	0.1201	0.0890	0.1018	0.2017
T_4	0.0801	0.1748	0.1329	0.0545	0.0910
F	1	1	1	1	1

From this fuzzy relation we now commence the successive phases of the Hungarian algorithm.

1ª) We subtract from *each column* the lowest value of each of them, placing the subtracted amounts below:

	P_1	P_2	P_3	P_4	P_5
T_1	0.0983	0.0036	0	0.0747	0.1418
T_2	0.1310	0.0001	0.1181	0.0530	0.1273
T_3	0.0854	0	0.0308	0.0473	0.1107
T_4	0	0.0547	0.0747	0	0
F	0.9199	0.8799	0.9418	0.9455	0.9090
	0.0801	0.1201	0.0582	0.0545	0.0910

With these operations what we arrive at is that there is a least one zero in each column.

2ª) We subtract from each row the lowest value of each one, placing the subtracted amount alongside:

	P_1	P_2	P_3	P_4	P_5	
T_1	0.0983	0.0036	0	0.0747	0.1418	0
T_2	0.1309	0	0.1180	0.0529	0.1272	0.0001
T_3	0.0854	0	0.0308	0.0473	0.1107	0
T_4	0	0.0547	0.0747	0	0	0
F	0.0400	0	0.0619	0.0656	0.0291	0.8799

We will see that in this matrix there is at least one zero in each row and one zero in each column.

3ª) We then considered, one after the other, the rows containing the least amount of zeros (if there were more than one zero, as was the case, these were selected arbitrarily). We drew a square around one of the zeros in each row and crossed out the zeros in the row and column in which there was a squared zero. The selected order of rows was T_1, T_2, T_3, F y T_4. We have shown the result below:

	P_1	P_2	P_3	P_4	P_5
T_1			[0]		
T_2		[0]			
T_3		0̸			
T_4	[0]			0̸	0̸
F		0̸			

For the sake of greater visual clarity we have eliminated all the values other than zero. We can see that in the matrix there are only *three zeros from the possible five in a square*. This tells us that that we had yet not found *an optimum assignment*. The algorithm must continue.

4ª) This phase was a little more complicated, so we decided to describe it in more detail, by means of the following sections:

a) We signal with a horizontal arrow ← those *rows* in wich there was not a zero in a square.

b) We indicate with a vertical arrow ↑ those *columns* in which t*here is* a crossed out zero in *a row signalled with an arrow.*

c) We mark with a horizontal arrow ← those *rows* in which *there was a zero in a square in a column marked with an arrow.*

d) We must repeat sections b) and c) until such time as no more rows or columns can be marked.

e) We draw a line through those *rows not* marked with arrows, and another line *through the columns that are* marked with arrows.

	P_1	P_2	P_3	P_4	P_5	
T_1	~~0.0983~~	~~0.0036~~	~~[0]~~	~~0.0747~~	~~0.1418~~	
T_2	0.1309	[0]	0.1180	0.0529	0.1272	←
T_3	0.0854	~~0~~	0.0308	0.0473	0.1107	←
T_4	~~[0]~~	~~0.0547~~	~~0.0747~~	~~0~~	~~0~~	
F	0.0400	~~0~~	0.0619	0.0656	0.0291	←
		↑				

5ª) With the boxes that are not crossed with a line we form a sub-matrix.

	P_1	P_3	P_4	P_5
T_2	0.1309	0.1180	0.0529	0.1272
T_3	0.0854	0.0308	0.0473	0.1107
F	0.0400	0.0619	0.0656	(0.0291)

We look for the lowest number of the elements of this sub-matrix, which is (F, P_5) with 0.0291.

6ª) This minimum value (0.0291), was subtracted from the elements of the *columns that were not crossed with a line and was added* to those of the *rows that were crossed with a line*. The new matrix arrived at was as follows:

	P_1	P_2	P_3	P_4	P_5
T_1	0.0983	0.0327	0	0.0747	0.1418
T_2	0.1018	0	0.0889	0.0238	0.0981
T_3	0.0563	0	0.0017	0.0182	0.0816
T_4	0	0.0838	0.0747	0	0
F	0.0109	0	0.0328	0.0365	0

7ª) With this new matrix in which there is at least one zero in each row and in each column, we commence the process of squares and crossing out once again The order to be followed would be T_1, T_2, T_3, F, T_4. We then arrived at the following:

	P_1	P_2	P_3	P_4	P_5
T_1			[0]		
T_2		[0]			
T_3		Ø			
T_4	[0]			Ø	Ø
F		Ø			[0]

8ª) Again on this occasion did we did not arrive at an optimum assignment as there were only four of the possible zeros boxed. We therefore had to pass on to the following phase, repeating what we had done in the fourth phase (placing arrows) and crossing out rows and columns:

	P_1	P_2	P_3	P_4	P_5	
T_1	~~0.0983~~	~~0.0827~~	[0]	~~0.0747~~	~~0.1418~~	
T_2	0.1018	[0]	0.0889	0.0238	0.0981	←
T_3	0.0563	Ø	0.0017	0.0182	0.0816	←
T_4	[0]	~~0.0838~~	~~0.0747~~	Ø	Ø	
F	~~0.0109~~	Ø	~~0.0328~~	~~0.0365~~	[0]	
		↑				

9ª) We formed the following sub-matrix:

	P_1	P_3	P_4	P_5
T_2	0.1018	0.0889	0.0238	0.0981
T_3	0.0563	(0.0017)	0.0182	0.0816

The element with the lowest value was (T_3, P_3) with 0.0017

10ª) This value, 0.0017, was subtracted from the elements of the non-lined columns and added to the rows that were lined. We then arrived at the following matrix:

	P_1	P_2	P_3	P_4	P_5
T_1	0.0983	0.0344	0	0.0747	0.1418
T_2	0.1001	0	0.0872	0.0221	0.0964
T_3	0.0546	0	0	0.0165	0.0799
T_4	0	0.0855	0.0747	0	0
F	0.0109	0.0017	0.0328	0.0365	0

11ª) Once again we proceeded to place squares and cross out the zeros, following the order T_1, T_2, T_3, F, T_4:

	P_1	P_2	P_3	P_4	P_5
T_1			[0]		
T_2		[0]			
T_3		~~0~~	~~0~~		
T_4	[0]			~~0~~	~~0~~
F					[0]

12ª) Again no optimum assignment, so we placed the arrows and crossed out rows and columns

	P_1	P_2	P_3	P_4	P_5	
T_1	0.0983	0.0344	[0]	0.0747	0.1418	←
T_2	0.1001	[0]	0.0872	0.0221	0.0964	←
T_3	0.0546	~~0~~	~~0~~	0.0165	0.0799	←
T_4	[0]	0.0855	0.0747	~~0~~	~~0~~	
F	0.0109	0.0017	0.0328	0.0365	[0]	
		↑	↑			

13ª) We formed the sub-matrix with the elements that were not crossed out:

	P_1	P_4	P_5
T_1	0.0983	0.0747	0.1418
T_2	0.1001	0.0221	0.0964
T_3	0.0546	(0.0165)	0.0799

The lowest value, 0.0165 corresponded to (T_3, P_4).

14ª) The value 0.0165 was subtracted and added to the elements of the non-lined columns and lined rows respectively. The new matrix arrived at was as follows:

	P_1	P_2	P_3	P_4	P_5
T_1	0.0818	0.0344	0	0.0582	0.1253
T_2	0.0836	0	0.0872	0.0056	0.0799
T_3	0.0381	0	0	0	0.0634
T_4	0	0.1020	0.0912	0	0
F	0.0109	0.0165	0.0493	0.0365	0

15ª) For the corresponding squares and crossings out we followed the order T_1, T_2, F, T_3 y T_4:

	P_1	P_2	P_3	P_4	P_5
T_1			[O]		
T_2		[O]			
T_3		~~O~~	~~O~~	[O]	
T_4	[O]			~~O~~	~~O~~
F					[O]

Now in this phase we had arrived at the optimum assignment on there being five zeros in a square corresponding to the five possible assignments, four real and one fictitious.

The result we had arrived at on using the "Hungarian algorithm" was *different* from the one we had found when using the "algorithm of assignment by elimination of rows and columns". In this case we had:

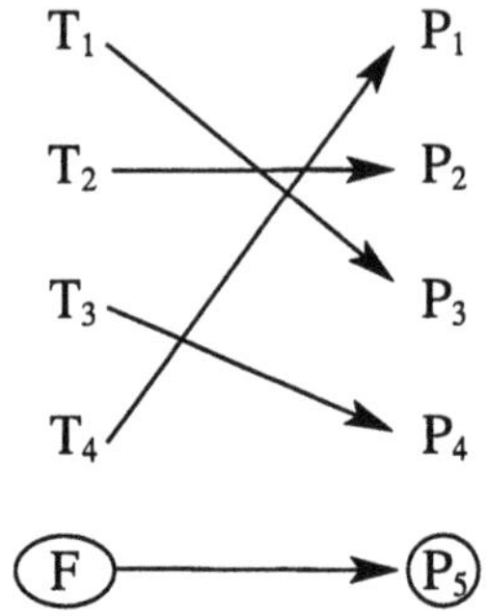

The difference consisted in the change of jobs for candidates P_3 and P_4 since in the first algorithm P_3 had been assigned T_1, with this method is was T_4; on the other hand then P_4 had been assigned T_4 now he was assigned T_1. What had occurred was a change over of jobs.

The Head Telegrapher, for along time removed from the academic world, shrugged his shoulders and said: What do we do now with these two employees (this was his administrative situation)? Although not necessary do to any calculations, since of the two algorithms, only the Hungarian ensures an optimum result, a simple comparative study would allow us to see this statement, and at the same time reassure my friend Pepe. What it meant was to compare *the sum of the closeness*, in one and the other algorithm, of the assignments we had made, or what should lead to the same result *the sum of the distances.*

In the first of these criteria (sum of closeness) we had seen that with the "algorithm of assignment by elimination of rows and columns" we had a total distance of :

$$S_1 = 0.9455 + 0.9418 + 0.8798 + 0.8345 = 3.6016$$

As the "Hungarian algorithm" had been used for the optimisation of minimums (distances) the complements to the unit should be calculated for the elements of fuzzy relation $[\underset{\sim}{R}]$.Therefore:

$$(T_1, P_3) = 0.0582 \quad ; \quad 1 - 0.0582 = 0.9418$$

$$(T_2, P_2) = 0.1202 \quad ; \quad 1 - 0.1202 = 0.8798$$

$$(T_3, P_4) = 0.1018 \quad ; \quad 1 - 0.1018 = 0.8982$$

$$(T_4, P_1) = 0.0801 \quad ; \quad 1 - 0.0801 = 0.9199$$

$$H = 0.9418 + 0.8798 + 0.8982 +.0.9199.= 3.6397$$

Although there was very little difference, it was clear that H was higher than S_1 so that optimisation by means of the Hungarian algorithm was proven.

The same conclusion was reached if, instead of using the criterion of closeness we used that of distance. In this case complementing to the unit should be done for the values of the closeness fuzzy relation $[\bar{\underset{\sim}{R}}]$ in the use of the algorithm for assignment by eliminations of rows and columns:

$$(P_1, T_3) = 0.8345 \quad ; \quad 1 - 0.8345 = 0.1655$$

$$(P_2, T_2) = 0.8798 \quad ; \quad 1 - 0.8798 = 0.1202$$

$$(P_3, T_1) = 0.9418 \quad ; \quad 1 - 0.9418 = 0.0582$$

$$(P_4, T_4) = 0.9455 \quad ; \quad 1 - 0.9455 = 0.0545$$

$$\bar{S}_1 = 0.1655 + 0.1202 + 0.0582 + 0.0545 = 0.3984$$

With the "Hungarian algorithm" we had:

$$\bar{H} = 0.0582 + 0.1202 + 0.1018 + 0.0801 = 0.3603$$

As $\bar{H} < \bar{S}_1$, it was once again proven that the best assignment, was an optimum, by means of the Hungarian algorithm. The very little difference existing in the use of either algorithm, showed that the result arrived at with the algorithm based on the elimination of rows and columns was a good approximation. The fact that so little difference existed between the assignment of P_1 and P_4 to jobs T_3 and T_4 made the Head of Telegraphers very happy since, independently of having made the best possible assignment, he now had available a certain interchangeability, which provided him with an amount of flexibility, very useful under certain circumstances such as sickness, accident, labour absences, among others. In this way the second of my friend Pepe's problems were resolved satisfactorily..

Interchangeability and grouping in human resources

The idea of making his employees interchangeable was linked, in a certain fashion, to the *training* of homogeneous groups, although with nuances that required a certain amount of thinking about. In fact, to be able to interchange two or more people in their respective jobs meant that all were able, more or less, to do the same task, both if they were occupied with one of them or if they did this with the rest. This would only be so if the employees possessed homogeneous characteristics relative to the necessary levels in the qualities, characteristics or peculiarities of the different tasks to be carried out. In a certain way this meant establishing homogeneous groups but only with regard to *labour aspects*. However what it appeared that Pepe wanted was not only this but something more: This homogeneity must also refer to aspects of *conviviality*. In this sense the formation of the groups was aimed at their being destined for geographical places separated from the official central office, with few personal contacts and with the permanent need of carrying out the work by using their own means and basically counting on the cooperation of their fellow workers. These groups were reduced (formed by two or a maximum of three people) as was small the group of people destined to these tasks. Actually the "particular offices" were *two* and the groups to be established could be comprised of *two* people each one, taking into account the availability for these tasks of *four* employees.

Once we had become aware of this fact we "got down to work". The first task to be done consisted in drawing up a list of the qualities, characteristics and peculiarities that were necessary both for doing the job at the designated offices as in order to make easy, or supportable if you like, coexistence with the companion. They would include, therefore, technical, cultural, psychological and even sociological aspects. After gathering a large number of "properties" we eliminated those we considered as less significant until they were reduced to the following:

C_1: Flexibility in relations with the public.

C_2: Dedication to the job.

C_3: Technical instrument training.

C_4: Integration into group work.

C_5: Experience in the entrusted functions.

C_6: Knowledge of computer operation.

C_7: Cooperative spirit.

C_8: Cultural level.

C_9: Stable married life.

C_{10}: Interest in sport

We were all conscious of the fact that there were other, up to a certain point, interesting qualities for working in common with others, but we felt that their inclusion in our work would provide no significant complementary element. Therefore we got down to describing each one of the four employees by means of a fuzzy sub-set of the referential of these qualities, characteristics and peculiarities. The result of our consultations, interviews and search provided us with the following descriptors:

	C_1	C_2	C_3	C_4	C_5	C_6	C_7	C_8	C_9	C_{10}
$\underset{\sim}{P}_1 =$	.6	1	.8	.4	.9	.7	.6	.3	1	.9

	C_1	C_2	C_3	C_4	C_5	C_6	C_7	C_8	C_9	C_{10}
$\underset{\sim}{P}_2 =$	.8	.6	.9	.7	.8	.5	.9	.8	.9	.7

	C_1	C_2	C_3	C_4	C_5	C_6	C_7	C_8	C_9	C_{10}
$\underset{\sim}{P}_3 =$	.9	.8	.6	.5	.9	.8	1	.7	.5	.9

	C_1	C_2	C_3	C_4	C_5	C_6	C_7	C_8	C_9	C_{10}
$\underset{\sim}{P}_4 =$	.7	.5	.8	.9	.8	.6	.7	.6	.8	.8

These fuzzy sub-sets brought to light the level possessed by each employee, of each one of the qualities, characteristics or peculiarities. What we had to do was to establish, as from which level, it could be considered that the employee possessed them at a sufficient degree to be able to satisfactorily comply with his labour responsibilities and with coexistence. This was equivalent to determining certain "thresholds" which we designated with the letter θ_j, j = 1, 2, ..., 10. The following were agreed upon:

$$\theta_1 = 0.8 \quad , \quad \theta_2 = 0.7 \quad , \quad \theta_3 = 0.8 \quad , \quad \theta_4 = 0.9 \quad , \quad \theta_5 = 0.8$$

$$\theta_6 = 0.7 \quad , \quad \theta_7 = 0.9 \quad , \quad \theta_8 = 0.7 \quad , \quad \theta_9 = 0.8 \quad , \quad \theta_{10} = 0.8$$

Once these thresholds had been established it was possible to convert each fuzzy sub-set, describing an employee, into a "Boolean sub-set"[11]. For this all that was required was to place, in the corresponding box, *a one if equal to or exceeding* the threshold and a *zero* if it were to be less than the threshold of the quality, characteristic or peculiarity. On doing these operations we arrived at the following Boolean sub-sets[12]:

	C_1	C_2	C_3	C_4	C_5	C_6	C_7	C_8	C_9	C_{10}
$P_1 =$		1	1		1	1			1	1

	C_1	C_2	C_3	C_4	C_5	C_6	C_7	C_8	C_9	C_{10}
$P_2 =$	1		1		1		1	1	1	

P_3 =

C_1	C_2	C_3	C_4	C_5	C_6	C_7	C_8	C_9	C_{10}
1	1			1	1	1	1		1

P_4 =

C_1	C_2	C_3	C_4	C_5	C_6	C_7	C_8	C_9	C_{10}
		1	1	1				1	1

We joined all these Boolean sub-sets into one table or matrix, and in this way arrived at:

[B] =

	C_1	C_2	C_3	C_4	C_5	C_6	C_7	C_8	C_9	C_{10}
P_1		1	1		1	1			1	1
P_2	1		1		1		1	1	1	
P_3	1	1			1	1	1	1		1
P_4			1	1	1				1	1

This Boolean matrix would constitute the basic support on which we would carry out the process leading to arriving at the sought after groups. Just as the Head Telegrapher described his requirements, we felt that the best means for solving these was to resort to a "theory" which has given us much satisfaction in the past. We are referring to the "theory of affinities"[13].From this theory certain algorithms had been drawn up which could be very useful for us[14]. From among these we chose the one called "algorithm for inverse maximum correspondence", the phases of which we have listed below:

1) We chose from among the *group of employees* and *the set of qualities, characteristics or peculiarities*, the one with the least amount of elements.

   The choice fell to the *employees*, as these were only four, while the qualities, characteristics or peculiarities were ten.

2) We listed all the possible groupings that could be formed with the four employees: no employee, taken one by one, two by two, three by three and lastly the four. This was as follows:

$$\{\emptyset, P_1, P_2, P_3, P_4, P_1P_2, P_1P_3, P_1P_4, P_2P_3, P_2P_4, P_3P_4,$$
$$P_1P_2P_3, P_1P_2P_4, P_1P_3P_4, P_2P_3P_4, P_1P_2P_3P_4\}$$

Mathematics calls this set the "power set".

3) *For each one of the groups* that constitute the elements of the "power set" a correspondence was made of the groupings of qualities, characteristics and peculiarities possessed by the employees of the group. From the Boolean matrix [B] the following were found:

$$\emptyset \rightarrow C_1\ C_2\ C_3\ C_4\ C_5\ C_6\ C_7\ C_8\ C_9\ C_{10}$$
$$P_1 \rightarrow C_2\ C_3\ C_5\ C_6\ C_9\ C_{10}$$
$$P_2 \rightarrow C_1\ C_3\ C_5\ C_7\ C_8\ C_9$$
$$P_3 \rightarrow C_1\ C_2\ C_5\ C_6\ C_7\ C_8\ C_{10}$$
$$P_4 \rightarrow C_3\ C_4\ C_5\ C_9\ C_{10}$$
$$.P_1\ P_2 \rightarrow C_3\ C_5\ C_9$$
$$P_1\ P_3 \rightarrow C_2\ C_5\ C_6\ C_{10}$$
$$P_1\ P_4 \rightarrow C_3\ C_5\ C_9\ C_{10}$$
$$P_2\ P_3 \rightarrow C_1\ C_5\ C_7\ C_8$$
$$.P_2\ P_4 \rightarrow C_3\ C_5\ C_9$$
$$..P_3\ P_4 \rightarrow C_5\ C_{10}$$
$$...P_1\ P_2\ P_3 \rightarrow C_5$$
$$.P_1\ P_2\ P_4 \rightarrow C_3\ C_5\ C_9$$
$$..P_1\ P_3\ P_4 \rightarrow C_5\ C_{10}$$
$$...P_2\ P_3\ P_4 \rightarrow C_5$$
$$...P_1\ P_2\ P_3\ P_4 \rightarrow C_5$$

4) For the same groupings of qualities, characteristics or peculiarities (right hand column) a choice was made of the group of employees (left hand column) comprising the greatest number of the same. The result was:

$$
\begin{array}{rcl}
C_5 & \rightarrow & P_1 P_2 P_3 P_4 \\
C_5 C_{10} & \rightarrow & P_1 P_3 P_4 \\
C_3 C_5 C_9 & \rightarrow & P_1 P_2 P_4 \\
C_1 C_5 C_7 C_8 & \rightarrow & P_2 P_3 \\
C_2 C_5 C_6 C_{10} & \rightarrow & P_1 P_3 \\
C_3 C_5 C_9 C_{10} & \rightarrow & P_1 P_4 \\
C_3 C_4 C_5 C_9 C_{10} & \rightarrow & P_4 \\
C_1 C_3 C_5 C_7 C_8 C_9 & \rightarrow & P_2 \\
C_2 C_3 C_5 C_6 C_9 C_{10} & \rightarrow & P_1 \\
C_1 C_2 C_5 C_6 C_7 C_8 C_{10} & \rightarrow & P_3 \\
C_1 C_2 C_3 C_4 C_5 C_6 C_7 C_8 C_9 C_{10} & \rightarrow & \emptyset
\end{array}
$$

These relations allowed us to establish, one by one, the affinities existing between the employees that formed each group and on having chosen for certain qualities, characteristics or peculiarities the largest group of people, these groups are considered as the maximum possible

5) It was quite easy to see that both the groupings of employees and that of qualities, characteristics or peculiarities could be structured in a coherent manner from the smallest groups to the largest, by placing after a certain group the group that possessed the highest number of elements plus one other. We began with the groups of employees.

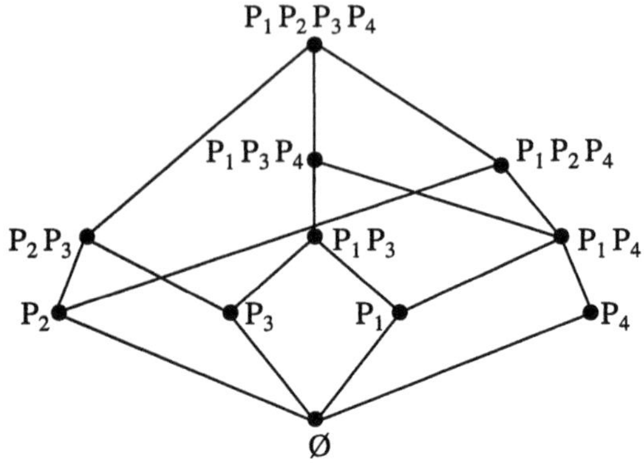

We continued after this with the groupings of qualities, characteristics and peculiarities:

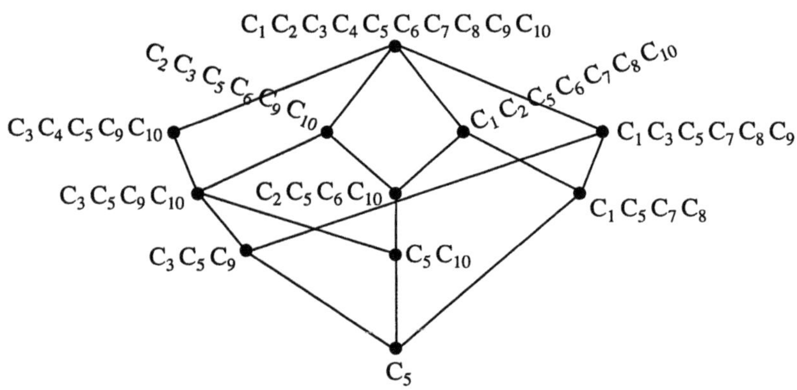

Having seen these two structured figures, known as lattices, Consol, women tend to be more observant, became aware of the fact that both structures had the same form, if one of them was rotated upside down. In technical words, what we had was two *isomorphic* lattices, so that it was also graphically possible to establish the relations, now correctly, called *of affinity*: A new representation will clarify this. We rotated the second of the graphs and arrived at the following result:

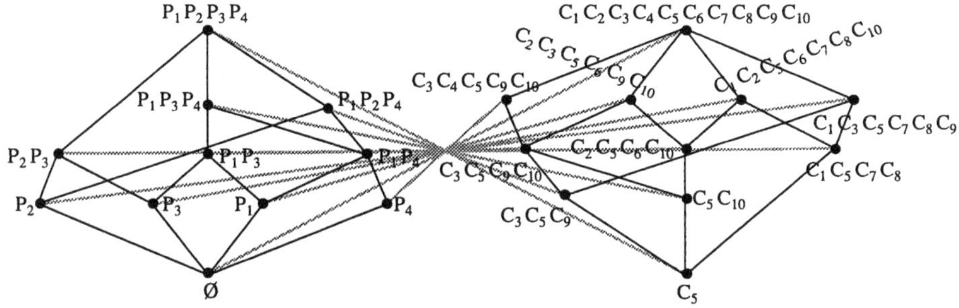

6) Making the two lattices coincide, taking into account the lines that join the vertices of one and the other, we arrived at a single lattice, in which appeared joined together the correspondences between the groups of employees and groups of qualities, characteristics or peculiarities possessed by all of them at the established level:

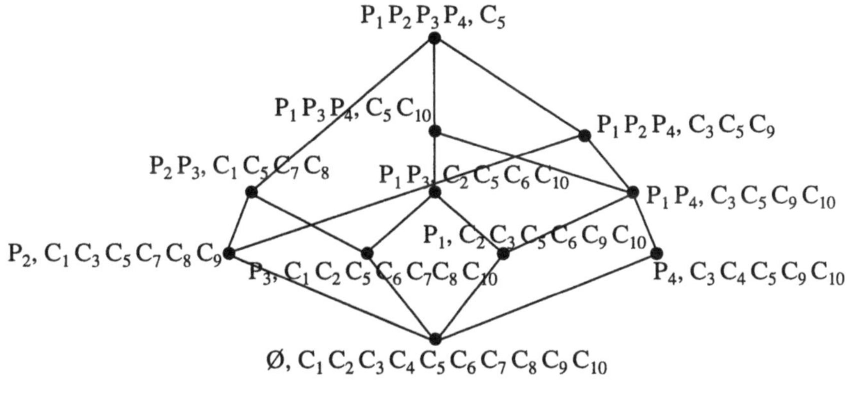

Here the Head Telegrapher, my friend Pepe, had all the data. Absolutely all of it. According to the qualities, characteristics or peculiarities that he wished to concur within the group, there automatically appeared in the lattice those employees possessing the same. This type of lattice is known as a "Galois lattice" in honour of the young mathematician who created it.

We could see that, as we moved within the lattice from the bottom to the top, the groupings of people became *larger,* but on the other hand the employees of the group possessed less qualities, characteristics and peculiarities. The opposite occurred when descending from the top to the bottom the number of qualities possessed by the group of people increased, but the latter decreased in number.

With the "Galois lattice" Pepe now had all the tools for carrying out the very best movement strategy. In fact, if for example, he wanted the qualities, characteristics or peculiarities to be C_1,C_5, C_7, C_8, that is flexibility in relations with the public, experience in the entrusted tasks, spirit of cooperation and cultural level, he should send employees P_2 and P_3, since they formed a more homogeneous group relative to the required "properties". If on the other hand what was desirable were qualities, characteristics or peculiarities C_3, C_5, C_9, and therefore: technical instrument training, experience in the entrusted tasks, and stable married life, then he could appoint for these jobs employees P_1, P_2 and P_4, or any combination of these three. While circumstances did not change excessively the Head Telegrapher had ensured an efficient management instrument.

We felt highly satisfied. What at the outset appeared to be complex problems had become converted in matters that were to prove normal action for the person responsible for a human group. From selection right through to assignment and homogeneous grouping we had tackled and resolved all the matters that had been raised. The outcome of events would corroborate the quality of the tools used but, above all else the correct action of my friend Pepe, who had overcome the enormous difficulty of continuing along this same line.

REFERENCES

[1] GIL ALUJA, J.: *La gestión interactiva de los recursos humanos en la incertidumbre*. publ. Ceura. Madrid, 1996, page 3.

[2] GIL ALUJA, J.: *La gestión interactiva de los recursos humanos en la incertidumbre*. publ. Ceura. Madrid, 1996, page 4.

[3] KAUFMANN, A. and GIL ALUJA, J.: *Introducción de la teoría de los subconjuntos borrosos a la gestión de empresas*. publ. Milladoiro. Santiago de Compostela, 1986, page. 138.

[4] GIL LAFUENTE, J.: *El "indice del máximo y mínimo nivel" en la optimización del fichaje de un deportista*. Proceeding of the AEDEM'01 Congress. Polytechnic University of Reggio, September 4-6, 2001.

[5] This index has been rounded off to thousandths so that the sum of all of them provided the unit.

[6] Let us recall that this is just one of the indexes used by Professor Jaime Gil Lafuente.

[7] The word algorithm stems from the Arab word Al-khuwarizmi, name of the mathematician to whom is attributed the commencement of these calculation procedures.

[8] GIL ALUJA, J.: *Elementos para una teoría de la decisión en la incertidumbre*. publ. Milladoiro, Vigo 1999, pages. 140-179.

[9] KAUFMANN, A. and GIL ALUJA, J.: *Introducción de la teoría de los subconjuntos borrosos a la gestión de las empresas*. publ. Milladoiro. Santiago de Compostela 1986, pages. 157-158.

GIL ALUJA, J.: *Selección de personal:el problema de la polivalencia y el de la uniformidad*. Cuadernos Ceura. Madrid, 1987, page. 8-10.

[10] KÖNIG, D.: *Théorie der endlichen und unendlichen graphen* (1916) reprinted later by Chelsea Publ. Cª. New York, 1950. This work was made known by Kuhn, H.W. in the article *"The hungarian method for the assignment problem"*. *Naval Research Logistics Quartely Vol. 2* Nos. 1-2, March-June 1955 pages 83-98. The algorithm was primanly made known through one of the works by Kaufmann, A.: *Méthodes et Modèles de la Recherche Opérationelle*. Volume I. Dunod Paris 1970, pages. 65-72.

[11] A Boolean sub-set is characterised by the fact that its levels are limited to *one* and *zero*, *yes* or *no*, *posses* or *does not posses*.

[12] We eliminate the zeros for greater clarity in the presentation.

[13] KAUFMANN, A. and GIL ALUJA, J.: *Selection of affinities by means of fuzzy relations and Galois lattices*. Proceedings of the Euro XI Congress O.R. Aachen, July 16-19, 1991.

[14] GIL ALUJA, J.: *Elementos para una teoría de la decisión en la incertidumbre*. publ. Milladoiro. Vigo 1999, pages. 215-222.

VII ACCOUNTING IN FINANCIAL ANALYSIS

Accounting as a support for financial analysis

Financial activity in businesses and institutions has traditionally been supported by three basic elements: *raising* monetary means, *placing* of these payment means and the *analysis* of the financial situation. Increasingly, the *analytical aspect* of financial management occupies the interest of scholars and professionals. The former for the challenges caused by changing realities and the latter by the pressing need of attempting to find a solution to emerging problems that arise from a society that is in continuous and accelerated movement.

The times have long gone by in which grasping financial phenomena was formalised with the aid of a mathematics of certainty, using precise data, nearly always of a historical nature, found in the financial statements provided by accounting. The brief immersion of stochastic techniques in this field, has been nothing more than a temporary dabbling into some of the aspects considered as essential, without their, up to now, having consolidated a vocation of permanence. It has been mainly in the last ten years when there have been considerable advances made in drawing up and using a good arsenal of operating instruments that are capable of allowing for adequate treatment of the problems arising in financial analysis[1] within a context of growing uncertainty. The path that has been taken has allowed for a new focus being given to financial analysis, which, nevertheless, has not closed the doors on the use of classical schemes, which, duly adapted, have been of unquestionable use for better understanding, conveniently explaining and carrying out an efficient treatment of the new financial situations, avoiding their simplification and deformation, which are joined one with the other when resorting to numerical precision.

We are convinced that the *number* constitutes a basic part of financial analysis. But this does not mean that the numerical study must of necessity be carried out in the field of certainty, frequently called, *determinism*, perhaps as a mere abuse of the language. That this should be so should not come as a surprise, if taking into account the fact that the source of the information from which the "data" flows for making an analysis was accounting, the desired faithful reflection of a supposedly precise reality. But not even the most refined *financial statements* have faithfully reflected the complex and involved current realities, neither are these realities assumable to absolute precision. It is absolutely certain that in one way or another the economic-financial situation of a business or institution must be *presented.* It is also certain that Accounting has played, in this sense, an important, essential we would say, historical role. Again it is certain to say that even today a coherent and structured amount of knowledge has not been found that is capable of providing even the same reply to the challenges posed. But it is also certain that more can be hoped for, much, much more, of the quasi infinite possibilities offered by this impressive scientific-technical body to which Fra Luca Pacciolo gave birth.

We have always asked ourselves why if we see economic-financial realities with a greater or lesser *vagueness* (which in no way whatsoever implies that this is erroneous) their formalization must be prevailingly precise. On acting in this way a serious dysfunction is introduced into the system right from the very outset. *Any distance* found between the imprecise reality and its noted down precise formalisation, constitutes an error. Later operatios, by mere arithmetical nature, will extend it. Is this serious? Supposedly not, since we have carried on for several centuries "living" from inherited and assumed principles without the occurrence of any large catastrophes, having extended and perfected a theoretical body that is generally considered as an important branch of Science. But the unquestionable predicament of Accounting and its efficiency is due, we feel, to two facts that in some manner are interrelated.

The first of these refers to *grasped realities.* At the very same time as the use commences of the instruments arising from the double entry, observation of the economic-financial phenomenology shows an acceptable stability and knowledge that is susceptible of making the bases of action of the "businessman" vary substantially, began to appear with a slow rhythm. The data and information from the past were a good source for understanding, explaining and treating the realities of the moment. Seen from current perspectives, the past appears to us as a very quiet backwater. The "Industrial Revolution" broke this supposed and idyllic business peace, the same way in which the more recent "Revolution of 68" broke down the ideas, although the former not sufficiently in order to shake the indomitable spirits of enterprising accountants. It was during the latter half of the last century, when profound changes commenced, developed and accelerated, which , in their acceleration, gave rise to very rapid changes, let us say continuous variations, in those *scenarios* in which men of business act. Today it is difficult for the data of the past to serve as the basis for the calculation of current data, and even less be the support for future estimates.

The second concerns the fundamental objects of Accounting. Without the desire whatsoever of stirring up any *controversy,* we feel it can be accepted that, up to not very long ago, accounting activity in its professional stream, was dedicated to drawing up, presenting and studying certain financial statements that should reflect in a suitable manner the economic-patrimonial situation of a business or institution; budget accounting was a magnificent and excellent exception. But the arising, or re-arising, with strength of the idea of *decisional accounting*, presented under different names but always leading to future actions, has given an about face to the traditional conception of looking at the past. And today, with increasing strength, can be seen a double stream, in the style of *Jano Bifronte*, one looking to the past and the other to the future.

It is possible that as a consequence of the coincidence between these two aspects, the one a member of the *real* field, the other of the *formal*, there may have arisen the incipient interest in seeking new ways that are susceptible of providing a methodology capable of finding a solution to the desire for a greater *representation of the complexity* of past and present economic-financial relations, as well as the creation of a basic set of instruments on which to support, suggest and limit the *capacity for decision* when faced with certain future realities, increasingly surrounded in the fog of uncertainty.

These thoughts intermittently came to our mind, whenever any event were to upset the relative calm of daily accounting mechanisms or we felt that available techniques were not being taken advantage of, which, easily adaptable, could become solid management instruments. These thoughts were commented upon with my companion, teacher and friend, Mario, when we met on a flight to Bilbao from Barcelona. At that time we were finishing writing, with Professor Kaufmann, our first joint book[2]. My teacher listened with interest to my explanations and we felt that he was receptive relative to the work we were doing on the creation of new techniques for improving management. Mario has been the leading professor of Accounting at the Spanish Universities and, under his inspiration or direction, the most eminent professors have developed their capacities. Thanks to his help and determination, new generatios of teachers and researchers were able to culminate their academic careers. The writer of these lines has the honour of being able to count himself among them. He attained the highest levels of glory and recognition. Today, in the peacefulness of his retirement, he continues to discretely exercise his teaching to those who have the wisdom of seeking his advise. His unquestionable authority was stimulus for entering, although briefly, into the task of incorporating the new techniques for the treatment of uncertainty in Accounting. And this was the result, fruit of the undertaking acquired on that flight, how we included a chapter in that book, first and very modest incursion into the theory of fuzzy sub-sets to the analysis of accounting statements. We deviated from this field of the study in order to continue our work in favour into the research of new models and algorithms, but we had the good fortune that a brilliant and young professor, Anna María, whose publications in the sphere of financial analysis in uncertainty have gained the highest praise for their originality, correct structuring and clarity of explanation, took over. The greater part of the contents of this chapter is consequent, directly or indirectly, upon her scientific works.

Provisional financial statements and uncertainty

During his active life, Mario combined his teaching activity and his professional activity successfully directing a consultancy office. The problems that arose there were of very diverse natures, but first among them were fiscal and economic problems. Thanks to the comments and conversations he held on these, it was possible to develop some of the techniques that are quite normal today. Even though it is some time ago I seem to recall that one of the first applications of fuzzy logic to Accounting was the consequence of a professional consultation relative to the possibility of making a complementary placement of monetary means in the business belonging to the person consulting. His indecision stemmed from *the lack of precision* in the consequences of said investment, as well as of the effects that would be caused not only in the *profitability* of his "business", but also in its *solvency*. He had three options open to him:

a) To renew existing fixed assets by acquiring new machinery.

b) Accumulate raw material stocks in view of possible price increases.

c) Continue with the then situation, waiting for successive economic accounting periods.

The approach for the solution of the problem could occur from several points of view, but everything led to the thought that the *balance sheet* and *the profit and loss account* should assume a certain prominence, implicitly required by the person responsible for the business.

We felt that the most important part of the consultation would arrive at a good solution if we managed to estimate, the *provisional balance sheets* and the *foreseen profit and loss accounts*, within the established economic horizon, which was two years ahead. The difficulty, on the other hand, was in arriving at these balance sheets, expressed in *exact numbers*, corresponding to a date located in the future. It appeared that the time had come to delve into the arithmetic of uncertainty in order to find a good substitute for the *certain number* and move on to find one from the sphere of fuzziness. For its simplicity we chose the *confidence interval*. Having taken this first decision, what remained was to establish the type of balance sheet and profit and loss account in which we wanted the required information to be presented. It should be in a very summarized format, in order for ease of reading and understanding for the company managers, who were not normally accustomed to the "details" of the accounts of the General Accounting Plan. What was required then, were certain, synthesised financial statements, but which must contain all the essential information. After due consultation we decided on the following:

BALANCE SHEET

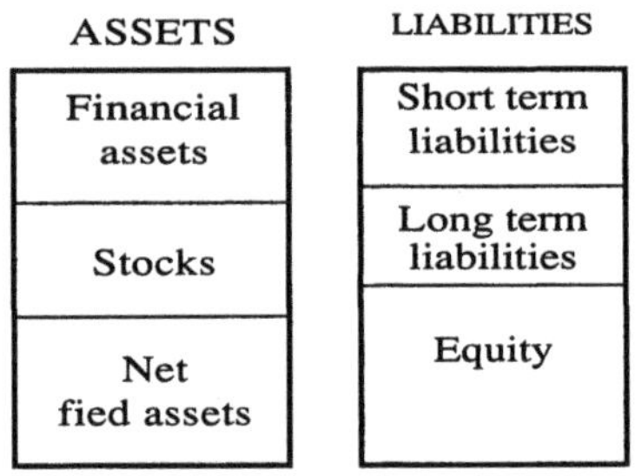

PROFIT AND LOSS ACCOUNT

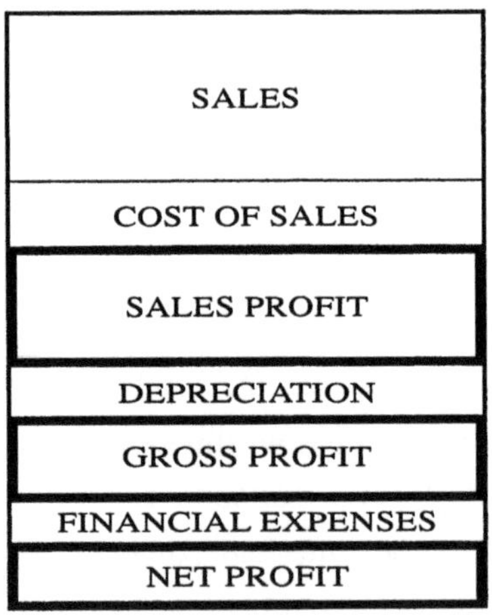

From this moment the principal task of gathering information was the responsibility of the business. The corresponding estimates should be expressed, just as we had decided, by means of *confidence intervals*, that is to say, bringing to light a *lower extreme* beneath which the hoped for reality would not be found and an *upper extreme*, over which this would also not occur. The phrase: "between such and such" was an exact reflection of the sense of the interval. Once we had available the estimates presented in this way, the process was fast and well known by us[3].

They did not take long to deliver a whole set of values, data, scenarios, presumptions and doubts that, slowly were put together until we had the following tables:

ITEM	FIXED ASSET MOVEMENT	STOCK ACCUMULATION	WAITING POSITION
Financial assets	[250, 280]	[300, 320]	[430, 450]
Stocks	[320, 340]	[800, 850]	[400, 430]
Net fixed assets	[1500, 1800]	[500, 540]	[500, 540]
Short term liabilities	[350, 380]	[550, 590]	[300, 330]
Long term liabilities	[820, 1220]	[150, 200]	[130, 170]
Equity	[900, 920]	[900, 920]	[900, 920]
Estimated sales	[2200, 2500]	[1500, 1600]	[1000, 1200]
Cost of sales	[1400, 1500]	[950, 1000]	[650, 700]
Depreciation	[220, 230]	[80, 100]	[80, 100]
Financial expenses	[30, 35]	[10, 20]	[0, 0]

From this set of estimates, we could immediately draw up the balance sheets and foreseen profit and loss accounts:

Decision a): Movements in fixed assets

BALANCE SHEET A

ASSETS		LIABILITIES	
Financial assets	[250, 280]	Short term liabilities	[350, 380]
Stocks	[320, 340]	Long term liabilities	[820, 1120]
Net fixed assets	[1500, 1800]	Equity	[900, 920]
Total	[2070, 2420]	Total	[2070, 2420]

PROFIT AND LOSS ACCOUNT A

Estimated sales		[2200, 2500]
Cost of sales	[1400, 1500]	
Depreciation	[220, 230]	[1620, 1730]
	Gross profit	[470, 880]
Financial expenses		[30, 35]
	Net profit	[435, 850]

Decision b): Stock accumulation

BALANCE SHEET B

ASSETS		LIABILITIES	
Financial assets	[300, 320]	Short term liabilities	[550, 590]
Stocks	[800, 850]	Long term liabilities	[150, 200]
Net fixed assets	[500, 540]	Equity	[900, 920]
Total	[1600, 1710]	Total	[1600, 1710]

PROFIT AND LOSS ACCOUNT B

Estimated sales		[1500, 1600]
Cost of sales	[950, 1000]	
Depreciation	[80, 100]	[1030, 1100]
	Gross profit	[400, 570]
Financial expenses		[10, 20]
	Net profit	[380, 560]

Decision c): Waiting position

BALANCE SHEET C

ASSETS		LIABILITIES	
Financial assets	[430, 450]	Short term liabilities	[300, 330]
Stocks	[400, 430]	Long term liabilities	[130, 170]
Net fixed assets	[500, 540]	Equity	[900, 920]
Total	[1330, 1420]	Total	[1330, 1420]

PROFIT AND LOSS ACCOUNT C

Estimated sales		[1000, 1200]
Cost of sales	[650, 700]	
Depreciation	[80, 100]	[730, 800]
	Gross profit	[200, 470]
Financial expenses		[0, 0]
	Net profit	[200, 470]

These three decisions meant, then, different future situations, reflected in the above financial statements. Taking into account the specific objectives of the person responsible for the business, a solution had to be found in two different areas: profitability and solvency.

Profitability in uncertainty

With regard to the first of these, that of *profitability*, we felt it would be very instructive to resort, at least initially, to arriving at certain ratios which in this case would be *uncertain ratios*. A prior question arose in all clarity and that was that which ever the path selected the profits ran no danger. What was being considered was their higher or lower level relative to other significant values.

In order to dispel any doubts in this sense, three ratios were taken into account:

$$\text{Net margin: } M_n = \frac{\text{Net profit}}{\text{Sales}}$$

$$\text{Financial profitability: } R_f = \frac{\text{Net profit}}{\text{Equity}}$$

$$\text{Economic profitability: } R_e = \frac{\text{Net profit}}{\text{Total assets}}$$

This calculation presented no problem whatsoever, as these were values, all of them (numerators and denominators), belonging to the positive field of real numbers. It is very well known that in order to divide intervals of confidence in R^+ normally, it is sufficient to arrive at the quotients "by crossing over the extremes". The results we arrived at were as follows:

Decision a):

$$\underset{\sim}{M}_n^{(A)} = \frac{[435, 850]}{[2200, 2500]} = \left[\frac{435}{2500}, \frac{850}{2200}\right] = [0.174, 0.386]$$

$$\underset{\sim}{R}_f^{(A)} = \frac{[435, 850]}{[900, 920]} = \left[\frac{435}{920}, \frac{850}{900}\right] = [0.473, 0.944]$$

$$\underset{\sim}{R}_e^{(A)} = \frac{[470, 880]}{[2070, 2420]} = \left[\frac{470}{2420}, \frac{880}{2070}\right] = [0.194, 0.425]$$

Decision b):

$$\underset{\sim}{M}_n^{(B)} = \frac{[380, 560]}{[1500, 1600]} = \left[\frac{380}{1600}, \frac{560}{1500}\right] = [0.238, 0.373]$$

$$\underset{\sim}{R}_f^{(B)} = \frac{[380, 560]}{[900, 920]} = \left[\frac{380}{920}, \frac{560}{900}\right] = [0.413, 0.622]$$

$$\underset{\sim}{R}_e^{(B)} = \frac{[400, 570]}{[1600, 1710]} = \left[\frac{400}{1710}, \frac{570}{1600}\right] = [0.234, 0.356]$$

Decision c):

$$\underset{\sim}{M}_n^{(C)} = \frac{[200, 470]}{[1000, 1200]} = \left[\frac{200}{1200}, \frac{470}{1000}\right] = [0.167, 0.470]$$

$$\underset{\sim}{R}_f^{(C)} = \frac{[200, 470]}{[900, 920]} = \left[\frac{200}{920}, \frac{470}{900}\right] = [0.217, 0.522]$$

$$\underset{\sim}{R}_e^{(C)} = \frac{[200, 470]}{[1330, 1420]} = \left[\frac{200}{1420}, \frac{470}{1330}\right] = [0.141, 0.353]$$

Once these uncertain ratios were arrived at it became necessary to establish a comparison for each type, with the object of finding an order from higher to lower. But this could not be done immediately, since on comparing on certain occasions it happened that the lower extreme of one ratio was higher than the lower extreme of another, the opposite occurred with the upper extremes. Faced with this difficulty we adopted the following criterion: "a confidence interval is higher than another when its average point is higher "[4].

Based on this criterion we continued with our work:

"Net margin" ratio:

$$\underset{\sim}{M}_n^{(A)} = [0.174, 0.386] \; ; \; \overline{\underset{\sim}{M}_n^{(A)}} = \frac{0.174 + 0.386}{2} = 0.280$$

$$\underset{\sim}{M}_n^{(B)} = [0.238, 0.373] \; ; \; \overline{\underset{\sim}{M}_n^{(B)}} = \frac{0.238 + 0.373}{2} = 0.306$$

$$\underset{\sim}{M}_n^{(C)} = [0.167, 0.470] \; ; \; \overline{\underset{\sim}{M}_n^{(C)}} = \frac{0.167 + 0.470}{2} = 0.319$$

The order found meant giving priority to decision c), followed by b) and finally a). Provisionally we had concluded by opting for the *waiting position* if we had the intention of attaining *the highest net profit per monetary unit sold* (per Euro sold). Now then one of the advantages of working with confidence intervals is that, with them, one has available more information in the results. And here we had a significant example. In fact, although the ratio $\underset{\sim}{M}_n^{(c)}$ was the greatest, its lower limit, 0.167, was the lowest of all. This indicated that, in the worst of circumstances (most pessimistic situation), the decision considered overall as the best would no longer be so, we repeat if circumstances were unfavourable. This position was compensated for by the fact that if things functioned "perfectly", the ratio was much more favourable than with the other options. With regard to the uncertainty inherent in the results, this also appears to a greater extent in $\underset{\sim}{M}_n^{(c)}$, on the difference between the extremes of the interval 0.470 – 0.167 = 0.303 is far higher than the other two possibilities. This mean an inconvenience, as the ratio can be placed from the worst situation to the best of the three. The election, relative to the significance of this ratio, was subject to controversy.

Ratio of financial profitability:

$$\underset{\sim}{R}_{(f)}^{(A)} = [0.473, 0.944] \; ; \; \ulcorner \underset{\sim}{R}_f^{(A)} \urcorner \; \frac{0.473 + 0.944}{2} = 0.709$$

$$\underset{\sim}{R}_f^{(B)} = [0.413, 0.622] \; ; \; \ulcorner \underset{\sim}{R}_f^{(B)} \urcorner \; \frac{0.413 + 0.622}{2} = 0.518$$

$$\underset{\sim}{R}_f^{(C)} = [0.217, 0.522] \; ; \; \ulcorner \underset{\sim}{R}_f^{(C)} \urcorner \; \frac{0.217 + 0.522}{2} = 0.370$$

The order of priority appeared here quite clearly commencing with decision a), followed by b) and finally c). We can see that in any of the three alternatives, when the lower extreme was greater, then the higher extreme was also greater. Based on this ratio *renovation and acquisition of fixed assets* would be the selection that would allow for optimisation relative to net profit per "own" *monetary unit placed within the business*, This was a ratio that was highly estimated by the owners of the company and which would play a leading role in the final decision. Nevertheless, in this ratio, uncertainty played against this decision as it was the highest in the best alternative 0.944 – 0.473 = 0.471. But this circumstance was toned down and buffered since, even the lower extreme of the interval (pessimistic position) gave a higher number than the lower extremes of the other intervals.

Ratio of economic profitability:

$$\underset{\sim}{R}_e^{(A)} = [0.194, 0.425] \ ; \ \overline{\underset{\sim}{R}_e^{(A)}} = \frac{0.194 + 0.425}{2} = 0.310$$

$$\underset{\sim}{R}_e^{(B)} = [0.234, 0.356] \ ; \ \overline{\underset{\sim}{R}_e^{(B)}} = \frac{0.234 + 0.356}{2} = 0.295$$

$$\underset{\sim}{R}_e^{(C)} = [0.141, 0.353] \ ; \ \overline{\underset{\sim}{R}_e^{(C)}} = \frac{0.141 + 0.353}{2} = 0.247$$

The same order as that for the ratio of financial profitability appeared again, a), b), c), but with less strength as the ratio for alternative b) was very close to ratio a). Further more, the lower extreme (pessimistic position) of the former was far in excess of that of the latter. The representativeness of this ratio, net profit by monetary unit (own and outside) placed in the business, did not appear to enthuse the businessman too much, with regard to this problem, obviously. Another thing altogether at the time of determining profitability of the business. But this was not at this time, the hub of his immediate concerns.

Throughout this study, one question awoke our interest, from a theoretical point of view: it appeared clear at first sight, that *as indebtedness increased, financial profit increased*. Could this be so in any event? What conditions were required for such a statement? When is it convenient to resort to outside financing? Could we formulate a law in this sense? We consulted with Anna María, who after a few days appeared with an interesting explanation which we reproduce below[5].

"When economic profitability is higher than the cost of the debts:

1ª) The greater the indebtedness, the higher the financial profitability.

2ª) The greater the indebtedness, the greater the uncertainty"

In her desire for a better presentation, she felt it was appropriate to use "confidence triplets" since they represented much appreciated further information. What was taken into account then were the extremes and the maximum presumption. She proposed as the initial source for the work the following provisional balance sheet, in which it can be seen that the business had no debts[6]:

BALANCE SHEET

ASSETS		LIABILITIES	
Financial assets	(430, 440, 450)		
Stocks	(400, 410, 430)		
Net fixed assets	(500, 530, 540)	Equity	(1330, 1380, 1420)
	Total (1330,1380,1420)		Total (1330, 1380, 1420)

Profit was estimated at:

$$\underset{\sim}{B} = (200, 400, 470)$$

"Gross Profit" was equal to "Net Profit" as there were no debts. For this reason, "Equity" would be equal to "Total Liabilities" (identical to "Total Assets") and, in this case, and only under these circumstances, "Economic Profitability" coincides with "Financial Profitability":

$$\underset{\sim}{R}_e = \underset{\sim}{R}_f = \frac{(200, 400, 470)}{(1330, 1380, 1420)} = \left(\frac{200}{1420}, \frac{400}{1380}, \frac{470}{1330}\right) = (0.14, 0.29, 0.33)$$

Form this starting out point, exempt from any debts and therefore, from financial expenses, the author established three different hypotheses, each one of which meant, at the same time, greater indebtedness and less equity, in such a way that neither the Economic Structure of the Balance Sheet (Assets) nor the Financial Structure (liabilities) would undergo any *overall variation.*

First hypothesis.

This consisted in a future reduction of Equity in 200 monetary units, compensated for exactly in the same amount by certain debts, the expected interest rates for which were expressed by the following confidence triplet:

$$\underset{\sim}{I} = (0.05, 0.06, 0.08)$$

This would cause the following changes in the economic figures which are listed:

Equity: $\underset{\sim}{C}_1 = (1330, 1380, 1420)\ (-)\ 200 = (1130, 1180, 1220)$

Liabilities: $\underset{\sim}{E}_1 = (200, 200, 200)$

Net profit: $\underset{\sim}{B}_1 = (200, 400, 470)\ (-)\ 200\ (.)\ (0.05, 0.06, 0.08) =$

$$= (200, 400, 470)\ (-)\ (10, 12, 16) = (184, 388, 460)$$

The "Financial profitability" would then be:

$$\underset{\sim}{R}_f^{(1)} = \frac{(184, 388, 460)}{(1130, 1180, 1220)} = \left(\frac{184}{1220}, \frac{388}{1180}, \frac{460}{1130}\right) = (0.15, 0.33, 0.41)$$

"Economic profitability" remained invariable by construction while the "Financial profitability" had increased, on passing from (0.14, 0.29, 0.33) to reach (0.15, 0.33, 0.41).

Second hypothesis.

The reduction in Equity and increase in Liabilities was in this case 400, a number which is higher than the one we considered before. Under these circumstances the variation in the economic values would be as follows:

Equity: $\underset{\sim}{C}_2 = (1330, 1380, 1420)\ (-)\ 400 = (930, 980, 1020)$

Liabilities: $\underset{\sim}{E}_2 = (400, 400, 400)$

Net profit: $\underset{\sim}{B}_2 = (200, 400, 470)\ (-)\ 400\ (.)\ (0.05, 0.06, 0.08) =$

$$= (200, 400, 470)\ (-)\ (20, 24, 32) = (168, 376, 450)$$

We would then have as "Financial profitability"

$$\underset{\sim}{R}_f^{(2)} = \frac{(168, 376, 450)}{(930, 980, 1020)} = \left(\frac{168}{1020}, \frac{376}{980}, \frac{450}{930}\right) = (0.16, 0.38, 0.48)$$

Once again the "Financial profitability" would be higher, passing from (0.15, 0.33, 0.41) to (0.16, 0.38, 0.48).

Third hypothesis

The further reduction in Equity and consequent increase in Liabilities was even higher and had been set at 600 monetary units. The resulting economic values for this change would have been those listed below:

Equity: $\underset{\sim}{C}_3 = (1330, 1380, 1420)\ (-)\ 600 = (730, 780, 820)$

Liabilities: $\underset{\sim}{E}_3 = (600,600, 600)$

Net profit: $\underset{\sim}{B}_3 = (200, 400, 470)\ (-)\ 600\ (.)\ (0.05, 0.06, 0.08) =$

$$= (200, 400, 470)\ (-)\ (30, 36, 48) = (152, 364, 440)$$

"Financial Profitability" would now be:

$$\underset{\sim}{R}_f^{(3)} = \frac{(152, 364, 440)}{(730, 780, 820)} = \left(\frac{152}{820}, \frac{364}{780}, \frac{440}{730}\right) = (0.19, 0.47, 0.60)$$

"Financial profitability" increases once again, even more vigorously passing from (0.16, 0.38, 0.48) to (0.19, 0.47, 0.60).

Having done the corresponding calculations for each one of the hypotheses, we now sum up the results that we arrived at:

$$\underset{\sim}{I} = (0.05, 0.06, 0.08)$$

$$\underset{\sim}{R}_e = (0.16, 0.29, 0.33)$$

It remained quite clear that the basic presumption was complied with: "economic profitability was higher than the cost of the debts". The successive financial profitabilities would be:

$$\underset{\sim}{R}_f^{(1)} = (0.15, 0.33, 0.41)$$

$$\underset{\sim}{R}_f^{(2)} = (0.16, 0.38, 0.48)$$

$$\underset{\sim}{R}_f^{(3)} = (0.19, 0.47, 0.60)$$

It was not too difficult to see the successive increase of the financial profitability as the indebtedness increased. Also, there could be no doubt of the fact that, in conjunction with these increases there was a "widening" of the triplets, in the sense of there being a greater distance between their extremes. This was equivalent to stating that uncertainty was also becoming greater. In fact:

$$\text{En } \underset{\sim}{R}_f^{(1)} : 0.41 - 0.15 = 0.26$$

$$\text{En } \underset{\sim}{R}_f^{(2)} : 0.48 - 0.16 = 0.32$$

$$\text{En } \underset{\sim}{R}_f^{(3)} : 0.60 - 0.19 = 0.41$$

Anna María had done a good job, which brought to light a law in uncertainty which could be capable of serving as a guide to business and institution executives.

Solvency in uncertainty

It appeared that there could be no further reasoning with regard to profitability. In this sense those responsible for the business who were requesting the work now had available all the elements required by them in order to be able to adopt the decision they considered to be more in line with their profit requirements. What remained was the second of the spheres they had pointed out as basic for the selection: solvency. It was very difficult to talk about *solvency* without expressly mentioning a concept that both accountants and financiers consider as essential in order for them to tackle the very minimum study. We are referring to "Working Capital". To be understood as Working Capital is *"that part of the permanent Capital that finances Current Assets"*. Therefore, this element of measurement, in conjunction with certain ratios selected to this effect, would be the instruments which we were going to use in order to come up with a reply to this part of the problem. Once again Anna María would be our "light and guide". Obviously Mario, teacher of teachers, was fully aware of all the ins and outs that this notion entails, used both in theory and in reality. But it was a novelty for him to see their transformation for normal use in a context of uncertainty. And it was this sense which justified my resorting to the young and brilliant researcher. Mario sustained, with all reason, the existence of a second definition of "Working Capital", arising from the "tautological" equality between Assets and Liabilities: *difference between current Assets and short term Liabilities"*. He justified this as follows, taking as a support one of the provisional balance sheets, expressed in intervals, which he converted into precise intervals "by making the entropy fall", that is to say, substituting each interval by its average point. From Balance sheet A, he arrived at the following representation by *patrimonial masses:*

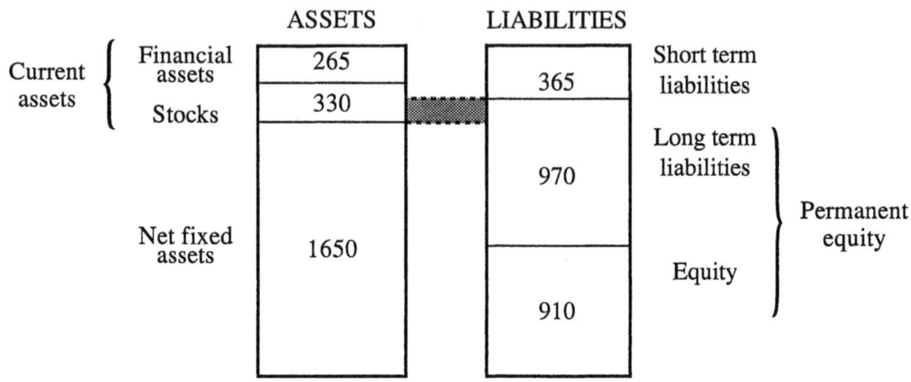

In fact the part of "Permanent equity" that financed "Current Assets" (first definition was:

$$FM_1 = (970 + 910) - 1650 = 230$$

And the difference between "Current Assets" and "Short term liabilities" gave us the following result:

$$FM_2 = (265 + 330) - 365 = 230$$

Indeed, Mario was right, if we adhered to the use of the arithmetic of certainty. But, what would happen in the transition to the arithmetic of uncertainty? We were soon going to find out by following the work of Anna María[7]. We continued with Balance sheet A, but now maintaining the patrimonial masses in the form of *confidence intervals,*

The first definition, "part of the permanent Equity that finances Current Assets" gave us the following result:

$$\underset{\sim}{FM}_1 = ([820, 1120] (+) [900, 920]) (-) [1500, 1800] =$$

$$= [1720, 2040] (-) [1500, 1800] = [-80, 540]$$

The second definition, "Current Assets less short term Liabilities", gave rise to the following:

$$\underset{\sim}{FM}_2 = ([250, 280] (+) [320, 340]) (-) [350, 380] =$$

$$= [570, 620] (-) [350, 380] = [190, 270]$$

The two definitions, *equivalent in the sphere of certainty were different when moving into uncertainty.* The first gave us [-80, 540] while the second was [190, 270]. Faced with these circumstances we were obliged to select one of the two as valid. The opinion of Anna María, ratified by all those present, was that the "original" could be none other than the first (part of the permanent Equity that finances current Assets) because the second was a consequence (effect) *of a tautological equality and* of an *arithmetic*. There was considerable surprise caused by this unusual fact which would require the rethinking of not a few classical Accounting concepts. But did this difference in results not invalidate the thesis sustained by us that the arithmetic of certainty is a special case of the arithmetic of uncertainty? Not only did this refute it but it once again verified it. In order to prove this statement it was sufficient to convert the uncertain Working Capitals, $\underset{\sim}{FM}_1$ and $\underset{\sim}{FM}_2$, into certain values, "making the entropy fall" (by finding the average point of the corresponding confidence intervals). This was easy and quick:

$$\overline{\underset{\sim}{FM}_1} = \frac{-80 + 540}{2} = 230$$

$$\overline{\underset{\sim}{FM}_2} = \frac{190 + 270}{2} = 230$$

As could be in no other way, one and the other result coincided and they were equal to the Working Capital Mario had found in the field of certainty.

Having made these theoretical considerations, which were not exempt from practical interest, we were now ready to redirect the work , having already agreed to use the first definition of Working Capital when finding ourselves immersed in the arithmetic of uncertainty. In this sense for *decision* a) we had as Working Capital:

$$\underset{\sim}{FM}^{(A)} = [-80, 540]$$

We now had to calculate the Working Capitals corresponding to the remaining alternatives.

For *decision b)* it would be:

$$\underset{\sim}{FM}^{(B)} = ([150, 200] (+) [900, 920]) (-) [500, 540] =$$

$$= [1050, 1120] (-) [500, 540] = [510, 620]$$

The result for *decision c*) it was:

$$\underset{\sim}{FM}^{(C)} = ([130, 170] (+) [900, 920]) (-) [500, 540] =$$

$$= [1030, 1090] (-) [500, 540] = [490, 590]$$

Also in this case the order appeared with total clarity:

$$\underset{\sim}{FM}^{(B)} > \underset{\sim}{FM}^{(C)} > \underset{\sim}{FM}^{(A)}$$

It is acknowledged that not always is the greater Working Capital the most advisable, since an excess of coverage could mean the existence of idle resources, which is not desirable for a business. In any event these results warranted being complemented by the use of some of the most significant ratios normally used as indices of solvency. For this work the following were chosen:

$$\text{Fixed asset financing} : F_i = \frac{\text{Permanent Equity}}{\text{Net Fixed Assets}}$$

$$\text{Total indebtedness: } E_t = \frac{\text{Total Liabilities}}{\text{Equity}}$$

$$\text{Long term indebtedness: } E_l = \frac{\text{Long term liabilities}}{\text{Equity}}$$

Having accepted these ratios as the most adaptable for the effects desired by the business, we prepared to do the calculation for each one of the alternatives.

Decision a):

$$\underset{\sim}{E}_i^{(A)} = \frac{[1720, 2040]}{[1500, 1800]} = \left[\frac{1720}{1800}, \frac{2040}{1500}\right] = [0.956, 1.360]$$

$$\underset{\sim}{E}_t^{(A)} = \frac{[1170, 1500]}{[900, 920]} = \left[\frac{1170}{920}, \frac{1500}{900}\right] = [1.272, 1.667]$$

$$\underset{\sim}{E}_l^{(A)} = \frac{[820, 1120]}{[900, 920]} = \left[\frac{820}{920}, \frac{1120}{900}\right] = [0.891, 1.244]$$

Decision b):

$$\underset{\sim}{E}_i^{(B)} = \frac{[1050, 1120]}{[500, 540]} = \left[\frac{1050}{540}, \frac{1120}{500}\right] = [1.944, 2.240]$$

$$\underset{\sim}{E}_t^{(B)} = \frac{[700, 790]}{[900, 920]} = \left[\frac{700}{920}, \frac{790}{900}\right] = [0.761, 0.878]$$

$$\underset{\sim}{E}_l^{(B)} = \frac{[150, 200]}{[900, 920]} = \left[\frac{150}{920}, \frac{200}{900}\right] = [0.163, 0.222]$$

Decision c):

$$\underset{\sim}{E}_i^{(C)} = \frac{[1030, 1090]}{[500, 540]} = \left[\frac{1030}{540}, \frac{1090}{500}\right] = [1.907, 2.180]$$

$$\underset{\sim}{E}_t^{(C)} = \frac{[430, 500]}{[900, 920]} = \left[\frac{430}{920}, \frac{500}{900}\right] = [0.467, 0.556]$$

$$\underset{\sim}{E}_l^{(C)} = \frac{[130, 170]}{[900, 920]} = \left[\frac{130}{920}, \frac{170}{900}\right] = [0.141, 0.189]$$

Lastly, we established the appropriate comparisons and consequent comments, for each type of ratio.

Ratio of "Fixed Asset Financing":

$$\underset{\sim}{F}_i^{(A)} = [0.956, 1.360]$$

$$\underset{\sim}{F}_i^{(B)} = [1.944, 2.240]$$

$$\underset{\sim}{F}_i^{(C)} = [1.907, 2.180]$$

This ratio is the one that best represents the *Working Capital* therefore it is recommended that it be *greater than the unit.* Now then, as we pointed out before, a value that is *excessively* distanced from one, would mean a risk of excessive financing of Current Assets by Permanent Equity, with the consequent decrease in the general profitability of the business. This comment led to the conclusion that *the most suitable ratio was not the highest*, but the one that was closest to the established ideal. Having made this comment it became evident that both decision b) and decision c) lacked a *certain excess of solvency* because even in the most pessimistic positions the ratio came close *to double* the value considered as minimum. Although it were to be by exclusion it was *decision* a) which was closest to the standards that were generally admitted for the activity exercised by the business under study. There was one inconvenience: in the worst of cases the index would not reach the unit (which would imply that the Working Capital, in its lower extreme would be negative). In spite of this it was felt that the risk of this likelihood was minimum.

Ratio of "Total Indebtedness":

$$\underset{\sim}{E}_t^{(A)} = [1.272, 1.667]$$

$$\underset{\sim}{E}_t^{(B)} = [0.761, 0.878]$$

$$\underset{\sim}{E}_t^{(C)} = [0.467, 0.556]$$

This ratio constitutes a comparison between total debts and equity. In general lines it is considered that solvency of a business or institution is not good when this ratio is much higher than one and a half. On the other hand values that are very separated bring to light a conservative policy. Leaving to one side the consequences of decision c) it would appear that: a) being relatively acceptable, was on the limit of assumable risk in its upper limit, being difficult to extend beyond that situation; with regard to b) in its upper limit its conservative position is hidden by the short term debts. This is very obvious in the following ratio.

Ratio of "Long Term Indebtedness".

$$\underset{\sim}{E}_1^{(A)} = [0.891, 1.244]$$

$$\underset{\sim}{E}_1^{(B)} = [0.163, 0.222]$$

$$\underset{\sim}{E}_1^{(C)} = [0.141, 0.189]$$

For this business, the unit appeared to be the ideal value for this ratio. From here, and even taking into account the suspicions raised by the pessimistic position of decision a) it appeared that outside all doubt that this index was the most correct index for the effects of a balanced solvency.

The joint consideration of the Working Capital and the three ratios that had been studied led us to the conviction that, from a solvency point of view, option a) "movements in fixed assets", in spite of the fact of having to resort to important outside financing did not infringe, in excess, against the solvency of the business. If to this was added the favourable level of the ratios for "Economic Profitability" and "Financial Profitability", *the recommendation should be aimed at renewal of existing fixed assets and acquisition of new equipment*, as foreseen in the project covered by decision a). And we have said "recommendation" since every executive should make his own decisions, in accordance with his manner of understanding the economic-financial life, with the freedom granted to him by the treatment of uncertainty by this method, but also with the responsibility which he is obliged to assume in his condition as the active subject of the financial management. Our objective was, and continues to be so today, to provide an effective aid, not only for getting to know the situation of businesses and institutions, but also to set down points of support for taking decisions in a context that is so involved and complex as the context in which businesses and institutions carry out their activities.

The use of fuzzy ratios

When Anna María showed us the relation there was between the increase in indebtedness and financial profitability she used confidence triplets situating between the extremes of the triplets a maximum presumption. As the arithmetic of triplets is simple, the increase in information relative to the interval constitutes an advantage not lacking in difficulties although non existent. But what happens if the patrimonial masses of the Balance Sheet and the values of the Profit and Loss Accounts are "described" by fuzzy numbers? Given the fact that the ratio is a quotient (non linear operation) fuzzy numbers could not be used in the "ternary" form but in the "form of α-cuts". For a suitable explanation of the circumstances we proposed setting out from Balance Sheet A and Profit and Loss Account A which, in short, showed the situation of the foreseen economic horizon according to the decision to be adopted. We made the transition of both financial statements expressed in confidence intervals, to the balance sheet and profit and loss account expressed in fuzzy numbers, situating between the two extremes a maximum presumption and assuming that the presumptions between this maximum and the extremes decreased proportionally (linearly), both to the right and to the left. In this way we had constructed "fuzzy triangular numbers", (F.T.N.).

In order to facilitate understanding, we reproduce the Balance Sheet and Profit and Loss Account in intervals and do the corresponding transformation:

BALANCE SHEET A

ASSETS		LIABILITIES	
Financial assets	[250, 280]	Short term liabilities	[350, 380]
Stocks	[320, 340]	Long term liabilities	[820, 1120]
Net Fixed Assets	[1500, 1800]	Equity	[900, 920]
Total	[2070, 2420]	Total	[2070, 2420]

PROFIT AND LOSS ACCOUNT A

Estimated sales		[2200, 2500]
Cost of sales	[1400, 1500]	
Depreciations	[220, 230]	[1620, 1730]
	Gross profit	[470, 880]
Finantial expenses		[30, 35]
	Net profit	[435, 850]

The Balance Sheet and Profit and Loss Account, now in F.T.N. were, after adding the information received on maximum presumptions:

BALANCE SHEET A

ASSETS		LIABILITIES	
Financial assets	(250, 260, 280)	Short term liabilities	(350, 370, 380)
Stocks	(320, 330, 340)	Long term Assets	(820, 900, 1120)
Net Fixed Assets	(1500, 1700,1800)	Equity (900, 910, 920)	
Total	(2070, 2290, 2420)	Total	(2070, 2180, 2420)

PROFIT AND LOSS ACCOUNT A

Estimated sales		(2200, 2300, 2500)
Cost of sales	(1400, 1450, 1500)	
Depreciations	(220, 225, 230)	(1620, 1675, 1730)
	Gross profit	(470, 625, 880)
Financial expenses		[30, 32, 35)
	Net profit	(435, 593, 850)

As is sufficiently well known, the difference between the confidence triplet and the fuzzy triangular number (F.T.N.) resides in the fact that while the triplet only posses information at three points (extremes and maximum presumption) the F.T.N. provides information uninterruptedly from the lower extreme to the upper extreme.

The name stems from the geometric figure of the triangle which is its graphical representation. Therefore the "drawn" version of the F.T.N. corresponding to estimated sales:

$$\underset{\sim}{V}_e = (2200, 2300, 2500)$$

would be as follows:

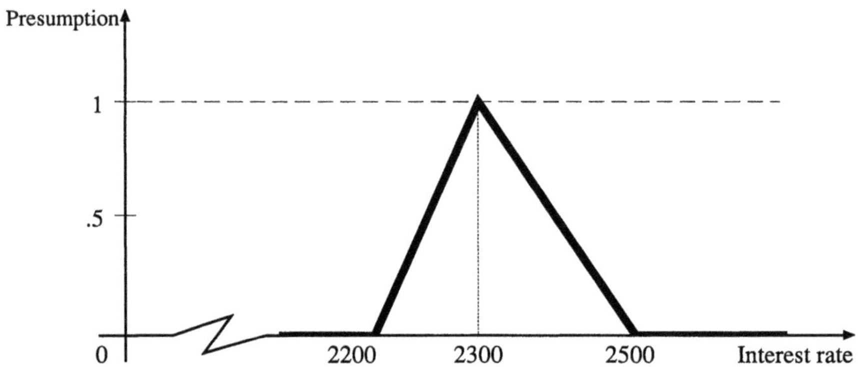

We could show each and every one of the patrimonial masses of the fuzzy Balance Sheet in the same way, as we could also the values that form the Profit and Loss Account.

Reiterating what we have said before, taking into account the fact that a ratio is a quotient, the ternary form is of no use, as shown up to now for doing this non linear operation. It is therefore necessary to transform the F.T.N. into α-cuts. The easiest way of doing this consists in forming an interval of functions, in such a way that we place:

As lower extreme: the *lower* extreme of the ternary form plus the subtraction of the maximum presumption and the lower extreme, multiplying this difference by α (parameter representing the level of presumption).

As upper extreme: the *upper* extreme of the ternary form less the subtraction between the upper extreme and maximum presumption, multiplying this difference by parameter α.

We could use as an example the F.T.N. representing "Estimated Sales" the ternary form of which was (2200,2300,2500) and its form in α-cuts:

$$\underset{\sim}{V}_e = [2200 + (2300 - 2200)\,\alpha,\ 2500 - (2500 - 2300)\,\alpha] =$$

$$= [2200 + 100\alpha,\ 2500 - 200\alpha]$$

The three ratios used as indicators of the profit were the ratio of "Net Margin", "Financial Profitability" and "Economic Profitability", which now moved on to become:

$$\underset{\sim}{M}_n = \frac{(435, 593, 850)}{(2200, 2300, 2500)} = \frac{[435 + (593 - 435)\alpha,\ 850 - (850 - 593)\alpha]}{[2200 + (2300 - 2200)\alpha,\ 2500 - (2500 - 2300)\alpha]}$$

$$= \frac{[435 + 158\alpha,\ 850 - 257\alpha]}{[2200 + 100\alpha,\ 2500 - 200\alpha]} = \left[\frac{435 + 158\alpha}{2500 - 200\alpha},\ \frac{850 - 257\alpha}{2200 + 100\alpha}\right]$$

For greater ease of reading we have shown this result in the hendecagonal form:

Net Margin

α	$\left[\frac{435 + 158\alpha}{2500 - 200\alpha}\right.$,	$\left.\frac{850 - 257\alpha}{2200 + 100\alpha}\right]$
1	0.258	
.9	0.249	0.270
.8	0.240	0.283
.7	0.231	0.295
.6	0.223	0.308
.5	0.214	0.321
.4	0.206	0.334
.3	0.198	0.347
.2	0.190	0.360
.1	0.182	0.373
0	0.174	0.386

$$\mathrm{R}_f=\frac{(435, 593, 850)}{(900, 910, 920)}=\frac{[465+158\alpha, 850-257\alpha]}{[900+10\alpha, 920-10\alpha]}=$$

$$=\left[\frac{435+158\alpha}{920-10\alpha}, \frac{850-257\alpha}{900+10\alpha}\right]$$

From this hendecagonal system we arrive at:

Financial Profitability

α	$\left[\frac{435+158\alpha}{920-10\alpha}\right.$,	$\left.\frac{850-257\alpha}{900+10\alpha}\right]$
1	0.652	
.9	0.634	0.681
.8	0.616	0.710
.7	0.598	0.739
.6	0.580	0.768
.5	0.562	0.797
.4	0.544	0.827
.3	0.526	0.856
.2	0.508	0.885
.1	0.491	0.915
0	0.473	0.944

$$\mathrm{R}_e=\frac{(470, 625, 880)}{(2070, 2290, 2420)}=\frac{[470+155\alpha, 880-255\alpha]}{[2070+220\alpha, 2420-130\alpha]}=$$

$$=\left[\frac{470+155\alpha}{2420-130\alpha}, \frac{880-255\alpha}{2070+220\alpha}\right]$$

The result in the hendecagonal system was:

Economic Profitability

α	$\left[\frac{470 + 155\alpha}{2420 - 130\alpha}\right.$,	$\left.\frac{880 - 255\alpha}{2070 + 220\alpha}\right]$
1	0.273	
.9	0.265	0.287
.8	0.256	0.301
.7	0.248	0.315
.6	0.240	0.330
.5	0.232	0.345
.4	0.225	0.361
.3	0.217	0.376
.2	0.209	0.392
.1	0.202	0.408
0	0.194	0.425

The informative advantage of these results could be seen at first glance, when comparing them to those arrived at when we used confidence intervals. Under those circumstances the only thing we knew was that the future reality could be found between two extremes. Now, further to this information we knew the maximum presumption and "all" the presumptions for the values located between the two extremes, in this case shown in a scale of eleven positions. Just a simple glance was enough for seeing that at each level of α, $\alpha = 1, 0.9, 0.8, \ldots, 0.2, 0.1, 0$, the F.T.N. becomes converted into a confidence interval and that, *when the higher the level, the lower the uncertainty* (the intervals were increasingly narrower as they went from zero to one). This meant that, although the person responsible for the decision could select any interval corresponding to one of the levels, he took a greater risk that *the reality would escape him* (it would move outside the limits) the higher the level selected. Only level zero (maximum uncertainty) "avoided", within the established hypotheses, falling into any error. This was the double sensation permanently felt by executives: "security in uncertainty" or "insecurity in certainty". In this sense it could be seen that the intervals of the fuzzy triangular number corresponding to level zero coincided, this could not be in any other way, with the results found with the example of confidence intervals.

The arithmetical operations destined to the solvency study followed the same path relative to ratios, but it was even easier to arrive at the Working Capital, since the only operations necessary were limited to the addition and subtraction, both linear in nature and therefore susceptible to being treated with F.T.N. expressed in the ternary form. We can see this below:

$$\underset{\sim}{FM} = ((820, 900, 1120) (+) (900, 910, 920)) (-) (1500, 1700, 1800) =$$
$$= (1720, 1810, 2040) (-) (1500, 1700, 1800) = (-80, 110, 240)$$

Here what we had was the sum and difference between the F.T.N, which always gives another F.T.N. In this way it could be said that the Working Capital was a fuzzy triangular number. In order to express the same in the hendecagonal system it was quite easy now, with the result obtained to pass over to the a α-cuts form. In accordance with previous mechanisms the transition to this form gave rise to:

$$\underset{\sim}{FM} = [-80 + 190\alpha, 240 - 130\alpha]$$

In this way we arrived at:

Working Capital.

α	[-80 + 190α, 240 - 130α]	
1	110	
.9	91	123
.8	72	136
.7	53	149
.6	34	162
.5	15	175
.4	- 4	188
.3	- 23	201
.2	- 42	214
.1	- 61	227
0	- 80	240

The linear form of the F.T.N. could be seen by observing how from *the maximum presumption* to the lower extreme the numbers decreased in *19 units* and to the *upper extreme* they increased in 13 units.

We now move on to find the three fuzzy solvency ratios, also used previously with the confidence intervals, that is the one for "Fixed Asset Financing", "Total Indebtedness" and "Long Term Indebtedness". These were found as follows:

$$F_i = \frac{(1720, 1810, 2040)}{(1500, 1700, 1800)} = \frac{[1720 + 90\alpha, 2040 - 230\alpha]}{[1500 + 200\alpha, 1800 - 100\alpha]} =$$

$$= \left[\frac{1720 + 90\alpha}{1800 - 100\alpha}, \frac{2040 - 230\alpha}{1500 + 200\alpha}\right]$$

With the hendecagonal form:

Fixed Asset Financing

α	$\left[\frac{1720 + 90\alpha}{1800 - 100\alpha}\right.$,	$\left.\frac{2040 - 230\alpha}{1500 + 200\alpha}\right]$
1	1.065	
.9	1.053	1.091
.8	1.042	1.118
.7	1.031	1.146
.6	1.020	1.174
.5	1.009	1.203
.4	0.998	1.233
.3	0.987	1.263
.2	0.976	1.295
.1	0.966	1.327
0	0.956	1.360

$$\underset{\sim}{E}_t = \frac{(1170, 1810, 1500)}{(900, 910, 920)} = \frac{[1170 + 100\alpha, 1500 - 230\alpha]}{[900 + 10\alpha, 920 - 10\alpha]} =$$

$$= \left[\frac{1170 + 100\alpha}{920 - 10\alpha}, \frac{1500 - 230\alpha}{900 + 10\alpha} \right]$$

The presentation by means of the hendecagonal system gave us:

Total Indebtedness

α	$\left[\frac{1170 + 100\alpha}{920 - 10\alpha}\right.$,	$\left.\frac{1500 - 230\alpha}{900 + 10\alpha} \right]$
1	1.396	
.9	1.383	1.422
.8	1.371	1.449
.7	1.358	1.476
.6	1.346	1.503
.5	1.333	1.530
.4	1.321	1.558
.3	1.309	1.585
.2	1.296	1.612
.1	1.284	1.639
0	1.272	1.667

$$\underset{\sim}{E}_l = \frac{(820, 900, 1120)}{(900, 910, 920)} = \frac{[820 + 80\alpha, 1120 - 220\alpha]}{[900 + 10\alpha, 920 - 10\alpha]} =$$

$$= \left[\frac{820 + 80\alpha}{920 - 10\alpha}, \frac{1120 - 220\alpha}{900 + 10\alpha} \right]$$

And we arrived at in the hendecagonal form:

Long Term Indebtedness

α	$\left[\frac{820 + 80\alpha}{920 - 10\alpha}\right.$,	$\left.\frac{1120 - 220\alpha}{900 + 10\alpha}\right]$
1	0.989	
.9	0.979	1.014
.8	0.969	1.040
.7	0.959	1.065
.6	0.950	1.091
.5	0.940	1.116
.4	0.930	1.142
.3	0.920	1.167
.2	0.911	1.193
.1	0.901	1.219
0	0.891	1.244

The general consideration made for the study of the profit were also technically valid in the case of solvency: more structured information; the possibility of the deciding person to be placed at a more or less high level of presumption, assuming a greater or lesser risk of error; and limitation of the uncertainty by the values of the extremes of the F.T.N.

The tasks had concluded and the results obtained were to the liking of the representatives of the requesting business. Time played in the favour of the adopted decision. The “modernisation of the industrial equipment, included in fixed assets”, brought greater competitiveness, which considerably increased invoicing to levels close to the upper extreme of the interval (and of the F.T.N.) they had estimated. Our satisfaction was enormous since we had been able to test a modest but innovative instrument in which at that time, few people believed.

General presentation of the instruments used

But three university professors, of nearly three different generations, could not conclude their task by arriving at a professional result. In some way what we had arrived at should be expressed in certain *general expressions* susceptible to being used over the widest possible range of future situations, extending perhaps, some element not used in the approach to this work in particular but maybe used for many others. Two were the instruments that had been used: the ratios and the Working Capital.

From a general point of view, the theoretical approach was simplified since all the values that intervened, were confidence intervals, triplets or fuzzy triangular numbers, members of the set R_0^+

With regard to the techniques used, the problems that arose required the solution to two problems:

1º Arriving at additions and subtractions of uncertain numbers for the Working Capital and quotients, also uncertain, for the ratios.

2º Comparison of uncertain values (confidence intervals, triplets and F.T.N.)

As already mentioned, for the normal approach, within the sphere of determinism, the amount of the *Working Capital*[8] could be arrived at by two different ways as a consequence of the equality between the Assets and Liabilities of the balance Sheet. In fact we call:

A_c = Current assets

A_f = Fixed assets (Net Fixed Assets)

P_c = Current liabilities (Short term liabilities)

C_p = Permanent equity

The following must always be complied with:

$$A_c + A_f = P_c + C_p$$

And, therefore:

$$C_p - A_f = A_c - P_c$$

The first member of this equality provides us with the definition of Working Capital as "the part of Permanent Equity that finances Current Assets". The second member, the other concept, brings to light "the difference between Current Assets less Current Liabilities".

When passing over to the sphere of uncertainty, if we adhere to the case of confidence intervals, we have:

$$\underset{\sim}{A}_c = [a_1, a_2] \qquad \underset{\sim}{P}_c = [c_1, c_2]$$

$$\underset{\sim}{A}_f = [b_1, b_2] \qquad \underset{\sim}{C}_p = [d_1, d_2]$$

In the first definition the following is written:

$$\underset{\sim}{FM}_1 = \underset{\sim}{C}_p (-) \underset{\sim}{A}_f = [d_1, d_2] (-) [b_1, b_2] = [d_1 - b_2, d_2 - b_1]$$

"the significance of which is as follows: if in the *pessimistic* position there is only available d_1 monetary units and it is necessary to invest in fixed assets the highest possible amount b_2, , then remaining as Working Capital will be $d_1 - b_2$ monetary units. In the *optimistic* position, if d_2 monetary units were to be available and it was sufficient to place in fixed assets b_1, then it would be possible to dedicate to Current Assets $d_2 - b_1$ monetary units"[9].

Let us move on to the second definition:

$$\underset{\sim}{FM}_2 = \underset{\sim}{A}_c (-) \underset{\sim}{P}_c = [a_1, a_2] (-) [c_1, c_2] = [a_1 - c_2, a_2 - c_1]$$

which brings to light the fact that in order to be able to make certain minimum payments a_1 there should be available at least a maximum of c_2 in the pessimistic position in order to be solvent, whilst in order to be able to make a maximum payment of a_2 it would only be necessary to posses the least possible monetary units c_1 in the optimistic position considering a situation of solvency.

On comparing both "Working Capitals" $\underset{\sim}{FM}_1$ and $\underset{\sim}{FM}_2$, it is easily seen that, *except for exceptional cases*, they provide different intervals:

$$\underset{\sim}{FM}_1 \neq \underset{\sim}{FM}_2$$

that is to say:

$$[d_1 - b_2, d_2 - b_1] \neq [a_1 - c_2, a_2 - c_1]$$

On the other hand, if the uncertainty is reduced to certainty finding, both for interval $\underset{\sim}{FM}_1$ as well as for interval $\underset{\sim}{FM}_2$, their average point, we arrive at:

$$\overline{\underset{\sim}{FM}}_1 = \frac{(d_1 - b_2) + (d_2 - b_1)}{2} = \frac{(d_1 + d_2) - (b_1 + b_2)}{2}$$

$$\overline{\underset{\sim}{FM}}_2 = \frac{(a_1 - c_2) + (a_2 - c_1)}{2} = \frac{(a_1 + a_2) - (c_1 + c_2)}{2}$$

In order for:

$$\overline{\underset{\sim}{FM}}_1 = \overline{\underset{\sim}{FM}}_2$$

it is necessary that:

$$\frac{(d_1 + d_2) - (b_1 + b_2)}{2} = \frac{(a_1 + a_2) - (c_1 + c_2)}{2}$$

and also:

$$(d_1 + d_2) - (b_1 + b_2) = (a_1 + a_2) - (c_1 + c_2)$$

$$(a_1 + a_2) + (b_1 + b_2) = (c_1 + c_2) + (d_1 + d_2)$$

but as:

$$a_1 + b_1 = c_1 + d_1$$

$$a_2 + b_2 = c_2 + d_2$$

their addition also complies with equality and, with it that of the two average points of the Working Capital $\underset{\sim}{\overline{FM}}_1$ and $\underset{\sim}{\overline{FM}}_2$. This indicates that, in spite of the uncertainty passing over to certainty, (the latter being a special case of the former) the values are merged into one.

All we have just formulated for the confidence intervals is extensible to the fuzzy numbers, expressed in the form of α-cuts. For this all that has to be done is to substitute the *values of the extremes* by the corresponding functions of α. The operation has no complications worthy of mention.

In order in a general way to present a ratio in the form of a confidence interval, the following terminology is normally used[10]:

Numerator: $\underset{\sim}{N} = [n_1, n_2]$

Denominator: $\underset{\sim}{D} = [d_1, d_2]$

Quotient: $\underset{\sim}{R} = [r_1, r_2]$

As a consequence of the well known arithmetic of confidence intervals and in the normal event that $n_1, n_2, d_1, d_2 > 0$, that is positive, the following is written:

$$\underset{\sim}{R} = \underset{\sim}{N}\ (:)\ \underset{\sim}{D} = [n_1, n_2]\ (:)\ [d_1, d_2] = \left[\frac{n_1}{d_2}, \frac{n_2}{d_1}\right]$$

Therefore:

$$[r_1, r_2] = \left[\frac{n_1}{d_2}, \frac{n_2}{d_1}\right]$$

The comparison between intervals can be done by following two paths. The one we used in the previous development consisting in the order from the *average point* of each interval. Let us assume three ratios in the form of intervals:

$$[r_1^{(1)}, r_2^{(1)}] = \left[\frac{n_1^{(1)}}{d_2^{(1)}}, \frac{n_2^{(1)}}{d_1^{(1)}}\right];\ [r_1^{(2)}, r_2^{(2)}] = \left[\frac{n_1^{(2)}}{d_2^{(2)}}, \frac{n_2^{(2)}}{d_1^{(2)}}\right];\ [r_1^{(3)}, r_2^{(3)}] = \left[\frac{n_1^{(3)}}{d_2^{(3)}}, \frac{n_2^{(3)}}{d_1^{(3)}}\right]$$

The average points are:

$$\overline{\underset{\sim}{R}}_1 = \frac{r_1^{(1)} + r_2^{(1)}}{2}; \quad \overline{\underset{\sim}{R}}_2 = \frac{r_1^{(2)} + r_2^{(2)}}{2}; \quad \overline{\underset{\sim}{R}}_3 = \frac{r_1^{(3)} + r_2^{(3)}}{2}$$

As $\overline{\underset{\sim}{R}}_1$, $\overline{\underset{\sim}{R}}_2$, $\overline{\underset{\sim}{R}}_3$ are exact numbers, their magnitude will automatically determine their order from greater to lesser, and from lesser to greater, as corresponds to the nature of the ratios under study.

There could appear a very special circumstance: that the average point of two or more of the ratios coincides, the extremes of one and the other being different., for example ratios [2, 6] and [3, 5], in which the average point is, for both, 4. When this occurs, and with the object of separating the two ratios, it is normal to adopt a complementary criterion which, *according to the nature of the ratio* can be, alternatively:

- The lower extreme determines precedence.
- The upper extreme determines precedence.
- The uncertainty (difference between the extremes) determines precedence.

Another way that is often used for comparing confidence intervals consists in finding the "maximum" or "supremum" , in some cases, or the "minimum" or "inferum" in others. Then the *Hamming Distance* is found between each ratio and the "supremum" or "inferum" as corresponds and the least difference determines precedence.

As is well known, the "maximum" of several confidence intervals in R_0^+ is arrived at by placing, in the result, as the lower extreme the greater of the lower extremes and as the upper extreme also the *greater* of the upper extremes. This is done as follows:

$$\underset{\sim}{R}_1 \,(\vee)\, \underset{\sim}{R}_2 \,(\vee)\, \underset{\sim}{R}_3 = [r_1^{(1)}, r_2^{(1)}] \,(\vee)\, [r_1^{(2)}, r_2^{(2)}] \,(\vee)\, [r_1^{(3)}, r_2^{(3)}] =$$

$$= [r_1^{(1)} \vee r_1^{(2)} \vee r_1^{(3)}, r_2^{(1)} \vee r_2^{(2)} \vee r_2^{(3)}] =$$

$$= \left[\frac{n_1^{(1)}}{d_2^{(1)}} \vee \frac{n_1^{(2)}}{d_2^{(2)}} \vee \frac{n_1^{(3)}}{d_2^{(3)}}, \frac{n_2^{(1)}}{d_1^{(1)}} \vee \frac{n_2^{(2)}}{d_1^{(2)}} \vee \frac{n_2^{(3)}}{d_1^{(3)}}\right]$$

The Hamming distances for each ratio with the "maximum" are:

$$d\,(\underset{\sim}{R}_1, \underset{\sim}{R}_1\,(\vee)\,\underset{\sim}{R}_2\,(\vee)\,\underset{\sim}{R}_3) = \left| \frac{n_1^{(1)}}{d_2^{(1)}} - \frac{n_1^{(1)}}{d_2^{(1)}} \vee \frac{n_1^{(2)}}{d_2^{(2)}} \vee \frac{n_1^{(3)}}{d_2^{(3)}} \right| + \left| \frac{n_2^{(1)}}{d_1^{(1)}} - \frac{n_2^{(1)}}{d_1^{(1)}} \vee \frac{n_2^{(2)}}{d_1^{(2)}} \vee \frac{n_2^{(3)}}{d_1^{(3)}} \right|$$

$$d\,(\underset{\sim}{R}_2, \underset{\sim}{R}_1\,(\vee)\,\underset{\sim}{R}_2\,(\vee)\,\underset{\sim}{R}_3) = \left| \frac{n_1^{(2)}}{d_2^{(2)}} - \frac{n_1^{(1)}}{d_2^{(1)}} \vee \frac{n_1^{(2)}}{d_2^{(2)}} \vee \frac{n_1^{(3)}}{d_2^{(3)}} \right| + \left| \frac{n_2^{(2)}}{d_1^{(2)}} - \frac{n_2^{(1)}}{d_1^{(1)}} \vee \frac{n_2^{(2)}}{d_2^{(2)}} \vee \frac{n_2^{(3)}}{d_2^{(3)}} \right|$$

$$d\,(\underset{\sim}{R}_3, \underset{\sim}{R}_1\,(\vee)\,\underset{\sim}{R}_2\,(\vee)\,\underset{\sim}{R}_3) = \left| \frac{n_1^{(3)}}{d_2^{(3)}} - \frac{n_1^{(1)}}{d_2^{(1)}} \vee \frac{n_1^{(2)}}{d_2^{(2)}} \vee \frac{n_1^{(3)}}{d_2^{(3)}} \right| + \left| \frac{n_2^{(3)}}{d_1^{(3)}} - \frac{n_2^{(1)}}{d_1^{(1)}} \vee \frac{n_2^{(2)}}{d_2^{(2)}} \vee \frac{n_2^{(3)}}{d_2^{(3)}} \right|$$

The shorter the distance, the greater the ratio. The order is established in this way from greater to lower or, if wanted, also from lower to greater.

It is also well known that the "minimum" of several intervals can be found by obtaining as the lower extreme the *smallest* of the lower extremes, and as the higher *the lowest* of the upper extremes. This can be written as follows:

$$\begin{aligned} \underset{\sim}{R}_1\,(\wedge)\,\underset{\sim}{R}_2\,(\wedge)\,\underset{\sim}{R}_3 &= [r_1^{(1)}, r_2^{(1)}]\,(\wedge)\,[r_1^{(2)}, r_2^{(2)}]\,(\wedge)\,[r_1^{(3)}, r_2^{(3)}] = \\ &= [r_1^{(1)} \wedge r_1^{(2)} \wedge r_1^{(3)}, r_2^{(1)} \wedge r_2^{(2)} \wedge r_2^{(3)}] = \\ &= \left[\frac{n_1^{(1)}}{d_2^{(1)}} \wedge \frac{n_1^{(2)}}{d_2^{(2)}} \wedge \frac{n_1^{(3)}}{d_2^{(3)}}, \frac{n_2^{(1)}}{d_1^{(1)}} \wedge \frac{n_2^{(2)}}{d_1^{(2)}} \wedge \frac{n_2^{(3)}}{d_1^{(3)}} \right] \end{aligned}$$

The Hamming distances of each ratio with the "minimum" are:

$$d\,(\underset{\sim}{R}_1, \underset{\sim}{R}_1\,(\wedge)\,\underset{\sim}{R}_2\,(\wedge)\,\underset{\sim}{R}_3) = \left| \frac{n_1^{(1)}}{d_2^{(1)}} - \frac{n_1^{(1)}}{d_2^{(1)}} \wedge \frac{n_1^{(2)}}{d_2^{(2)}} \wedge \frac{n_1^{(3)}}{d_2^{(3)}} \right| + \left| \frac{n_2^{(1)}}{d_1^{(1)}} - \frac{n_2^{(1)}}{d_1^{(1)}} \wedge \frac{n_2^{(2)}}{d_1^{(2)}} \wedge \frac{n_2^{(3)}}{d_1^{(3)}} \right|$$

$$d\,(\underset{\sim}{R}_2, \underset{\sim}{R}_1\,(\wedge)\,\underset{\sim}{R}_2\,(\wedge)\,\underset{\sim}{R}_3) = \left| \frac{n_1^{(2)}}{d_2^{(2)}} - \frac{n_1^{(1)}}{d_2^{(1)}} \wedge \frac{n_1^{(2)}}{d_2^{(2)}} \wedge \frac{n_1^{(3)}}{d_2^{(3)}} \right| + \left| \frac{n_2^{(2)}}{d_1^{(2)}} - \frac{n_2^{(1)}}{d_1^{(1)}} \wedge \frac{n_2^{(2)}}{d_2^{(2)}} \wedge \frac{n_2^{(3)}}{d_2^{(3)}} \right|$$

$$d\,(\underset{\sim}{R}_3, \underset{\sim}{R}_1\,(\wedge)\,\underset{\sim}{R}_2\,(\wedge)\,\underset{\sim}{R}_3) = \left| \frac{n_1^{(3)}}{d_2^{(3)}} - \frac{n_1^{(1)}}{d_2^{(1)}} \wedge \frac{n_1^{(2)}}{d_2^{(2)}} \wedge \frac{n_1^{(3)}}{d_2^{(3)}} \right| + \left| \frac{n_2^{(3)}}{d_1^{(3)}} - \frac{n_2^{(1)}}{d_1^{(1)}} \wedge \frac{n_2^{(2)}}{d_2^{(2)}} \wedge \frac{n_2^{(3)}}{d_2^{(3)}} \right|$$

Likewise, the least distance determines the lowest ratio and from this the order from least to greater, or, why not, from greater to least can be found.

When the information of the Balance Sheet and the Profit and Loss Account art expressed in *F.T.N.*, the ratios, as a quotient to the F.T.N. do not remain as fuzzy triangular numbers, but normally undergo "deformations" of their sides. Even so, there is no problem whatsoever in establishing an *order of precedence* by following the same path as we did for the confidence intervals, that is, using the *Hamming Distance* between the ratio and the "supremum" or "inferum", as is considered suitable, of those that are being compared. In order to make it easier to follow this process we propose making reference to the estimates we made of the "Ratio for economic profitability" for the three possible options a), b) and c), by converting the intervals into F.T.N.

With regard to decision a) we already had the ration represented by the following fuzzy number:

Economic Profitability A

α	$\left[\frac{470 + 155\alpha}{2420 - 130\alpha}\right.$,	$\left.\frac{880 - 255\alpha}{2070 + 220\alpha}\right]$
1	0.273	
.9	0.265	0.287
.8	0.256	0.301
.7	0.248	0.315
.6	0.240	0.330
.5	0.232	0.345
.4	0.225	0.361
.3	0.217	0.376
.2	0.209	0.392
.1	0.202	0.408
0	0.194	0.425

For obtaining the fuzzy ratio consequent upon decision b) we did as follows:

The following are estimated as F.T.N.:

Gross profit: (400, 500, 570)

Total assets: (1600,1650, 1710))

Therefore the sought after ratio was:

$$\underset{\sim}{R}_e^{(B)} = \frac{(400, 500, 570)}{(1600, 1650, 1710)} = \frac{[400 + 100\alpha, 570 - 70\alpha]}{[1600 + 50\alpha, 1710 - 60\alpha]} = \left[\frac{400 + 100\alpha}{1710 - 60\alpha}, \frac{570 - 70\alpha}{1600 + 50\alpha}\right]$$

and their hendecagonal representation:

Economic Profitability B

α	$\left[\frac{400 + 100\alpha}{1710 - 60\alpha}\right.$,	$\left.\frac{570 - 70\alpha}{1600 + 50\alpha}\right]$
1	0.303	
.9	0.296	0.308
.8	0.289	0.313
.7	0.282	0.319
.6	0.275	0.324
.5	0.268	0.329
.4	0.261	0.335
.3	0.254	0.340
.2	0.247	0.345
.1	0.241	0.351
0	0.234	0.356

Finally we moved on and found the ratio for "Economic Profitability" corresponding to decision c):

We estimated the following F.T.N.

Gross Profit: (200, 300, 470)

Total Assets: (1330, 1400, 1420)

in such a way that the fuzzy ratio was:

$$\underset{\sim}{R}_e^{(C)} = \frac{(200, 300, 470)}{(1330, 1400, 1420)} = \frac{[200 + 100\alpha, 470 - 170\alpha]}{[1330 + 70\alpha, 1420 - 20\alpha]} = \left[\frac{200 + 100\alpha}{1420 - 20\alpha}, \frac{470 - 170\alpha}{1330 + 70\alpha}\right]$$

with the following hendecagonal representation:

Economic Profitability C

α	$\left[\frac{200 + 100\alpha}{1420 - 20\alpha}\right.$,	$\left.\frac{470 - 170\alpha}{1330 + 70\alpha}\right]$
1	0.214	
.9	0.207	0.228
.8	0.199	0.241
.7	0.192	0.255
.6	0.185	0.268
.5	0.177	0.282
.4	0.170	0.296
.3	0.163	0.310
.2	0.155	0.324
.1	0.148	0.339
0	0.141	0.353

We then moved on to find the "maximum" or "supremum" of these three fuzzy ratios, choosing level by level, the *highest* values for the lower extremes and for the upper extremes

α	$R_e^{(A)}$		$R_e^{(B)}$		$R_e^{(C)}$		$R_e^{(A)} (\vee) R_e^{(B)} (\vee) R_e^{(C)}$	
1	0.273		0.303		0.214		0.303	
.9	0.265	0.287	0.296	0.308	0.207	0.228	0.296	0.308
.8	0.256	0.301	0.289	0.313	0.199	0.241	0.289	0.313
.7	0.248	0.315	0.282	0.319	0.192	0.255	0.282	0.319
.6	0.240	0.330	0.275	0.324	0.185	0.268	0.275	0.330
.5	0.232	0.345	0.268	0.329	0.177	0.282	0.268	0.345
.4	0.225	0.361	0.261	0.335	0.170	0.296	0.261	0.361
.3	0.217	0.376	0.254	0.340	0.163	0.310	0.254	0.376
.2	0.209	0.392	0.247	0.345	0.155	0.324	0.247	0.392
.1	0.202	0.408	0.241	0.351	0.148	0.339	0.241	0.408
0	0.194	0.425	0.234	0.356	0.141	0.353	0.234	0.425

We felt at that time that there might be an interest in a graphical representation of each one of the fuzzy numbers representing the "Financial Profitability" (the triangles are an approximation) in a thin line and the "maximum" or "supremum" of all in a thick line. The following was the result:

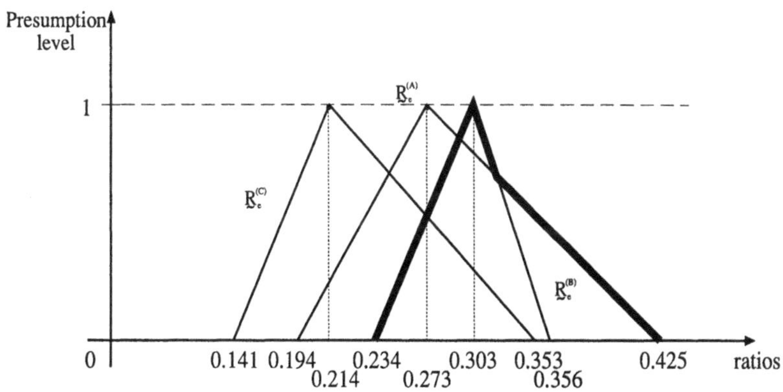

Once having arrived at the "supremum" for the three ratios of Economic Profitability we were ready to calculate the Hamming *distance* for each one of them from this "supremum". The results have been detailed below:

α	$d(\underset{\sim}{R}_e^{(A)}, \underset{\sim}{R}_e^{(A)} (\vee) \underset{\sim}{R}_e^{(B)} (\vee) \underset{\sim}{R}_e^{(C)})$	
1	|0.273 - 0.303| + |0.273 - 0.303|	0.060
.9	|0.265 - 0.296| + |0.287 - 0.308|	0.052
.8	|0.256 - 0.289| + |0.301 - 0.313|	0.045
.7	|0.248 - 0.282| + |0.315 - 0.319|	0.038
.6	|0.240 - 0.275| + |0.330 - 0.330|	0.035
.5	|0.232 - 0.268| + |0.345 - 0.345|	0.036
.4	|0.225 - 0.261| + |0.361 - 0.361|	0.036
.3	|0.217 - 0.254| + |0.376 - 0.376|	0.037
.2	|0.209 - 0.247| + |0.392 - 0.392|	0.038
.1	|0.202 - 0.241| + |0.408 - 0.408|	0.039
0	|0.194 - 0.234| + |0.425 - 0.425|	0.040
	Total........	0.456

α	$d(\underset{\sim}{R}_e^{(B)}, \underset{\sim}{R}_e^{(A)} (\vee) \underset{\sim}{R}_e^{(B)} (\vee) \underset{\sim}{R}_e^{(C)})$	
1	|0.303 - 0.303| + |0.303 - 0.303|	0.000
.9	|0.296 - 0.296| + |0.308 - 0.308|	0.000
.8	|0.289 - 0.289| + |0.313 - 0.313|	0.000
.7	|0.282 - 0.282| + |0.319 - 0.319|	0.000
.6	|0.275 - 0.275| + |0.324 - 0.330|	0.006
.5	|0.268 - 0.268| + |0.329 - 0.345|	0.016
.4	|0.261 - 0.261| + |0.335 - 0.361|	0.026
.3	|0.254 - 0.254| + |0.340 - 0.376|	0.036
.2	|0.247 - 0.247| + |0.345 - 0.392|	0.047
.1	|0.241 - 0.241| + |0.351 - 0.408|	0.057
0	|0.234 - 0.234| + |0.356 - 0.425|	0.069
	Total........	0.257

α	$d(\underset{\sim}{R}_e^{(C)}, \underset{\sim}{R}_e^{(A)} (\vee) \underset{\sim}{R}_e^{(B)} (\vee) \underset{\sim}{R}_e^{(C)})$	
1	$\lvert 0.214 - 0.303\rvert + \lvert 0.214 - 0.303\rvert$	0.178
.9	$\lvert 0.207 - 0.296\rvert + \lvert 0.228 - 0.308\rvert$	0.169
.8	$\lvert 0.199 - 0.289\rvert + \lvert 0.241 - 0.313\rvert$	0.162
.7	$\lvert 0.192 - 0.282\rvert + \lvert 0.255 - 0.319\rvert$	0.154
.6	$\lvert 0.185 - 0.275\rvert + \lvert 0.268 - 0.330\rvert$	0.152
.5	$\lvert 0.177 - 0.268\rvert + \lvert 0.282 - 0.345\rvert$	0.154
.4	$\lvert 0.170 - 0.261\rvert + \lvert 0.296 - 0.361\rvert$	0.156
.3	$\lvert 0.163 - 0.254\rvert + \lvert 0.310 - 0.376\rvert$	0.157
.2	$\lvert 0.155 - 0.247\rvert + \lvert 0.324 - 0.392\rvert$	0.160
.1	$\lvert 0.148 - 0.241\rvert + \lvert 0.339 - 0.408\rvert$	0.162
0	$\lvert 0.141 - 0.234\rvert + \lvert 0.353 - 0.425\rvert$	0.165
		Total........ 1.769

These distances from the "maximum" or "supremum" gave us a good order among the fuzzy ratios for "Economic Profitability" estimated as a consequence of each one of the decisions to be adopted. As in the case of the *confidence intervals* ratios $\underset{\sim}{R}_e^{(A)}$ and $\underset{\sim}{R}_e^{(B)}$ stand out over ratio $\underset{\sim}{R}_e^{(C)}$, although with a notable difference, since the order taking into account the intervals was:

$$\underset{\sim}{R}_e^{(A)} > \underset{\sim}{R}_e^{(B)} > \underset{\sim}{R}_e^{(C)}$$

with the greater information of the *fuzzy numbers* it was:

$$\underset{\sim}{R}_e^{(B)} > \underset{\sim}{R}_e^{(A)} > \underset{\sim}{R}_e^{(C)}$$

as:

$$d(\underset{\sim}{R}_e^{(B)}, \underset{\sim}{R}_e^{(A)} (\vee) \underset{\sim}{R}_e^{(B)} (\vee) \underset{\sim}{R}_e^{(C)}) > d(\underset{\sim}{R}_e^{(A)}, \underset{\sim}{R}_e^{(A)} (\vee) \underset{\sim}{R}_e^{(B)} (\vee) \underset{\sim}{R}_e^{(C)}) >$$

$$d(\underset{\sim}{R}_e^{(C)}, \underset{\sim}{R}_e^{(A)} (\vee) \underset{\sim}{R}_e^{(B)} (\vee) \underset{\sim}{R}_e^{(C)})$$

The existence of a maximum presumption in the fuzzy numbers makes the difference and this acts as a "strengthener" of ratio $\underset{\sim}{R}_e^{(B)}$ tive to ratio $\underset{\sim}{R}_e^{(A)}$. The same would occur if we were dealing with confidence triplets, since on making the "entropy fall" , starting out from:

$$\underset{\sim}{R}_e^{(A)} = (0.194, 0.273, 0.425), \qquad \underset{\sim}{R}_e^{(B)} = (0.234, 0.303, 0.356)$$

we would arrive at:

$$\overline{\underset{\sim}{R}}_e^{(A)} = 1/4(0.194 + (2) . (0.273) + 0.425) = 0.291$$

$$\overline{\underset{\sim}{R}}_e^{(B)} = 1/4(0.234 + (2) . (0.303) + 0.356) = 0.299$$

This consideration on the ratio for "Economic Profitability", which we included under this section as an example, did not change the adopted decision based on *confidence intervals* taking into account the results and reasoning there had been with the other analysis instruments, above all considering the ratios for "Financial Profitability" and those for "Solvency". But it could have been very different, if other elements had concurred with different mechanisms. From here stems the importance acquired, normally, by the possibility of being able to include, in the studies, all the available information, even when this to a certain degree is tainted with a degree of subjectivity.

These applications of the Theory of fuzzy sub-sets in Accounting and Financial Analysis were the commencement of a new line of research, which, very skilfully and with magnificent achievements, was followed by Anna María. Her last work[11] testifies to this. Other works from a select group of professors have been added to her investigations in order today to be able to offer a compact group of new knowledge of great value in order to be able to attain the objective of re-conducting and limiting financial uncertainty, that is consubstantial with our day and age. But is was Mario, with his serene wisdom, who brought up the first concern, from which stemmed a whole range of theoretical and technical elements, forming a homogeneous body which, increasingly grows with the efforts of the components of those groups of professors trained on the five continents, with the idea a greater service to society by the universities. This service vocation was a constant during the life of Mario, who instilled it into numerous disciples. As always happens, by the very nature of things, some assumed it and other abused it. From his privileged social and academic position he offered a hand to all those who came to him for help or advice. Today in his situation of well earned rest, he contemplates from the distance the results of his efforts. His very high concept of friendship has not suffered at all, in spite of treachery and disappointments. After so much ungratefulness I feel it is justifiable to ask myself; what remains after glory?... fortunate is he who can answer: *a friend.*

REFERENCES

[1] The most interesting contributions have been made by GIL LAFUENTE, A.M. in her works.; El análisis financiero en la incertidumbre, Publ. Ariel Barcelona, 1990; Fundamentos de análisis financiero. Publ. Ariel Barcelona 1993; Nuevas estrategias para la gestión de las empresas, Publ. Ariel Barcelona 2001 which have served us as the basis and orientation for the development of this chapter.

[2] KAUFMANN, A and GIL ALUJA, J.: *Introducción de la teoría de los subconjuntos borrosos a la gestión de las empresas*. Publ. Malladoiro Santiago de Compostela 1986.

[3] GIL LAFUENTE, A.M.: *El análisis financiero en la incertidumbre*. Publ.. Ariel. Barcelona 1990, pages. 81-87.

GIL LAFUENTE, A.M.: *Fundamentos de análisis financiero*. Publ.. Ariel. Barcelona 1993, pages 119-126.

[4] KAUFMANN, A. and GIL ALUJA, J.: *Técnicas operativas de gestión para el tratamiento de la incertidumbre*. Publ.. Hispano Europea. Barcelona 1987, pages 39.

[5] GIL LAFUENTE, A.M.: *El análisis financiero en la incertidumbre*. Publ.. Ariel. Barcelona 1990, pages 91-99.

[6] See that this refers to the same balance sheet as for the option "waiting position" in which the values of the patrimony have each been completed with a maximum presumption. In Liabilities only Equity appears.

[7] GIL LAFUENTE, A.M.: *El análisis financiero en la incertidumbre*. Publ.. Ariel, Barcelona 1990, pages 41-45.

[8] GIL LAFUENTE, A.M.: *El análisis financiero en la incertidumbre*. Publ.. Ariel Barcelona 1990, pages. 40-45.

[9] GIL LAFUENTE, A.M.: *El análisis financiero en la incertidumbre*. Publ.. Ariel Barcelona 1990, pages 43.

[10] KAUFMANN, A. and GIL ALUJA, J.: *Introducción de la teoría de los subconjuntos borrosos a la gestión de las empresas*. Publ..Milladoiro. Santiago de Compostela 1986, pages 168-172.

[11] GIL LAFUENTE, A.M.: *Nuevas estrategias para la gestión de las empresas*. Publ.. Ariel. Barcelona 2001.

VIII DISTRIBUTION IN COMMERCIAL ACTIVITY

The search for distribution channels

The relationship existing between a business that manufactures certain products and its eventual clients includes a whole build-up of problems that commercial studies approach from different angles. From one point of view, an attempt has been made to give a single sense to the elements created for their adequate treatment. This is how the term "marketing-mix" is born and extended as agglutinative for four words which Anglo Saxon literature has divulged from four words beginning with P: "product", "place", "promotion" and "price". With all the usual exceptions, the simplicity and elementary nature of this approach has made their divulging easy up to the point that it is difficult to find a work on commercial studies that does not make a more or less extensive reference to it. In an effort to summarise this concept it could be accepted to state as a general objective of *marketing-mix* that it is the strengthening of competitiveness, in an effort to attain the commercial and general aims.

In this chapter, we are going to give our attention to one of these four terms: *distribution.* Its study is aimed at achieving that the "products" of a business, institution and, why not, also from an individual producer, are to be found in the precise place and time in order to satisfy a need or desire. For this it is necessary to carry out a whole lot of activities that include tasks from finishing the product right up to its arrival in the hands of the client. *Distribution forms a part* of a nucleus of knowledge that specialised literature normally designates with the name of *"logistics"*. Its objective is the study and management of all the activities that are necessary for material and product handling from purchase right up to sale. A small illustration will clearly show these concepts:

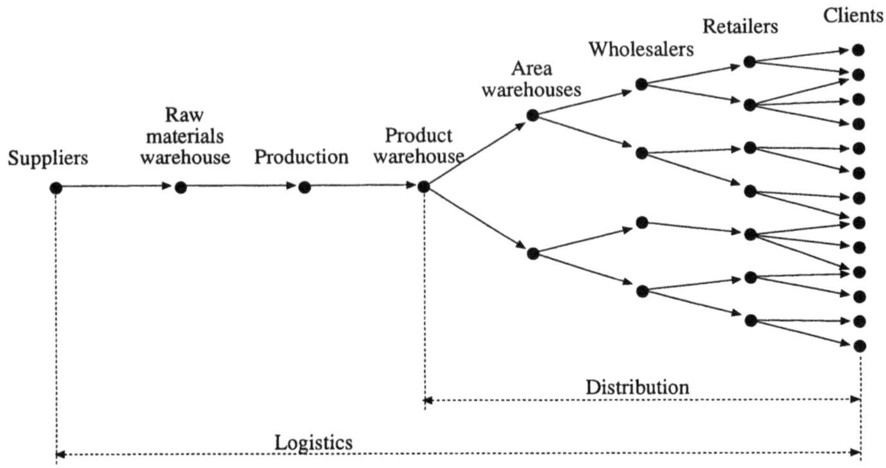

Finally, another concept linked to the previous one stands out in commercial practice. We are referring to *distribution channels*. These are the channels by which the products circulate. They are formed by those human and material means that are capable of facilitating their movement and sale. The search for the best distribution channels constitutes one of the aspects that are most studied both the academic and professional fields.

In the life of businesses and institutions, a constant is constituted by the need to resolve situations in which distribution, in one of its variations, acquires a leading part. But perhaps what is not quite so normal is that it arises in all its extension in the economic activity of individuals. This exception in no way means straying from the general studies, neither does it even require deviating from the models that are useable in large organisations. We would like to show palpable proof of the universality of the instruments that are day by day attracting followers.

It is generally known that people whose life has been lived along the byways of art in any of its variations, are not accustomed to possessing the "virtue" of order, neither do they stand out as having a mechanist and structured vision of their activity. Materialism, which occasionally emerges in their professional relations, is no more than a fleeting murmur in the midst of a certain and necessary mental disorder, source of their creativity. True artists are the possessors of a divine gift, of this "golden box" which is none other than *imagination*. Creation, in any of its variations, is the fruit of three phenomena[1]: *mental entropy, alternation* of the *entropy,* and the *frequency* of the alternation[2]. Entropy is an estimate (measurement or valuation) of the disorder existing in human thought. An excess of order is equivalent to robotization, an excess of disorder the absence of ideas. Therefore, the human mind fluctuates, fortunately, between order and disorder, between structuring and de-structuring. When this alternation process takes place with adequate frequency, creative imagination arises.

Innovations are not normally the fruit of walking the paths of classicism. Artists like researchers seek a discovery. And this is done with solitary work always accompanied by doubt and supported, on many occasions, by hypotheses that escape the eyes of the majority of mortals. Their findings are the result of a mixture of observations, presumptions and audacious adventures, which lead to their breaking down the barriers of the prejudices that their formation has erected. It seems like a miracle that these sublime beings are capable of surviving in the social and economic system that is so complex and de-humanised here in the beginning of this millennium. For them they exist, triumph or fail, walk or amble, pulling behind them, either knowingly or unknowingly, that they have similar problems to the remaining citizens of the world. My love of art has allowed me to know quite a few of them, good creators and false geniuses, nearly always not knowing where the limit is, if it exists, that separates the truth from the lie, the authentic from the false, the original or from the copy. Art has been, in a certain way, the counter weight to academic rigidity, the *disorder* that compensates the *excess of order* which has allowed for a certain balance in our thoughts.

Among those artists who in a more or less fleeting manner I crossed paths with, I should like, on this occasion, to mention an excellent ceramic artist, Neus, whose work stands out as excellently made and due to her sensitivity, and originality in tones and forms. The fame and recognition for this creator of lines and masses has transcended the narrow limits of our country and her works, her production, is requested by different segments of the population, through different distribution channels all over the world. Her creative capacity, which is very important, does not reach sufficient volume in order to please all those desirous of possessing one of her clay marvels. What to do faced the alternatives that arise? How to select points of sale? Which type of work was the most suitable for each area? How to add homogeneous products in order not to confuse collectors? And finally, Which is the best route for the delivery of the works of art? These were just some of the questions that Neus brought up, when thinking about her immediate future.

From the opposite side of thought, one could perceive that this plethora of questions were interconnected and some of them even overlapped. Therefore it would be a convenient move to bring some order to these concerns and see if we could come up with some solutions by seeking the most suitable techniques for their treatment

Preference between points of sale

We proposed commencing with an attempt to establish an *order of preference* between the possible *distribution channels* by which her "production" of works of art could be distributed. Obviously, this ordering should take place according to certain predetermined *criteria*. In order to arrive at this objective it was necessary to become aware of two sets of elements: possible distribution channels and the criteria considered as important for the preference of one channel over another. Little by little we accumulated all the information.

With regard to the first set, "distribution channels" the following were accepted as possible:

A: Art galleries.

B: Art and antique fairs.

C: Auction halls.

D: Airport shops

E: Internet sales.

F: Exclusive art dealer.

G: Direct sales

The second of the sets, included the *criteria* to be used for making an order. In this respect it should be pointed out that the criteria adopted at the time cannot be considered of general use in other cases nor at different times. They possessed a high degree of subjectivity, although they were highly suggestive for the study of the world of art. The following were taken into account:

a: A good position with private collectors.

b : Contacts with public institutions.

c : Possibility of placing works in museums.

d : Suitable programming of prices for the works.

e : Sales level.

f : Net profits from sales.

g : Stability and permanence in the market.

h : Strengthening of artist image.

i : Presence in the media.

j : Possibility of passing into posterity.

In this way, we had sufficient information to commence the process that would lead to the desired order of preferences. In this sense we drew up a "table" in which we placed the *channels* in the rows and the *criteria* in the columns. In each box we set down the "degree" or "level" of the established criteria presumably possessed by each channel, on the understanding that the higher the level a value was assigned the nearer to the unit (maximum value) and therefore further distanced from zero (minimum value) according to the well known hendecagonal semantic scale which we reproduce below:

1: Maximum level

.9: Very, very high level

.8: Very high level

.7: High level

.6: Quite high level

.5: Normal level

.4: Quite low level

.3: Low level

.2: Very low level

.1: Very, very low level

0: Minimum level

A meeting with Neus with collecting experts was sufficient to reach an agreement on the valuation to be assigned to each one of the criteria of each channel, which we have placed in the respective boxes. The result is as presented below:

[R]=

	a	b	c	d	e	f	g	h	i	j
A	.9	.4	.2	.7	.8	.7	1	.6	.6	.3
B	.6	.3	.2	.1	.6	.8	.2	.8	.7	.3
C	.7	.3	.1	.2	.5	.7	.4	.7	.4	.4
D	.4	0	0	.2	.7	.7	.5	.4	0	.1
E	.8	.1	.2	.6	.8	.9	.7	.7	.1	.4
F	.8	.9	.8	1	.7	.4	.8	.6	.8	.5
G	.9	.3	.4	.6	.6	1	.7	.3	.1	.2

It goes without saying that this table, which we call a *fuzzy relation*, corresponded to the artist Neus, with her possibilities of the moment, her relationships with people and institutions, her extraordinary worth and the unquestionable quality of her works. Every artist (painter, sculptor, engraver, ceramist, photographer, etc..) possess their own fuzzy relation which can vary as time goes by.

Having found this fuzzy relation, it was now necessary make a point that could have important consequences: the adopted criteria did not have the same importance for the aspirations, desires and needs of the ceramicist. In some way or another a different *weight* should be given to each one of them. And this is what we were going to do after an "interrogation" to which we submitted our friend Neus. We felt we had interpreted her thoughts adequately on establishing the following values, in which to the greatest weight we assigned a value of *one* and the lightest a *zero*.

a: good position with private collectors. 0.7

b: contacts with public institutions 0.6

c: possibility of placing works in museums 0.8

d: suitable programming of prices for the works 0.9

e: sales level... 0.6

f: net profit from sales 0.5

g: stability and permanence in the market 0.6

h: strengthening of artist image 0.7

i: presence in the media 0.4

j: possibility of passing into posterity 1

We were conscious of the fact of the inter-relationship existing between some of these criteria and although this fact was underlying in the assigned weights, we felt that the task would become too complicated if we included them in an explicit manner, without our acting in this way signifying any radical variation in the result obtained. We opted therefore for simplicity and speed in the operations to be done. In order to make later calculation easier we considered that it would be useful to convert the weights into values the *sum of which would be the unit*. For this it was sufficient to add all the weights and then divide each individual weight by the sum we arrived at. The result was the following:

$$S = 0.7 + 0.6 + 0.8 + 0.9 + 0.6 + 0.5 + 0.6 + 0.7 + 0.4 + 1 = 6.8$$

The new "normalised" numbers are:

$$w_a = \frac{0.7}{6.8} = 0.103 \;; \quad w_b = \frac{0.6}{6.8} = 0.088 \;; \quad w_c = \frac{0.8}{6.8} = 0.118$$

$$w_d = \frac{0.9}{6.8} = 0.132 \;; \quad w_e = \frac{0.6}{6.8} = 0.088 \;; \quad w_f = \frac{0.5}{6.8} = 0.074$$

$$w_g = \frac{0.6}{6.8} = 0.088 \;; \quad w_h = \frac{0.7}{6.8} = 0.103 \;; \quad w_i = \frac{0.4}{6.8} = 0.059$$

$$w_j = \frac{1}{6.8} = 0.147$$

It is customary to present these "normalised " weights by means of a fuzzy subset, *descriptor* of the importance of the criteria, as follows:

	a	b	c	d	e	f	g	h	i	j
$\underset{\sim}{W}$ =	.103	.088	.118	.132	.088	.074	.088	.103	.059	.147

Once the necessary information is available, we can do the calculations, leading to arriving at an order of preference of the suggested channels. For this we suggest affecting *each channel* possessing the criteria, with *the corresponding weight* representing its importance, in order later to do the addition. The result was as expressed below:

Channel A

(0.9) . (0.103) + (0.4) . (0.088) + (0.2) . (0.118) + (0.7) . (0.132) + (0.8) . (0.088) +

+ (0.7) . (0.074) + (1) . (0.088) + (0.6) . (0.103) + (0.6) . (0.059) + (0.3) . (0.147) =

= 0.5954

Channel B

(0.6) . (0.103) + (0.3) . (0.088) + (0.2) . (0.118) + (0.1) . (0.132) + (0.6) . (0.088) +
+ (0.8) . (0.074) + (0.2) . (0.088) + (0.8) . (0.103) + (0.7) . (0.059) + (0.3) . (0.147) =
= 0.4224

Channel C

(0.7) . (0.103) + (0.3) . (0.088) + (0.1) . (0.118) + (0.2) . (0.132) + (0.5) . (0.088) +
+ (0.7) . (0.074) + (0.4) . (0.088) + (0.7) . (0.103) + (0.4) . (0.059) + (0.4) . (0.147) =
= 0.4222

Channel D

(0.4) . (0.103) + (0) . (0.088) + (0) . (0.118) + (0.2) . (0.132) + (0.7) . (0.088) +
+ (0.7) . (0.074) + (0.5) . (0.088) + (0.4) . (0.103) + (0) . (0.059) + (0.1) . (0.147) =
= 0.2809

Channel E

(0.8) . (0.103) + (0.1) . (0.088) + (0.2) . (0.118) + (0.6) . (0.132) + (0.8) . (0.088) +
+ (0.9) . (0.074) + (0.7) . (0.088) + (0.7) . (0.103) + (0.1) . (0.059) + (0.4) . (0.147) =
= 0.5294

Channel F

(0.8) . (0.103) + (0.9) . (0.088) + (0.8) . (0.118) + (1) . (0.132) + (0.7) . (0.088) +
+ (0.4) . (0.074) + (0.8) . (0.088) + (0.6) . (0.103) + (0.8) . (0.059) + (0.5) . (0.147) =
= 0.7321

Channel G

(0.9) . (0.103) + (0.3) . (0.088) + (0.4) . (0.118) + (0.6) . (0.132) + (0.6) . (0.088) +
+ (1) . (0.074) + (0.7) . (0.088) + (0.3) . (0.103) + (0.1) . (0.059) + (0.2) . (0.147) =
= 0.5001

From these calculations we arrived at certain values which brought to light the level of *"suitability"* of the corresponding channel, which would allow us to establish an order of preference among them. Thus ordering them from greater to least we obtained:

1) Channel F: Exclusivity of an "art dealer".

2) Channel A: Art galleries.

3) Channel E: Internet sales.

4) Channel G: Direct sales.

5)Channel B: Art and antique fairs

6) Channel C: Auction rooms.

7) Channel D: Airport shops.

This result did not appear to surprise Neus too much. Nevertheless, a question remained pending: although there were not too many doubts relative to the assigned weights, which made the importance of each one of the criteria intervene, there was a certain reticence relative to the valuations given to the levels of each criterion for the different channels. Thus, for example, it was established that an "art dealer" was perfect (value one) for programming the prices of the works of art. It could be argued that this was only so for a good "art dealer" but not all of then were or are. Indeed this is true, but is it so for all the channels and all the criteria. In some way we had to find a *representative unit* of the corresponding channel, that was reasonable for the magnificent artist for whom we were carrying out the study. The fuzzy relation did not have a universal nature nor the property of permanence. Each artist has his or her own which can vary over time.

We had established the operating process in such a way that following it up would be easy for the non-initiated into the mathematics of uncertainty. But, perhaps, it would be a good thing to present the general formula to facilitate the use of this tool in other cases and under other circumstances. This is really a very simple scheme: Given $[\underset{\sim}{R}]$ and $[\underset{\sim}{W}]$, it is sufficient to calculate the addition-product:

$$[\underset{\sim}{R}] \circ [\underset{\sim}{W}]$$

which in our case is:

[R] • [W] =

	a	b	c	d	e	f	g	h	i	j
A	.9	.4	.2	.7	.8	.7	1	.6	.6	.3
B	.6	.3	.2	.1	.6	.8	.2	.8	.7	.3
C	.7	.3	.1	.2	.5	.7	.4	.7	.4	.4
D	.4	0	0	.2	.7	.7	.5	.4	0	.1
E	.8	.1	.2	.6	.8	.9	.7	.7	.1	.4
F	.8	.9	.8	1	.7	.4	.8	.6	.8	.5
G	.9	.3	.4	.6	.6	1	.7	.3	.1	.2

•

a	.103
b	.088
c	.118
d	.132
e	.088
f	.074
g	.088
h	.103
i	.059
j	.147

=

A	B	C	D	E	F	G
.5954	.4224	.4222	.2809	.5294	.7321	.5001

Having reached this point, we feel it is necessary to bring to light the fact that in the preceding development we had transgressed, perhaps too much, the normally accepted rules relative to the use of operators when considering fuzzy information. In fact the composition operator addition-product is not precisely the most adequate from a theoretical point of view. But as this is a very specific application we felt that we could incur in this "formal sin" in the interest of simplicity, speed of comprehension and operation. Under other circumstances we would have been inclined to use another calculation procedure[3]. In any event we now had *an* order of preference of channels, in which the importance of the "art dealer" and "art galleries" stood out in order for us to arrive at the sought after objectives represented by the *criteria.* It goes without saying that for different *criteria* (or objectives if you like) a different order of preference between the channels could occur.

The use of several channels

Now, artists know perfectly well that the exclusivity of an "art dealer" does not allow then to resort to other channels. This then perhaps would make it necessary to establish a type of comparison in which were *included* those channels that were compatible. It was in this sense that we considered jointly:

A Art Galleries
B Art and antique fairs
C Auction rooms

D Airport shops
E Internet sales
G Direct sales

F Exclusivity of an "art dealer"

It was very obvious that with these groupings of channels, fuzzy relation [$\underset{\sim}{R}$]) would undergo important modifications. Taking into account the nature of the three sets we had established and, without varying the levels given in the aforementioned matrix, we came to the conclusion that the operator of *union or reunion* ($\cup$) would be the most suitable for resolving our problems. We are now going to include our reasoning of that time.

Each one of the *channels* can be described by means of a fuzzy sub-set (row of the fuzzy relation [$\underset{\sim}{R}$]). Therefore:

	a	b	c	d	e	f	g	h	i	j
$\underset{\sim}{A}$ =	.9	.4	.2	.7	.8	.7	1	.6	.6	.3

	a	b	c	d	e	f	g	h	i	j
$\underset{\sim}{B}$ =	.6	.3	.2	.1	.6	.8	.2	.8	.7	.3

	a	b	c	d	e	f	g	h	i	j
$\underset{\sim}{C}$ =	.7	.3	.1	.2	.5	.7	.4	.7	.4	.4

	a	b	c	d	e	f	g	h	i	j
$\underset{\sim}{D}$ =	.4	0	0	.2	.7	.7	.5	.4	0	.1

	a	b	c	d	e	f	g	h	i	j
$\underset{\sim}{E}$ =	.8	.1	.2	.6	.8	.9	.7	.7	.1	.4

	a	b	c	d	e	f	g	h	i	j
$\underset{\sim}{F}$ =	.8	.9	.8	1	.7	.4	.8	.6	.8	.5

	a	b	c	d	e	f	g	h	i	j
$\underset{\sim}{G}$ =	.9	.3	.4	.6	.6	1	.7	.3	.1	.2

Now then, if we resort to the first of the sets A, B, C the consequent level for each of the criteria would be the *highest* of the three. In this way for criterion a it would be 0.9, as this is the highest value of 0.9, 0.6 and 0.7. Acting in this way we arrived at:

	a	b	c	d	e	f	g	h	i	j
$\underset{\sim}{A} (\cup) \underset{\sim}{B} (\cup) \underset{\sim}{C}$ =	.9	.4	.2	.7	.8	.8	1	.8	.7	.4

	a	b	c	d	e	f	g	h	i	j
$\underset{\sim}{D} (\cup) \underset{\sim}{E} (\cup) \underset{\sim}{G}$ =	.9	.3	.4	.6	.8	1	.7	.7	.1	.4

	a	b	c	d	e	f	g	h	i	j
$\underset{\sim}{F}$ =	.8	.9	.8	1	.7	.4	.8	.6	.8	.5

The new fuzzy relation that was formed then only had three rows, corresponding to the new groups of channels we had established. It was as follows:

	a	b	c	d	e	f	g	h	i	j
$\underset{\sim}{A} (\cup) \underset{\sim}{B} (\cup) \underset{\sim}{C}$	.9	.4	.2	.7	.8	.8	1	.8	.7	.4
$[\underset{\sim}{R}'] = \underset{\sim}{D} (\cup) \underset{\sim}{E} (\cup) \underset{\sim}{G}$	.9	.3	.4	.6	.8	1	.7	.7	.1	.4
$\underset{\sim}{F}$	.8	.9	.8	1	.7	.4	.8	.6	.8	.5

We continue with the same process in order to do the corresponding calculations, arriving at the following result:

	a	b	c	d	e	f	g	h	i	j
$\underset{\sim}{A} (\cup) \underset{\sim}{B} (\cup) \underset{\sim}{C}$	.9	.4	.2	.7	.8	.8	1	.8	.7	.4
$[\underset{\sim}{R}] \cdot [\underset{\sim}{W}] = \underset{\sim}{D} (\cup) \underset{\sim}{E} (\cup) \underset{\sim}{G}$	.9	.3	.4	.6	.8	1	.7	.7	.1	.4
$\underset{\sim}{F}$	.8	.9	.8	1	.7	.4	.8	.6	.8	.5

•

a	.103
b	.088
c	.118
d	.132
e	.088
f	.074
g	.088
h	.103
i	.059
j	.147

=

$\underset{\sim}{A} (\cup) \underset{\sim}{B} (\cup) \underset{\sim}{C}$	$\underset{\sim}{D} (\cup) \underset{\sim}{E} (\cup) \underset{\sim}{G}$	$\underset{\sim}{F}$
.6440	.5883	.7321

Continuing at the head of the order was the exclusivity of an "art dealer" followed, very closely by, art galleries, in conjunction with fairs and auctions, and in last place was direct sales, complemented by sales by Internet and at airports. Let us remember, however, that the latter two channels are undergoing rapid progress and it should not come as a surprise that in the not too distant future, they could become highly, we could even say vitally, important. We should also point out that the grouping made allowed for a significant increase in the "suitability" of the jointly used channels.

We now had available a whole arsenal of techniques, corresponding to the sphere of uncertainty that were capable of providing solutions that were scientifically more valued and technically better placed, but we felt that Neus wanted no further "ramblings" from her *mad friend,* but on the other had she was concerned about other problems that were troubling her artistic mind. Therefore we centred our attention on another aspect of her promotional objective.

Distribution taking homogenous groups as a starting point

The works of artists, in general, and those of our illustrious friend, in particular, even when they possess certain characteristics in common that define their personality and make them recognisable from the works done by other artists, are not normally totally homogeneous relative to the segments of the population who are eventual purchasers, be this because of their size, by the singularity of the piece (unique or multiple piece), pigmentation (type of colouring), form or even the subject matter. *A point of sale* can have the whole range of works or only a part of them. Too much diversity may disconcert the collector-investor, a little varied offer runs the risk of "weakening" the image of the artist and "frighten off" the client, because of not being able to find what was being sought. Apart from the "testimonial" exhibitions, or those exhibitions called "retrospective", in which are included works from different "periods" of an artist, *normal exhibitions* require a *coherence* of the works presented, in such a way that viewers capture, with a certain ease, the fundamental characteristics of the artist and there occurs a "flow" of *attraction,* let us say nearly of love, between the object (artistic work) and subject (eventual purchaser). Therefore it should not come as a surprise that the preparation of a "vernissage" (placing of the works) causes in the artist great concern and not a few doubts and uncertainties in those few hours prior to "the moment of truth", to triumph or failure, which for all mortals is always being played down, but for the sensitive soul of the artist acquires all that is categorical of what is Boolean.

Be it either for a suitable geographical distribution (in one point of sale eventual clients could accept certain works and not others), or in order to attain coherence in a certain artistic manifestation (certain works should not be in the way of others), it is vitally important to make *homogeneous groupings* relative to certain previously established criteria. We felt then and still feel today that it would be of enormous assistance if we separated the collection of the works of Neus into *groups*, each one suitable for sending to an art gallery, to form, or not a monographic exhibition, in such a way that viewers could admire, in all its splendour, the beauty emerging from the works of this famous ceramist. For this it is necessary to establish on the one hand, a collection by the *different types of works* normally done by the artist and, on the other hand, a collection made up of the criteria used for their differentiation.

With regard to the first we reached an agreement to adopt the ones listed below:

C_1: Large size ceramics.

C_2: Singular polychrome plates.

C_3: Multiple edition plates.

C_4: Singular polychrome vases.

C_5: Multiple edition vases.

C_6: Singular works to order

C_7: Multiple edition works to order.

C_8: Large edition small pieces.

The criteria considered important for the desired homogenisation not always coincide with those used in the previous section, when the objective was different, consisted in establishing an order of preference of the distribution channels although they were not, in our case, total disassociation with the same. We have listed them below:

a: Interest for private collectors

b: Interest for public institutions

c: Adequate for purchase by museums

d: Price level

e: Sales level

f: Capacity for generating future sales

g: Improve the image of the artist

h: Make the work of the artist popular

Having made this list we now moved on to describe each type of work by means of a fuzzy sub-set of the referential of the criteria. As was usual for us, we used the semantic hendecagonal scale which, if we recall assigns eleven values from zero to one, in such a way that when the more the type of work possess the corresponding criterion, the nearer the value will be to the unit.

This approach allowed us to establish the following fuzzy sub-sets:

	a	b	c	d	e	f	g	h
$\underset{\sim}{C}_1$=	.2	1	.8	1	.3	.8	.9	.4

	a	b	c	d	e	f	g	h
$\underset{\sim}{C}_2$=	1	.6	.8	.7	.7	.3	.8	.3

	a	b	c	d	e	f	g	h
$\underset{\sim}{C}_3$=	.5	.4	.1	.4	.8	.6	.4	.9

	a	b	c	d	e	f	g	h
$\underset{\sim}{C}_4$=	1	.5	.8	.8	.7	.3	.8	.3

	a	b	c	d	e	f	g	h
$\underset{\sim}{C}_5$=	.5	.2	.1	.5	.7	.6	.4	.8

	a	b	c	d	e	f	g	h
$\underset{\sim}{C}_6$=	.1	.9	.6	1	.3	.4	.6	.3

	a	b	c	d	e	f	g	h
$\underset{\sim}{C}_7$=	.3	.2	.1	.4	.6	.7	.3	.9

	a	b	c	d	e	f	g	h
$\underset{\sim}{C}_8$=	.1	.3	.1	.1	1	.9	.3	1

These eight descriptors constituted all the available information from which we had to make the homogeneous grouping of the types of works normally done by Neus. The first step to be taken, in order for us to attain this objective, was to see the *degree of distance* of each work of art relative to the remainder. The further the distance, the less homogeneous they were. In order to find this degree of distance we resorted to a very well known concept: the relative Hamming distance. Its simplicity and ease of operation made it highly recommendable. As is so well known, to find the relative *Hamming distance* between two *fuzzy* sub-sets, all that needs to be done is to subtract the levels existing in each criterion, add these differences without considering the negative signs (sum of absolute values) and divide the sum by the number of criteria. In this way we found the following distances:

$$\delta(\underset{\sim}{C}_1, \underset{\sim}{C}_2) = 1/8\,(|.2 - 1| + |1 - .6| + |.8 - .8| + |1 - .7| + |.3 - .7| + |.8 - .3| + |.9 - .8| + |.4 - .3|)$$
$$= 1/8\,(.8 + .4 + 0 + .3 + .4 + .5 + .1 + .1) = 2.6/8 = 0.325$$

$$\delta(\underset{\sim}{C}_1, \underset{\sim}{C}_3) = 1/8\,(|.2 - .5| + |1 - .4| + |.8 - .1| + |1 - .4| + |.3 - .8| + |.8 - .6| + |.9 - .4| + |.4 - .9|)$$
$$= 1/8\,(.3 + .6 + .7 + .6 + .5 + .2 + .5 + .5) = 3.9/8 = 0.488$$

$$\delta(\underset{\sim}{C}_1, \underset{\sim}{C}_4) = 1/8\,(|.2 - 1| + |1 - .5| + |.8 - .8| + |1 - .8| + |.3 - .7| + |.8 - .3| + |.9 - .8| + |.4 - .3|)$$
$$= 1/8\,(.8 + .5 + 0 + .2 + .4 + .5 + .1 + .1) = 2.6/8 = 0.325$$

$$\delta(\underset{\sim}{C}_1, \underset{\sim}{C}_5) = 1/8\,(|.2 - .5| + |1 - .2| + |.8 - .1| + |1 - .5| + |.3 - .7| + |.8 - .6| + |.9 - .4| + |.4 - .8|)$$
$$= 1/8\,(.3 + .8 + .7 + .5 + .4 + .2 + .5 + .4) = 3.8/8 = 0.475$$

For the remaining distances we have just shown the results:

$\delta(\underset{\sim}{C}_1, \underset{\sim}{C}_6) = 0.150$	$\delta(\underset{\sim}{C}_1, \underset{\sim}{C}_7) = 0.463$	$\delta(\underset{\sim}{C}_1, \underset{\sim}{C}_8) = 0.550$
$\delta(\underset{\sim}{C}_2, \underset{\sim}{C}_3) = 0.388$	$\delta(\underset{\sim}{C}_2, \underset{\sim}{C}_4) = 0.025$	$\delta(\underset{\sim}{C}_2, \underset{\sim}{C}_5) = 0.375$
$\delta(\underset{\sim}{C}_2, \underset{\sim}{C}_6) = 0.300$	$\delta(\underset{\sim}{C}_2, \underset{\sim}{C}_7) = 0.463$	$\delta(\underset{\sim}{C}_2, \underset{\sim}{C}_8) = 0.575$
$\delta(\underset{\sim}{C}_3, \underset{\sim}{C}_4) = 0.388$	$\delta(\underset{\sim}{C}_3, \underset{\sim}{C}_5) = 0.063$	$\delta(\underset{\sim}{C}_3, \underset{\sim}{C}_6) = 0.438$
$\delta(\underset{\sim}{C}_3, \underset{\sim}{C}_7) = 0.100$	$\delta(\underset{\sim}{C}_3, \underset{\sim}{C}_8) = 0.188$	$\delta(\underset{\sim}{C}_4, \underset{\sim}{C}_5) = 0.375$
$\delta(\underset{\sim}{C}_4, \underset{\sim}{C}_6) = 0.300$	$\delta(\underset{\sim}{C}_4, \underset{\sim}{C}_7) = 0.463$	$\delta(\underset{\sim}{C}_4, \underset{\sim}{C}_8) = 0.575$
$\delta(\underset{\sim}{C}_5, \underset{\sim}{C}_6) = 0.425$	$\delta(\underset{\sim}{C}_5, \underset{\sim}{C}_7) = 0.088$	$\delta(\underset{\sim}{C}_5, \underset{\sim}{C}_8) = 0.225$
$\delta(\underset{\sim}{C}_6, \underset{\sim}{C}_7) = 0.438$	$\delta(\underset{\sim}{C}_6, \underset{\sim}{C}_8) = 0.525$	$\delta(\underset{\sim}{C}_7, \underset{\sim}{C}_8) = 0.163$

With all these distances which, we insist on very much were indices of separation between types of works of art, we could form a table which, visibly, included in a unitary manner, all this important information. This is normally called a "matrix of dissimilitude" or "fuzzy relation of dissimilitude".

$[\underset{\sim}{\delta}]=$

	C_1	C_2	C_3	C_4	C_5	C_6	C_7	C_8
C_1	0	.325	.488	.325	.475	.150	.463	.550
C_2	.325	0	.388	.025	.375	.300	.463	.575
C_3	.488	.388	0	.388	.063	.438	.100	.188
C_4	.325	.025	.388	0	.375	.300	.463	.575
C_5	.475	.375	.063	.375	0	.425	.088	.225
C_6	.150	.300	.438	.300	.425	0	.438	.525
C_7	.463	.463	.100	.463	.088	.438	0	.163
C_8	.550	.575	.188	.575	.225	.525	.163	0

It could be seen that in order to fill in the boxes of the above table it was necessary to calculate the distance of one type of work from another and the distance of the other from the first, since both were the same. In fact the distance between "large size ceramics" and "multiple edition plate" is the same as the distance between "multiple edition plate" and "large size ceramics". In our case both were 0.488. When this occurs it is said that the matrix or fuzzy relation possess the *property of symmetry*. This was particularly interesting because it allowed us to apply an *algorithm* that would make later calculations far easier.

From the fuzzy relation of distances $[\underset{\sim}{\delta}]$ we could move over to a relation of *closeness* known as a "relation of similitude". For this it was sufficient to place, in each box, the complement of the unit. This we did:

$[\tilde{S}]=$

	C_1	C_2	C_3	C_4	C_5	C_6	C_7	C_8
C_1	1	.675	.512	.675	.525	.850	.537	.450
C_2	.675	1	.612	.975	.625	.700	.537	.425
C_3	.512	.612	1	.612	.937	.562	.900	.812
C_4	.675	.975	.612	1	.625	.700	.537	.425
C_5	.525	.625	.937	.625	1	.575	.912	.775
C_6	.850	.700	.562	.700	.575	1	.562	.475
C_7	.537	.537	.900	.537	.912	.562	1	.837
C_8	.450	.425	.812	.425	.775	.475	.837	1

This matrix or fuzzy relation constituted the support for making homogeneous groupings. Actually, in itself it already provided groups, although two by two. In fact, it can be seen that C_1 and C_2 form a group with a degree of homogeneity of 0.675: C_1 and C_3 another group with a degree of homogeneity of 0.512 (that is to say slightly less); and so on successively. But what we were looking for was not to obtain sets formed by two types of works of art but made up of the largest number possible of types of works. This led us to consider a mathematical element which we had used on other occasions: the *maximum sub-relations of similarity*. In order to arrive at these it was necessary to take an important prior decision: which group or level of homogeneity did we consider sufficient for our requirements. The greater the level required the less number of works of art would form the group. On the other hand if we accepted a lower level of similarity the groups would be more numerous. Faced with this circumstance we tested out the groupings for two different minimum levels: the level that was equal to or greater than 0.700 and the level that was greater or equal to 0.800. In both cases we acted as follows: we would for a relation such that in those boxes where there was a value equal to or greater than the required threshold we would place a one, and if not a zero (or whatever was easier, we would leave it blank) In this way we converted a *fuzzy matrix* into a *Boolean matrix* We began with the degree or level $\alpha \geq 0.700$:

$[\beta_{0.700}]=$

	C_1	C_2	C_3	C_4	C_5	C_6	C_7	C_8
C_1	1					1		
C_2		1		1		1		
C_3			1		1		1	1
C_4		1		1		1		
C_5			1		1		1	1
C_6	1	1		1		1		
C_7			1		1		1	1
C_8			1		1		1	1

If we had selected level $\alpha \geq 0.800$ the corresponding Boolean matrix would be:

$[\beta_{0.800}]=$

	C_1	C_2	C_3	C_4	C_5	C_6	C_7	C_8
C_1	1					1		
C_2		1		1				
C_3			1		1		1	1
C_4		1		1				
C_5			1		1		1	
C_6	1					1		
C_7			1		1		1	1
C_8			1				1	1

Having reached this point we now had to "exploit" these matrices. For this we selected a very interesting algorithm[4], the automatisms of which make it very desirable, particularly if wanting to use computers.

We will explain the stages to follow for the use of the Pichat[5] algorithm in some detail:

a) For each one of the rows of the Boolean matrix *that are above* the main diagonal we take into consideration the 1.

b) In those boxes where ones do not exist we look at the column to which they correspond, assigning them a Boolean variable expressed by means of a sub-index of the same. (Therefore if we are talking about column C_5 we assign the "symbolised" variable by means of a 5).

c) We then do the *addition*, $(\dot{+})$, of the expressed Boolean variable by the sub-index of the element of each row (for example, for row C_2 it would be 2) with the Boolean product of all the variables found in the previous section. When in a row there are no vacant boxes (nil values) to the right of the principal diagonal, this row is not considered.

d) We calculate the Boolean product, (.) of all the sums obtained in the preceding section. In this sense it should be remembered that:

$$x\ (\cdot)\ x = x$$

$$x\ (\dot{+})\ x = x$$

$$x\ (\dot{+})\ x\ (\cdot)\ y = x$$

As a result of this product we arrive at a polynomial S.

e) Each of the monomials of polynomial S is transformed into its complement (values that are *not* found in monomials). In this way we arrive at a new polynomial S'. It should be taken into account that the sub-index is *not* a number but a symbol representing a Boolean variable.

f) Each of the monomials of polynomial S' is a maximum sub-relation of similarity. With this we arrive at the grouping we were looking for.

Although we have attempted to make the description comprehensible to non experts, some of the words we have used (monomial, polynomial, sum and Boolean products,...) are sufficient to create a certain rejection by those, such as Neus, who are not used to this vocabulary. For this reason we desisted from following the path of theory to immediately move on to their practical use, specifically to our problem. That is following the steps we set out. We begin with the Boolean matrix $[\beta_{0.700}]$:

a) Consideration of the 1 existing above the principal diagonal.

	C_1	C_2	C_3	C_4	C_5	C_6	C_7	C_8
C_1	1					1		
C_2		1		1		1		
C_3			1		1		1	1
C_4				1		1		
C_5					1		1	1
C_6						1		
C_7							1	1
C_8								1

b) Assignment of the Boolean variables by means of sub-indexes.

	C_1	C_2	C_3	C_4	C_5	C_6	C_7	C_8
C_1	1	2	3	4	5		7	8
C_2		2	3		5		7	8
C_3			3	4		6		
C_4				4	5		7	8
C_5					5	6		
C_6						6	7	8
C_7							7	
C_8								8

c) Boolean sum of the sub-index of each row by the product of the Boolean variables of the previous section.

Row C_1: $1 \dotplus 234578$

Row C_2: $2 \dotplus 3578$

Row C_3: $3 \dotplus 46$

Row C_4: $4 \dotplus 578$

Row C_5: 5 ∔ 6

Row C_6: 6 ∔ 78

Row C_7: —

Row C_8: —

d) Arriving at the Boolean products.

Product of row C_1 by row C_2:

(1 ∔ 234578) . (2 ∔ 3578) = 12 ∔ 13578 ∔ 234578 ∔ ~~234578~~

Product of the previous result by row C_3:

(12 ∔ 13578 ∔ 234578) . (3 ∔ 46) =
= 123 ∔ 1246 ∔ 13578 ∔ ~~1345678~~ ∔ 234578 ∔ ~~2345678~~

Product of the previous result by row C_4:

(123 ∔ 1246 ∔ 13578 ∔ 234578) . (4 ∔ 578) =
= 1234 ∔ ~~123578~~ ∔ 1246 ∔ ~~1245678~~ ∔ ~~134578~~ ∔ 13578 ∔ 234578 ∔ ~~234578~~

Product of the previous result by row C_5:

(1234 ∔ 1246 ∔ 13578 ∔ 234578) . (5 ∔ 6) =
= 12345 ∔ ~~12346~~ ∔ ~~12456~~ ∔ 1246 ∔ 13578 ∔ ~~135678~~ ∔ 234578 ∔ ~~2345678~~

Product of the previous result by row C_6:

(12345 ∔ 1246 ∔ 13578 ∔ 234578) . (6 ∔ 78) =
~~123456~~ ∔ ~~1234578~~ ∔ 1246 ∔ ~~124678~~ ∔ ~~135678~~ ∔ 13578 ∔ ~~2345678~~ ∔ 234578

The polynomial found was:

S = 1246 ∔ 13578 ∔ 234578

e) Obtaining the complement of polynomial S:

S' = 3578 ∔ 246 ∔ 16

f) Presentation of the maximum sub-relations of similarity.

The three monomials found determined the three maximum homogeneous groupings. These were as follows:

Group I
- C_3: Multiple edition plates
- C_5: Multiple edition vases
- C_7: Multiple edition works to order
- C_8: Large edition small pieces

Group II
- C_2: Singular polychrome plates
- C_4: Singular polychrome vases
- C_6: Singular works to order

Group III
- C_1: Large size ceramics
- C_6: Singular works to order

The results we had arrived at appeared reasonable, if what was wanted was a homogeneous presentation of the work of the artist. But the importance of this approach appeared to require a "technical" verification of the groupings. For this we formed sub-relations, giving them the form of *sub-matrices*, in such a way that we assigned to each "box" of the table, the number representing the closeness between types of works which were included in the fuzzy relation $[\underset{\sim}{\delta}]$. In this way we arrived at:

	C_3	C_5	C_7	C_8
C_3	1	.937	.900	.812
C_5	.937	1	.912	.775
C_7	.900	.912	1	.837
C_8	.812	.775	.837	1

	C_2	C_4	C_6
C_2	1	.937	.700
C_4	.975	1	.700
C_6	.700	.700	1

	C_1	C_6
C_1	1	.850
C_6	.850	1

The value of the result was obvious since we could see that all the *closeness indexes* were equal to or higher than 0.700, the threshold we had allowed as the lower limit for homogeneity.

Neus now had before her the evidence of what she was doing intuitively, perhaps not in the strictest manner. Now, all of what we had done was correct on one condition: the criteria we had established should, for the artist, have the same importance. If this were not so, we would have modified the terms of this study by incorporating to each difference between absolute values of the Hamming distance, a multiplier representing the relative weight of each criterion. We knew how to do this[6] but did not feel it was necessary to redo all the calculation with this variation.

On the other hand, artistic manifestations do not always seek a homogeneous presentation. On certain occasions the objective to be attained is to give an overall view of the work of the artist. But, then, the problem disappears, since the approach itself provides the solution: only one grouping exists formed by all and each one of the types of work of the artist. We now believed that the subject was closed for the most important of its aspects. All that was left then was to think about what grouping would have been made if we had required a *greater degree of homogeneity*. We could do a test by taking $\alpha \geq 0.800$ as the threshold. Intuitively we saw that some of the types of works were going to disappear from the sub-relations we had found before. Verification was fast by starting out from the Boolean matrix $[\beta_{0.800}]$ we had found before, which we have reproduced below, placing only the 1 existing above the principal diagonal:

	C_1	C_2	C_3	C_4	C_5	C_6	C_7	C_8
C_1	1					1		
C_2		1		1				
C_3			1		1		1	1
C_4				1				
C_5					1		1	
C_6						1		
C_7							1	1
C_8								1

We assign the Boolean variables:

	C_1	C_2	C_3	C_4	C_5	C_6	C_7	C_8
C_1	1	2	3	4	5		7	8
C_2		2	3		5	6	7	8
C_3			3	4		6		
C_4				4	5	6	7	8
C_5					5	6		8
C_6						6	7	8
C_7							7	
C_8								8

We do the Boolean sum:

Row C_1: $1 \dotplus 234578$

Row C_2: $2 \dotplus 35678$

Row C_3: $3 \dotplus 46$

Row C_4: $4 \dotplus 5678$

Row C_5: $5 \dotplus 68$

Row C_6: $6 \dotplus 78$

We arrived at the Boolean products:

Product of row C_1 by row C_2:

$(1 \dotplus 234578) . (2 \dotplus 35678) =$

$= 12 \dotplus 135678 \dotplus 234578 \dotplus$ ~~2345678~~

Product of the previous result by row C_3:

$(12 \dotplus 135678 \dotplus 234578) . (3 \dotplus 46) =$

$= 123 \dotplus 1246 \dotplus 135678 \dotplus$ ~~1345678~~ $\dotplus 234578 \dotplus$ ~~2345678~~

Product of the previous result by row C_4:

$(123 \dotplus 1246 \dotplus 135678 \dotplus 234578) \,.\, (4 \dotplus 5678) =$

$= 1234 \dotplus \cancel{1235678} \dotplus 1246 \dotplus \cancel{1245678} \dotplus \cancel{1345678} \dotplus 135678 \dotplus$

$\dotplus 234578 \dotplus \cancel{2345678}$

Product of the previous result by row C_5:

$(1234 \dotplus 1246 \dotplus 135678 \dotplus 234578) \,.\, (5 \dotplus 68) =$

$= 12345 \dotplus \cancel{123468} \dotplus 12456 \dotplus 12468 \dotplus$

$\dotplus 135678 \dotplus \cancel{135678} \dotplus 234578 \dotplus \cancel{2345678}$

Product of the previous result by row C_6:

$(12345 \dotplus 12456 \dotplus 12468 \dotplus 135678 \dotplus 234578) \,.\, (6 \dotplus 78) =$

$\cancel{123456} \dotplus \cancel{1234578} \dotplus 12456 \dotplus \cancel{1245678} \dotplus$

$\dotplus 12468 \dotplus \cancel{124678} \dotplus 135678 \dotplus \cancel{135678} \dotplus \cancel{2345678} \dotplus 234578$

From here we arrived at the following polynomial:

$$S = 12456 \dotplus 12468 \dotplus 135678 \dotplus 234578$$

Its complement was:

$$S' = 378 \dotplus 357 \dotplus 24 \dotplus 16$$

Therefore at the required level $\alpha \geq 0.800$, the homogenous groupings increased in number (four in stead of three), but on the other hand, the number of components was reduced. We now had:

Group I { C_3: Multiple edition plates; C_7: Multiple edition ordered works; C_8: Large edition small pieces }

The *multiple edition vases* "had broken away" from this group as there existed a certain degree of incompatibility (the similarity of 0.775 did not reach the desired 0.800) with the *large edition small pieces* but, on the other hand, a new group was formed, in which the type of work that was absent from the group was precisely the latter :

Group II $\begin{cases} C_3\text{: Multiple edition plates} \\ C_5\text{: Multiple edition vases} \\ C_7\text{: Multiple edition ordered works} \end{cases}$

Now multiple edition plates and vases, once again formed a homogeneous group:

Group III $\begin{cases} C_2\text{: Singular polychrome plates} \\ C_4\text{: Singular polychrome vases} \end{cases}$

Included in this group was singular ordered works when the requirement for homogeneity was less, but in the exact degree it disappears as obviously for this type of work there has to be a greater similarity.

Group IV $\begin{cases} C_1\text{: Large size ceramics} \\ C_6\text{: Singular ordered works} \end{cases}$

Finally this group remained intact, even though the level required was higher.

Once in possession of all these results, we proceeded to do the "technical" comparison by constructing the sub-matrices corresponding to the maximum sub-relations of similarity.

	C_3	C_7	C_8
C_3	1	.900	.812
C_7	.900	1	.837
C_8	.812	.837	1

	C_3	C_5	C_7
C_3	1	.937	.900
C_5	.937	1	.912
C_7	.900	.912	1

	C_2	C_4
C_2	1	.975
C_4	.975	1

	C_1	C_6
C_1	1	.850
C_6	.850	1

A simple glance at these sub-matrices was sufficient to see that all their values complied with the condition of being equal to or higher than 0.800, the threshold established on this occasion.

In this way, our friend, the well known ceramist Neus, had a technical support for adopting, at each time, the most adequate decision for specific circumstances and for the objectives sought after when approaching the problem of grouping her work with a specific exhibition or artistic manifestation in view.

Movement of the products

Normally artists carry out their work and execute their works of art in a workshop located geographically next to a warehouse that constitutes the neuralgic centre of all their activities. This is no obstacle for them to go out into open air or go to closed places where they become inspired, have ideas and, at times, make drawings in order later to convert them into the works of art they qualify as definite. Our ceramist was no exception. This manner of operating was the reason that in her workshop she concentrated and even today concentrates the artistic objects created by her. These objects had to be transferred to the different points of sale (art galleries, auction rooms, etc...), mainly scattered all over the European Union. As is only logical, not all these points of sale were supplied regularly, but some of them only received the works sporadically, nearly always by order of the art dealers, as a consequence of orders from collectors. But, on the other hand, there were certain art galleries, to which she had to send the works of art with a certain regularity, duly packed and with all security devices and, in any event, under the watchful eye of Jesús, her intelligent, cultured and faithful co-operator, whose seriousness and efficiency was the guarantee for a job well done.

Neus was concerned by the high cost of transport for her works, particularly with regard to six galleries to which she sent products frequently, *following different circuits*, that Jesús tried out in an attempt to contain expenses at reasonable levels. Jesús placed special emphasis on the fact that the circuit found must posses a basic characteristic: *it should not pass twice running through the same city* in which a gallery was located (art dealers are very jealous of their clientele and they do not like to see the representatives of their artists too frequently). But as he rightly said, there existed just so many possibilities that "even if she lived for two hundred years" she could not exhaust them. When she told me what her problems were, we became aware immediately that her case was reduced to a question that was brought up many years ago, which was known as the "salesman problem" the solution to which was possible thanks to the "branch and bound" algorithm[7].We set out then on the task of gathering all the information necessary for using this interesting instrument. In this sense we had a great deal of data as a consequence of all the trips done by Jesús in the past, although it should not be forgotten that the evolution of the economic system included a future immersed in a thick cloud of uncertainty, to such an extent that the estimates of transport costs could not be foreseen with any certainty, but estimated by means of an *uncertain number.* For this we chose the *confidence interval.* With these premises the first information had to make reference to the location of the points of sale to which the works of art had to arrive. Everyone was in agreement with the following:

A: Reus (Central warehouse)

B: Rome

C: Grenoble

D: Paris

E: Frankfurt

F: Brussels

Having revised all the justifying documents of previous remittances and finding out what could occur over the next few years, an estimate was made for each trip *from one city to another,* of a minimum cost (in no event would it be lower) and maximum cost (it would never be higher). One special characteristic was that not always was the cost from one city to another the same as from that other city to the first, as a consequence of timetables, means of transport, etcetera...

After many discussions and comments a general consensus was reached. The result was presented by means of a table formed by six rows and six columns. The rows represented the start out cities and the columns the arrival cities. The intervals placed in each box brought to light the minimum costs and maximum costs, expressed in monetary units. In this way the following ϕ-fuzzy relation (fuzzy relation expressed in confidence intervals) was arrived at:

↗	A	B	C	D	E	F
A	∞	[30,35]	[15,17]	[20,25]	[34,36]	[25,28]
B	[32,38]	∞	[20,25]	[24,28]	[25,29]	[28,32]
C	[16,19]	[19,22]	∞	[6 , 8]	[12,14]	[10,14]
D	[24,30]	[22,24]	[7 , 10]	∞	[15,19]	[6 , 8]
E	[35,38]	[26,28]	[12,14]	[16,20]	∞	[17,21]
F	[30,35]	[27,30]	[12,15]	[7 ,10]	[18,22]	∞

We pointed out that in the principal diagonal of the ϕ-fuzzy relation we had placed the symbol ∞. With this what we desired was to remove from the calculations the fact of going f*rom one city to the same city*, a matter that had to be rejected as not realistic, but above all by the clause imposed, very rightly, by Jesús.

Having made this clarification we commenced our process by *separating*, for each interval, the lower extreme and the upper extreme, in order to form two fuzzy relations: one for the "optimistic position" in which would be incurred in every case certain *minimum costs*, and the other for the "pessimistic conception" in which for every journey certain *maximum costs would* have to be borne. The two matrices were as follows:

For the lower extreme or limit:

↗	A	B	C	D	E	F
A	∞	30	15	20	34	25
B	32	∞	20	24	25	28
C	16	19	∞	6	12	10
D	24	22	7	∞	15	6
E	35	26	12	16	∞	17
F	30	27	12	7	18	∞

For the upper extreme or limit:

↗	A	B	C	D	E	F
A	∞	35	17	25	36	28
B	38	∞	25	28	29	32
C	19	22	∞	8	14	14
D	30	24	10	∞	19	8
E	38	28	14	20	∞	21
F	35	30	15	10	22	∞

Our work now consisted of exploiting each one of these matrices by means of the mechanism of an algorithm, in which an arborescence (tree shaped graph) from bi-partitions is built. The stages of the "branch and bound" algorithm could be summarised as follows[8]:

a) Subtract from each row and each column of the matrix the smallest element of the corresponding row and column. From this a new matrix will be arrived at in which there will be, at least, one zero in each row and one zero in each column.

b) Calculate the *sum* of all the values subtracted from the rows and columns of the initial matrix. The result obtained provides the root of a tree *shaped graph.*

c) Make a *bi-partition*, ramifying in such a way that you *choose the greatest value* of the *sum* of the *lowest* value of the row, and column of which are members *all the zeros* of the matrix (without considering, obviously, the zero itself).

d) Place a vertex and an arc in the tree graph the significance of which is to *make the path* between the two points of sale corresponding to that vertex. Assign to this vertex a "limit" equal to the sum of the limit of the previous vertex plus the value found in stage c).

e) Place a vertex and an arc in the tree graph bringing to light the fact that the path must be taken between the points of sale expressed by that vertex.

Eliminate from the matrix, the row and column corresponding to this element. A symbol ∞ is placed in all those boxes representing an arc giving rise to a shorter circuit than the one sought or which prevents the formation of the desired circuit.

f) In the matrix found in the previous stage, subtract from each row and each column the smallest element of one and the others. With this we arrive at a matrix in which there exists at least one zero in each row and each column.

g) Calculate the sum of all the subtracted values of the rows and columns of the matrix. Add this amount to the limit obtained in the previous vertex of the tree graph. In this way what is obtained is the limit corresponding to the vertex of the tree graph which expresses that the trip represented by that vertex should not be done.

h) Compare the value of the limits found for all the *vertices that hang* and select the smallest (should there be two or more that are equally small then choose one arbitrarily).

i) The limit selected in the previous stage can be, either the vertex that expresses that that path must be taken, or that this path should not be taken. In the first case continue the algorithm from section c). In the second case move on to the following stage.

j) In the corresponding matrix, place the symbol ∞ in the box expressing *not to pass* between two points of sale. Subtract from its row and column the *smallest values* in each of them. Once again return to c).

I was conscious of the difficulties entailed by the correct assimilation of the contents of the algorithm, but I hoped that with the numerical development we were going to do its mechanisms would be fully understood. We "got down to work" with the matrix representing the *optimistic position*, as shown below:

↗	A	B	C	D	E	F
A	∞	30	15	20	34	25
B	32	∞	20	24	25	28
C	16	19	∞	6	12	10
D	24	22	7	∞	15	6
E	35	26	12	16	∞	17
F	30	27	12	7	18	∞

We followed the following steps:

We subtracted from the elements of each row the lowest value of each one of them and placed the result on the right of the resulting ,atrix. We arribed at the following:

↗	A	B	C	D	E	F	
A	∞	15	0	5	19	10	15
B	12	∞	0	4	5	8	20
C	10	13	∞	0	6	4	6
D	18	16	1	∞	9	0	6
E	23	14	0	4	∞	5	12
F	23	20	5	0	11	∞	7

We next deducted from the values of each column the lowest value, placing these minimum values below the resulting matrix:

↗	A	B	C	D	E	F
A	∞	2	0	5	14	10
B	2	∞	0	4	0	8
C	0	0	∞	0	1	4
D	8	3	1	∞	4	0
E	13	1	0	4	∞	5
F	13	7	5	0	6	∞

We added the amounts subtracted from all the rows and columns:

$$15 + 20 + 6 + 6 + 12 + 7 + 10 + 13 + 0 + 0 + 5 + 0 = 94$$

We commenced the tree graph by placing as the limit for the root the value we had found, 94:

E 94

We reproduce the previous matrix placing in the boxes where *the value zero* existed an additional box in which we place the sum of the least value of the row and the least value of the column of which each zero is a member (without taking into account the zero that was being treated at each moment). We then arrived at:

↗	A	B	C	D	E	F
A	∞	2	0 [2]	5	14	10
B	2	∞	0 [0]	4	0 [1]	8
C	0 [2]	0 [1]	∞	0 [0]	1	4
D	8	3	1	∞	4	0 [5]
E	13	1	0 [1]	4	∞	5
F	13	7	5	0 [5]	6	∞

We now had to choose from all the values placed within the squares one of the highest. We opted for the one corresponding to element (F, D).

The tree graph was commenced from root E, by placing an "arc" joined to a "vertex" $(\overline{F}, \overline{D})$, assigning an initial level of 94 + 5 = 99:

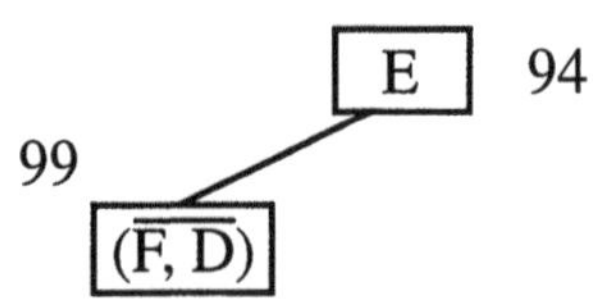

From the previous matrix row F and column D were eliminated and we now arrived at a new matrix:

→	A	B	C	E	F
A	∞	2	0	14	10
B	2	∞	0	0	8
C	0	0	∞	1	4
D	8	3	1	4	∞
E	13	1	0	∞	5

We placed the symbol ∞ in box (D, F) since this route was not possible if route (F, D) was taken. In this way it was eliminated from the calculation process.

The route we had found (F, D) that is to say (Brussels, Paris) we moved onto a graph which we commenced with this arc:

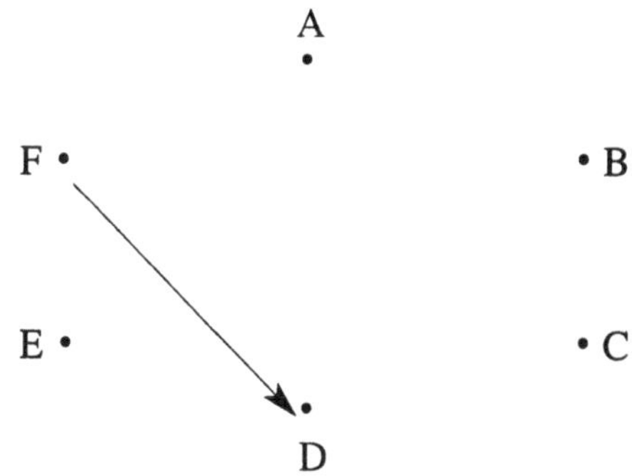

It was then necessary to modify the previous matrix so that there was at least one zero in each row and each column:

→	A	B	C	E	F	
A	∞	2	0	14	10	
B	2	∞	0	0	8	
C	0	0	∞	1	4	
D	7	2	0	3	∞	1
E	13	1	0	∞	5	

→	A	B	C	E	F
A	∞	2	0	14	6
B	2	∞	0	0	4
C	0	0	∞	1	0
D	7	2	0	3	∞
E	13	1	0	∞	1
					4

We then added to the tree graph vertex (F, D) with a limit equal to 94 + 5 = 99:

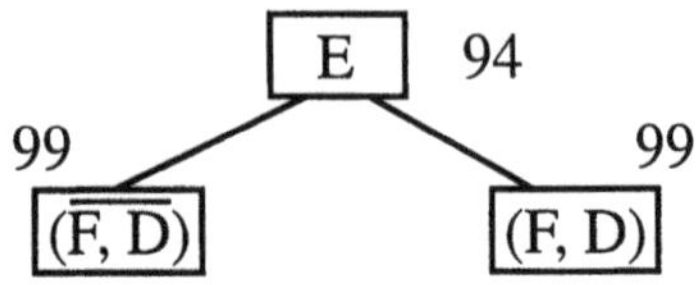

In the new matrix we placed squares within the boxes with zeros. These were filled in with the sum of the least values of the row and column of which they were members. We chose one of the vertices with the highest value. Specifically element (D, C):

↗	A	B	C	E	F
A	∞	2	0 [2]	14	6
B	2	∞	0 [0]	0 [1]	4
C	0 [2]	0 [1]	∞	1	0 [1]
D	7	2	0 [2]	3	∞
E	13	1	0 [1]	∞	1

We continued to build the tree graph, always starting out from one of the lowest limits of the hanging vertices. We decided on vertex (F, D) to which we assigned a limit equal to 99 + 2 = 101:

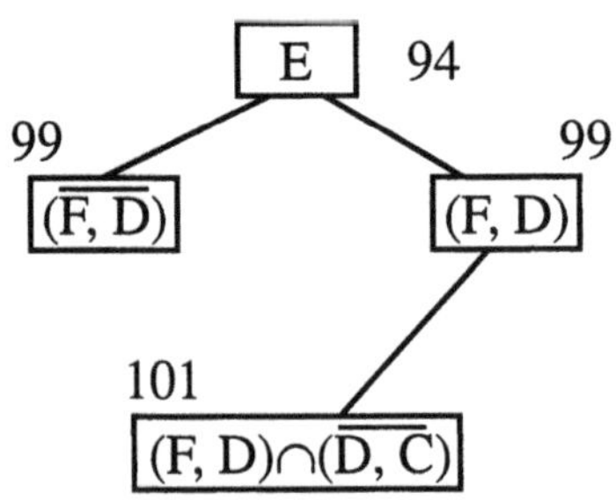

As we did before, we visualised the new route (D, C), Paris-Grenoble, in the proposed graph:

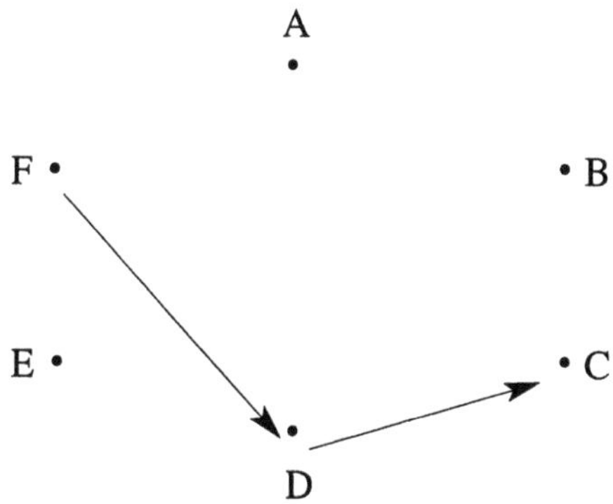

In the previous matrix we eliminated row D and column C, placing the symbol ∞ in box (C, F), for the reasons already stated and known of avoiding closing the circuit prematurely:

→	A	B	E	F
A	∞	2	14	6
B	2	∞	0	4
C	0	0	1	∞
E	13	1	∞	1

The corresponding deductions were made so that there remained at least one zero in each row and each column:

→	A	B	E	F	
A	∞	0	12	4	2
B	2	∞	0	4	
C	0	0	1	∞	
E	12	0	∞	0	1

We added a new vertex in the tree graph (F, D) ∩ (D, C) with a valuation of 99 + 3 = 102:

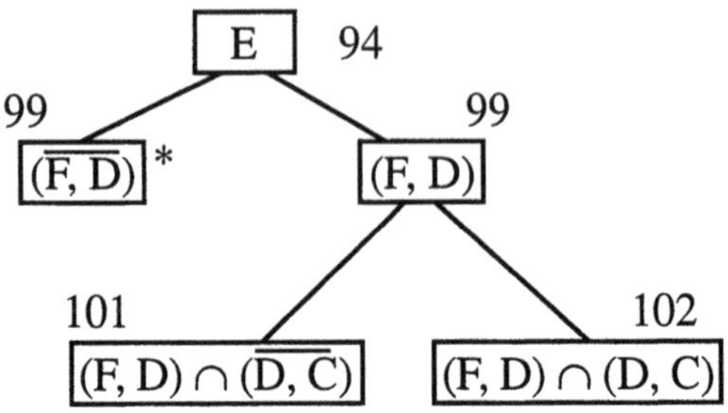

The lower of the limits of the vertices that hung corresponded to $\overline{(F, D)}$ with 99. From it the tree graph should continue. But as a consequence of it being a vertex that expressed that the route should not be done, to the matrix that we had obtained initially with zeros in all the rows and columns, should be added the symbol ∞ in box (F, D):

↗	A	B	C	D	E	F
A	∞	2	0	5	14	10
B	2	∞	0	4	0	8
C	0	0	∞	0	1	4
D	8	3	1	∞	4	0
E	13	1	0	4	∞	5
F	13	7	5	∞	6	∞

We now had to deduct from the row and column corresponding to (F, D) the lower of the respective values. This resulted in:

↗	A	B	C	D	E	F	
A	∞	2	0	5	14	10	
B	2	∞	0	4	0	8	
C	0	0	∞	0	1	4	
D	8	3	1	∞	4	0	
E	13	1	0	4	∞	5	
F	8	2	0	∞	1	∞	5

We continued with the algorithm normally, placing squares in the boxes with ceros:

↗	A	B	C	D	E	F
A	∞	2	0 [2]	5	14	10
B	2	∞	0 [0]	4	0 [1]	8
C	0 [2]	0 [1]	∞	0 [4]	1	4
D	8	3	1	∞	4	0 [5]
E	13	1	0 [1]	4	∞	5
F	8	2	0 [1]	∞	1	∞

The limit of the new vertex $(\overline{F, D}) \cap (\overline{D, F})$ was 99 + 5 = 104:

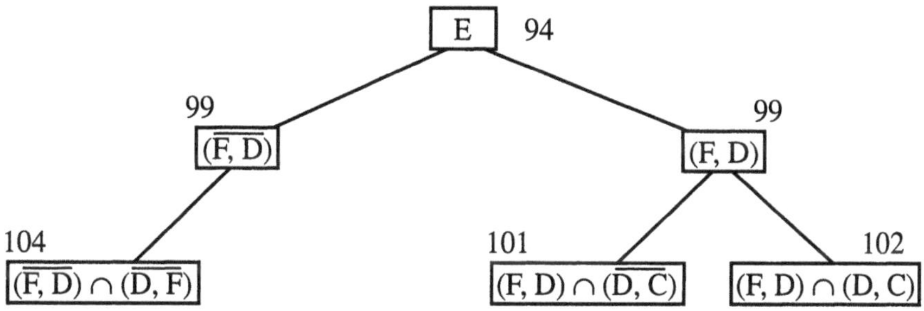

We eliminate row D and column F from the previous matrix:

↗	A	B	C	D	E
A	∞	2	0	5	14
B	2	∞	0	4	0
C	0	0	∞	0	1
E	13	1	0	4	∞
F	8	2	0	∞	1

As there were zeros in all the rows and columns, the limit of the new vertex $(\overline{F, D}) \cap (D, F)$ was equal to the vertex that preceded it, that is 99:

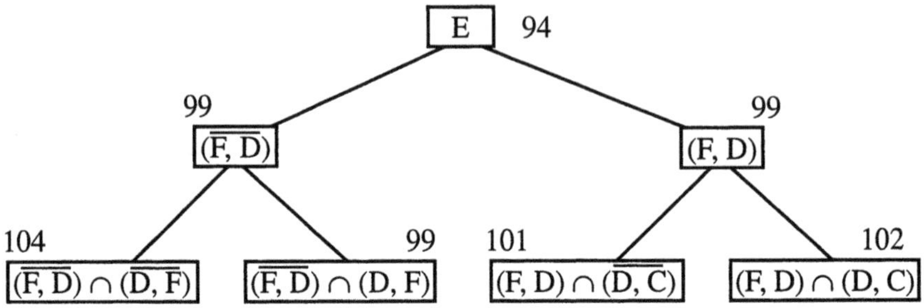

It was then necessary to redraw the graph from the beginning changing the direction of the initial arc:

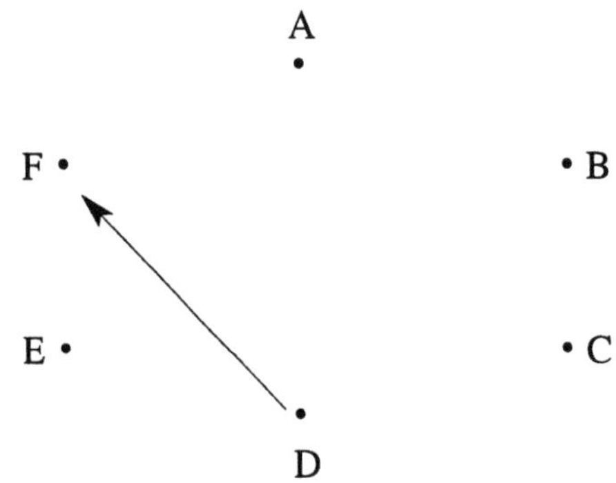

Once again we filled the squares in the boxes with zeros:

→	A	B	C	D	E
A	∞	2	0 [2]	5	14
B	2	∞	0 [0]	4	0 [1]
C	0 [2]	0 [1]	∞	0 [4]	1
E	13	1	0 [1]	4	∞
F	8	2	0 [1]	∞	1

The lowest limit of the hanging vertices corresponded to vertex $(\overline{F, D}) \cap (D, F)$ with a value of 99, From here the new bi-partition should commence:

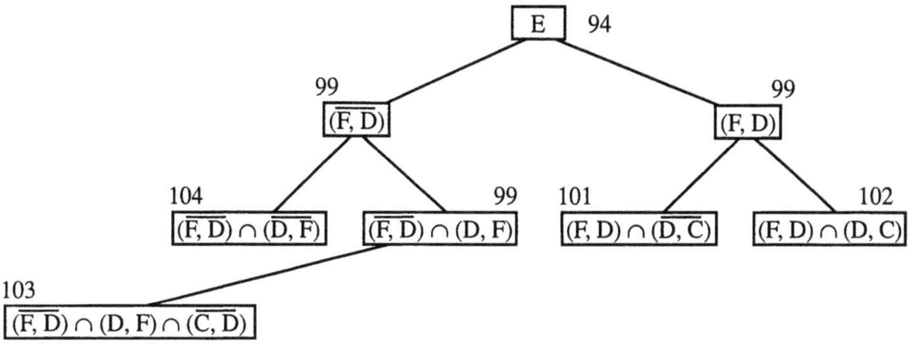

We continue constructing the graph:

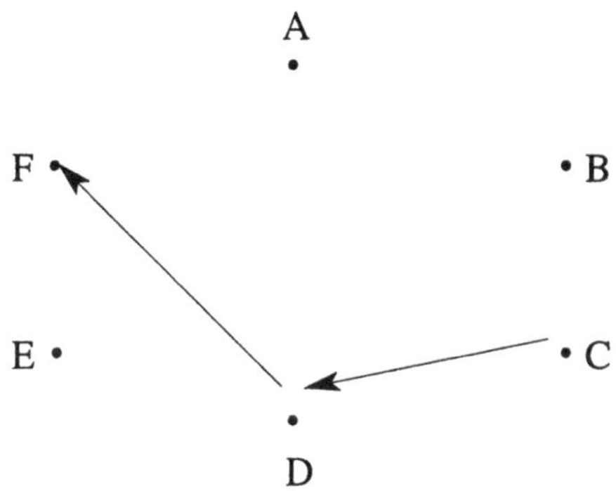

We eliminate row C and column D from the previous matrix placing a symbol ∞ in box (F, C) with the object of avoiding a non desirable circuit, (D, F, C):

→	A	B	C	E
A	∞	2	0	14
B	2	∞	0	0
E	13	1	0	∞
F	8	2	∞	1

We do the necessary adjustments in the matrix so that at least one zero exists in each row and column:

↗	A	B	C	E	
A	∞	2	0	14	
B	2	∞	0	0	
E	13	1	0	∞	
F	7	1	∞	0	1

↗	A	B	C	E
A	∞	1	0	14
B	0	∞	0	0
E	11	0	0	∞
F	5	0	∞	0
	2	1		

We add a vertex to the tree graph with a limit equal to 99 + 4 = 103:

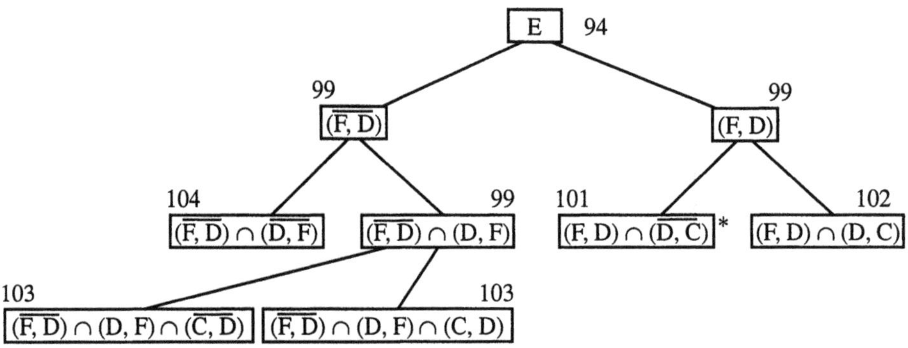

Between all the hanging vertices the one with the lowest level was $(F, D) \cap \overline{(D, C)}$ with a value equal to 101. We had therefore to go back in order to find the corresponding matrix in which to place a ∞ in box (D, C) as this route was negative:

↗	A	B	C	E	F
A	∞	2	0	14	6
B	2	∞	0	0	4
C	0	0	∞	1	0
D	7	2	∞	3	∞
E	13	1	0	∞	1

On the other hand we had to abandon the new graph illustrating the routes in order to continue with the one we had begun before:

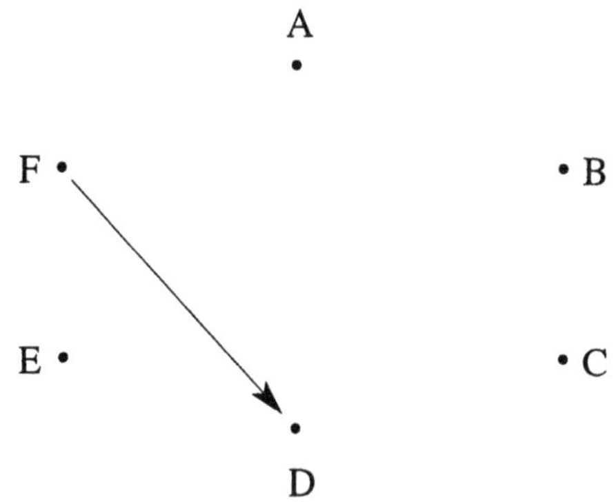

We then deducted the lowest values from the row and column corresponding to box (D, C) in which a symbol ∞. had been placed. We then arrived at:

↗	A	B	C	E	F	
A	∞	2	0	14	6	
B	2	∞	0	0	4	
C	0	0	∞	1	0	
D	5	0	∞	1	∞	2
E	13	1	0	∞	1	

We then immediately placed the corresponding numbers in the squares of the boxes with zeros:

↗	A	B	C	D	E
A	∞	2	0 [2]	14	6
B	2	∞	0 [0]	0	4
C	0 [2]	0 [0]	∞	1	0 [1]
E	5	0 [1]	∞	1	∞
F	13	1	0 [1]	∞	1

We chose anyone of the boxes from the ones with the highest values in squares. On this occasion the choice fell to (A, C). The limit if the new vertex was 101 + 2 = 103:

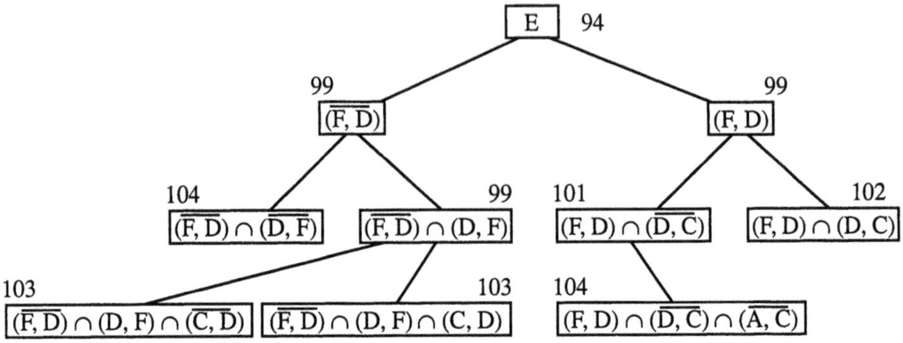

The new arc (A, C) was added to the graph:

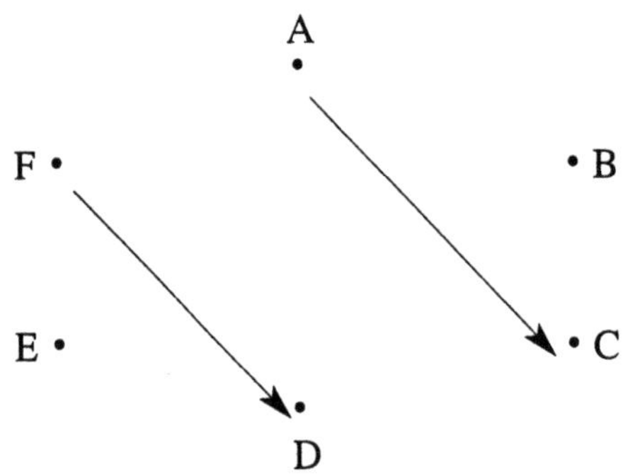

We then proceeded to eliminate row A and column C from the previous matrix and placed a symbol ∞ in box (C, A):

↗	A	B	E	F
B	2	∞	0	4
C	∞	0	1	0
E	5	0	1	∞
F	13	1	∞	1

We deducted the necessary amounts from the elements of this matrix so that there remained zeros in all rows and columns:

→	A	B	E	F	
B	2	∞	0	4	
C	∞	0	1	0	
E	5	0	1	∞	
F	12	0	∞	0	1

→	A	B	E	F
B	0	∞	0	4
C	∞	0	1	0
E	3	0	1	∞
F	10	0	∞	0
	2			

The new vertex would , in this way have an initial limit equal to 101 + 3 = 104:

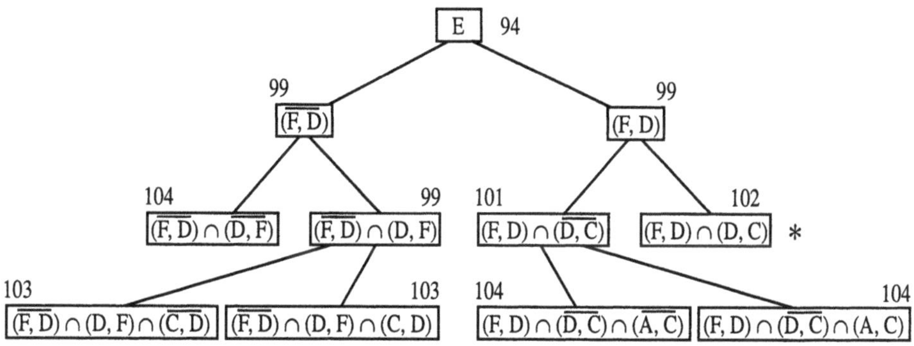

The lowest limit of the hanging vertices corresponded to vertex (F, D) ∩ (D, C) with a value of 102. Therefore we had to continue the tree graph through this vertex. Once again we had to resort to a previous matrix as follows:

→	A	B	E	F
A	∞	0	12	4
B	2	∞	0	4
C	0	0	1	∞
E	12	0	∞	0

With the corresponding graph:

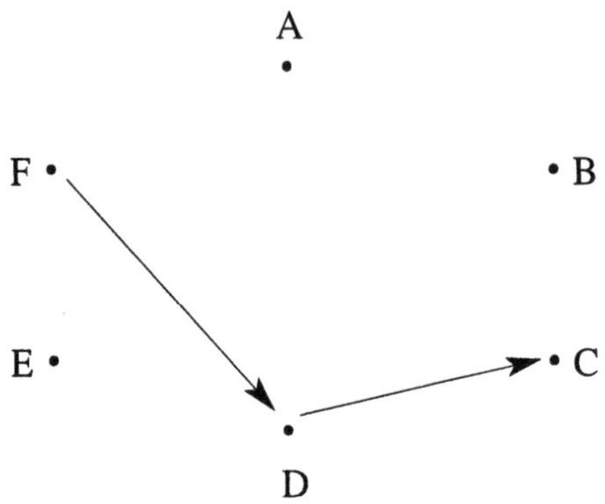

We then did the calculation leading to filling in the squares of the boxes with zero:

↗	A	B	E	F
A	∞	0 [4]	12	4
B	2	∞	0 [3]	4
C	0 [2]	0 [0]	1	∞
E	12	0 [0]	∞	0 [4]

We arbitrarily selected box (E, F) for the bi-partition, one of the two with the highest value in the square. The route would be from E, Frankfurt, to F, Brussels:

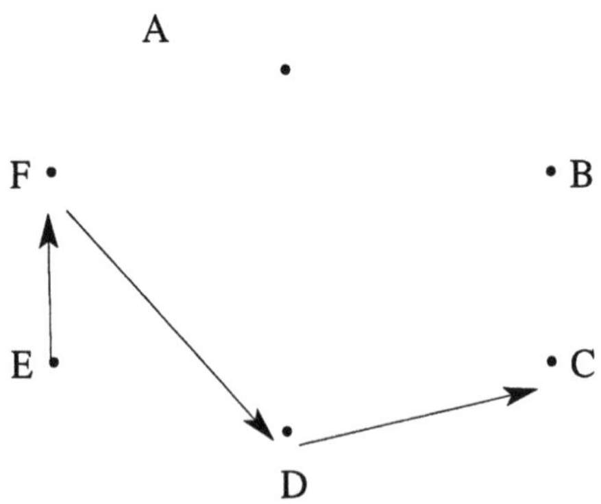

We continued the tree graph from vertex (F, D) $\cap$ (D, C). The new vertex would be assigned a limit if 102 + 4 = 106:

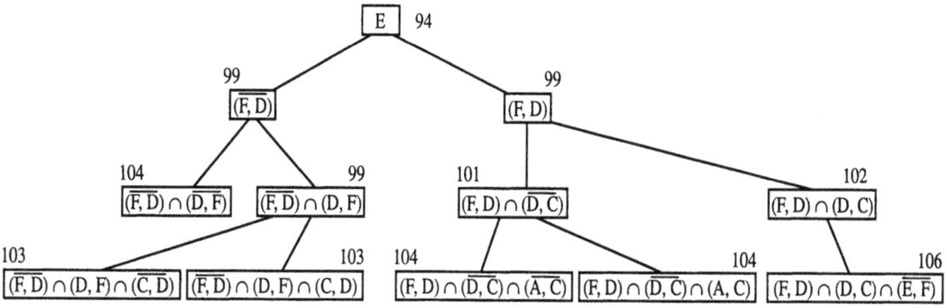

Row E and column F were eliminated from the matrix placing a symbol ∞ in box (C, E) in order to avoid it continuing in the circuit prematurely. We arrived at:

↗	A	B	E
A	∞	0	12
B	2	∞	0
C	0	0	∞

The new vertex would have a limit equal to the preceding vertex that is, 102 + 0 = 102:

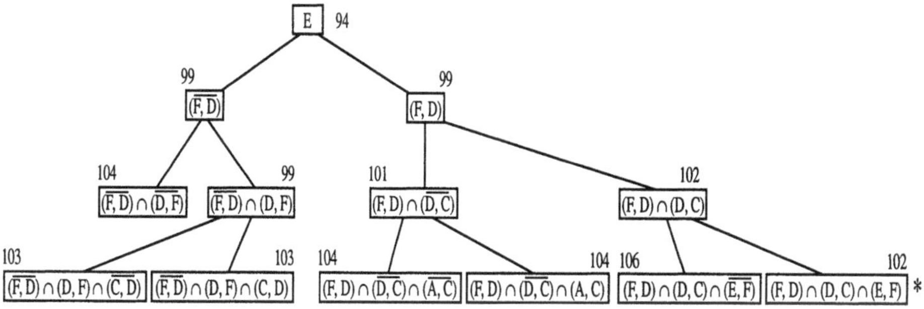

We then added a new arc to the graph:

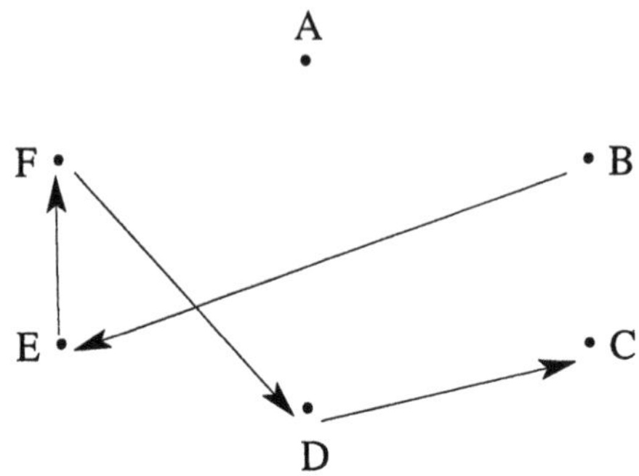

We could see that there already existed a zero in each row and each column. For this reason we moved on to assign values to the squares in the elements with zero:

→	A	B	E
A	∞	0 [12]	12
B	2	∞	0 [14]
C	0 [2]	0 [0]	∞

We obviously then chose element (B, E) in order to do the new partition, through vertex (F, D) ∩ (D, C) ∩ (E, F) with the lowest limit.

The limit of the new vertex would be 102 + 14 = 116:

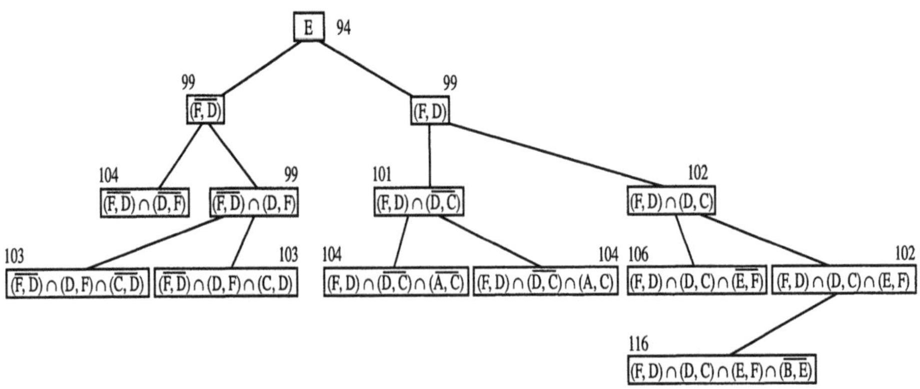

We eliminated row B and column E, placing symbol ∞ in box (C, B) for the reasons we have mentioned previously. We then arrived at the following matrix:

↗	A	B
A	∞	0
C	0	∞

It was quite clear that there was a zero in each row and each column. For this reason the new vertex would have the same value as the preceding vertex. We would then have:

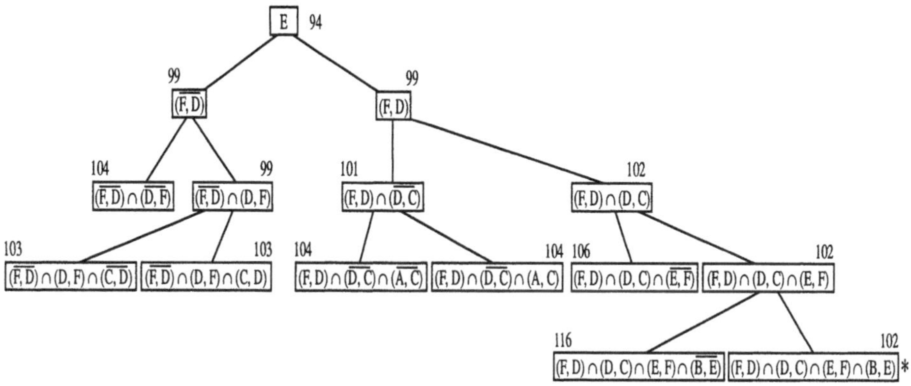

We assigned values to the squares with zeros:

↗	A	B
A	∞	0 ∞
C	0 ∞	∞

We selected one of the elements of the matrix with value zero, since both of them possessed the same value ∞, in the square. In any event in the next stage of the algorithm the value relegated at that time would appear. We took (C, A) as the vertex for incorporating into the graph:

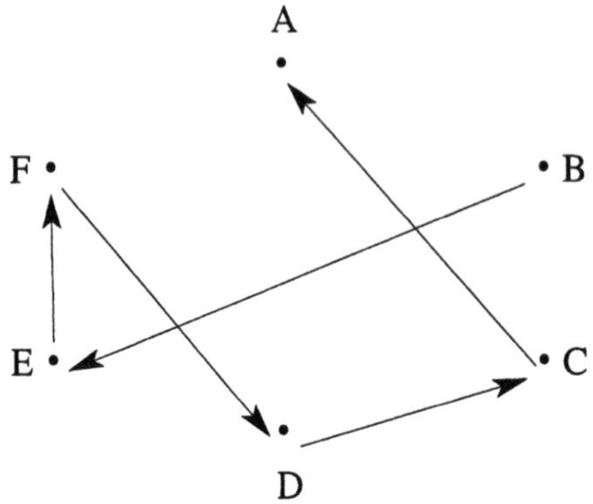

To the tree graph we would add the vertex corresponding to the limit of

$102 + \infty = \infty$:

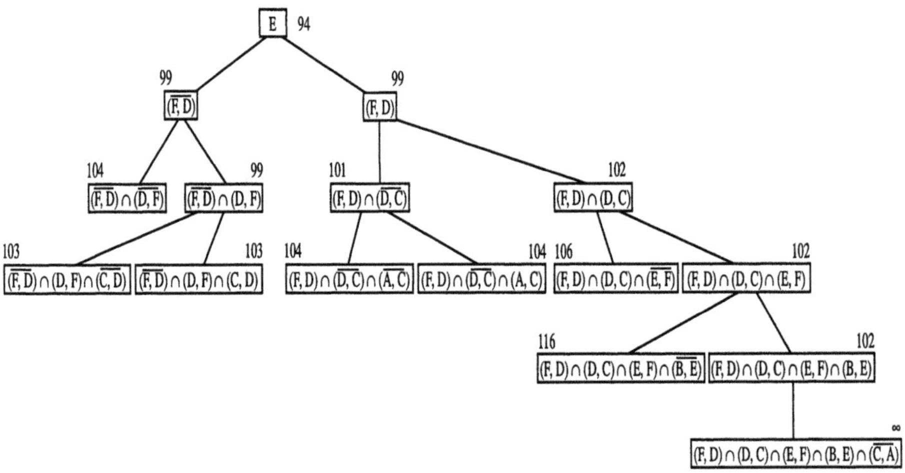

On eliminating row C and column A from the matrix the following matrix of the order of 1 x 1 remains, in which obviously there exists a zero in its row and column:

↗	B
A	0

The tree graph we now show is the final one, since we incorporate the last vertices, from the one possessing the lower limit, that is (F, D) ∩ (D, C) ∩ (E, F) ∩ (B, E) ∩ (C, A):

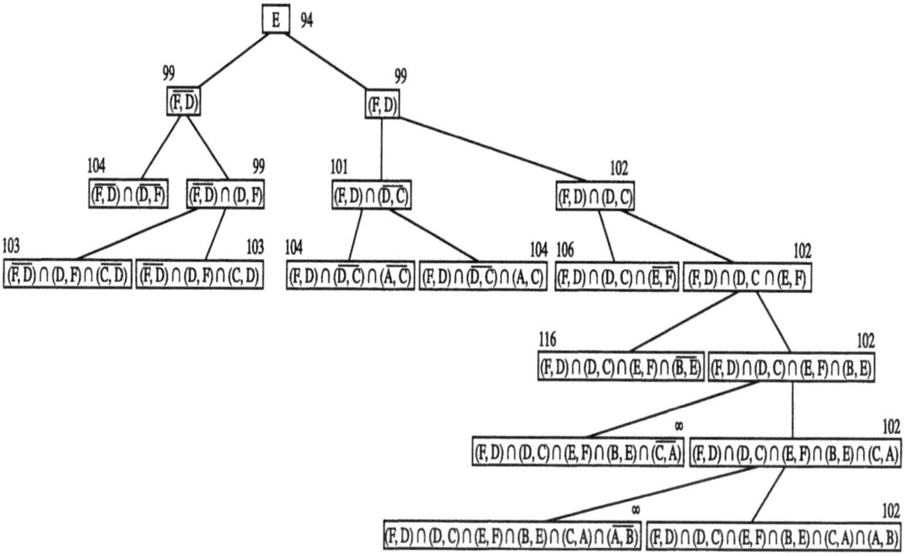

Finally we added to the previous graph the two arcs found, in this way showing the complete optimum circuit:

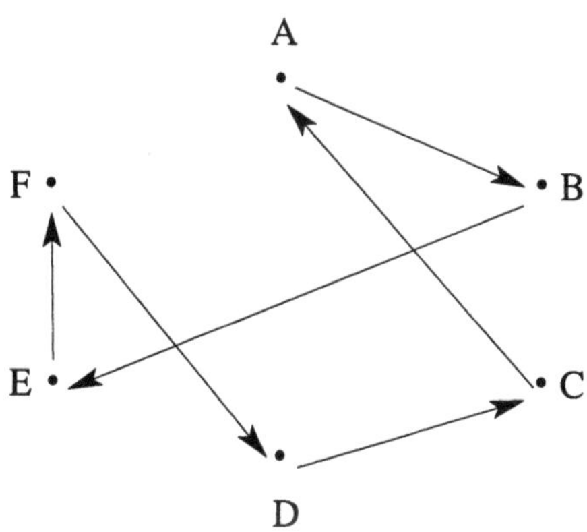

Therefore the route to be followed with the minimum cost was formed by the following stages:

Reus → Rome → Frankfurt → Brussels → Paris → Grenoble → Reus

And the known cost of the 102 monetary units was broken down as follows:

(A, B) : Reus → Rome	30	monetary units
(B, E) : Rome → Frankfurt	25	" "
(E, F) : Frankfurt → Brussels	17	" "
(F, D) : Brussels → Paris	7	" "
(D, C) : Paris → Grenoble	7	" "
(C, A) : Grenoble → Reus	16	" "
	Total 102	monetary units

All of those involved in this subject expressed their favourable opinion on the proposed circuit, not yet used, although as Jesús rightly stated, the reverse circuit would have a slightly higher cost. In fact assuming certain optimum quantity (as was the case according to the adopted data) the reverse circuit, commencing with the stage Reus → Grenoble and ending with Rome → Reus cost 121 monetary units. And this was so because the same cost between two cities did not exist when the direction was not the same.

Now then, all we had done up to then, meant that the cost to be borne would be the *minimum possible*. But what would happen if the cost were to be the *maximum admitted?* In order to answer this question it was essential to repeat the same process followed with the aid of the same algorithm, but starting out from the other matrix which in short, is the one we have reproduced below:

	A	B	C	D	E	F
A	∞	35	17	25	36	28
B	38	∞	25	28	29	32
C	19	22	∞	8	14	14
D	30	24	10	∞	19	8
E	38	28	14	20	∞	21
F	35	30	15	10	22	∞

We are not going to develop the same calculations we did before. We will limit ourselves to showing the results with the corresponding comments as complements.

The tree graph we found was the one shown below. In it we have put an asterisk at the vertices from which we made the bi-partition:

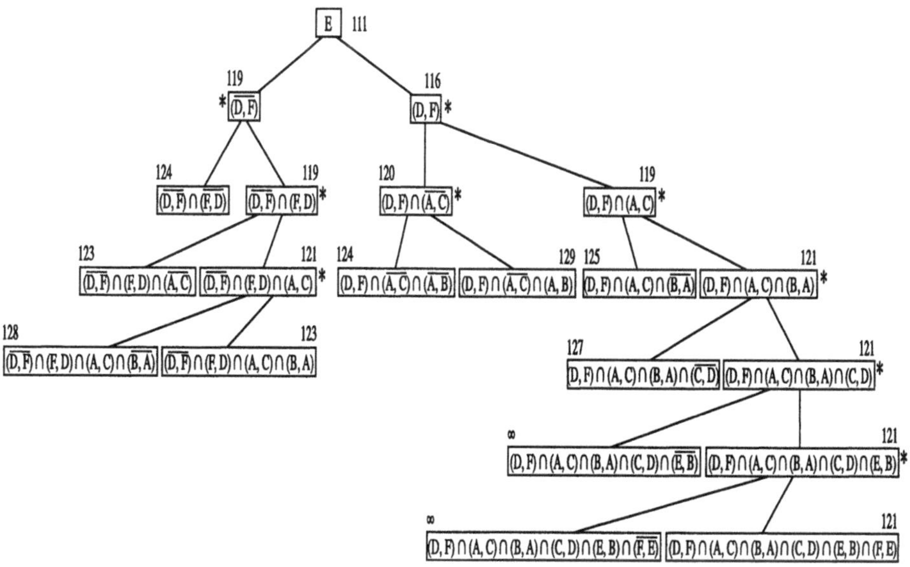

The following was the resulting optimum circuit in the event of bearing the higher costs:

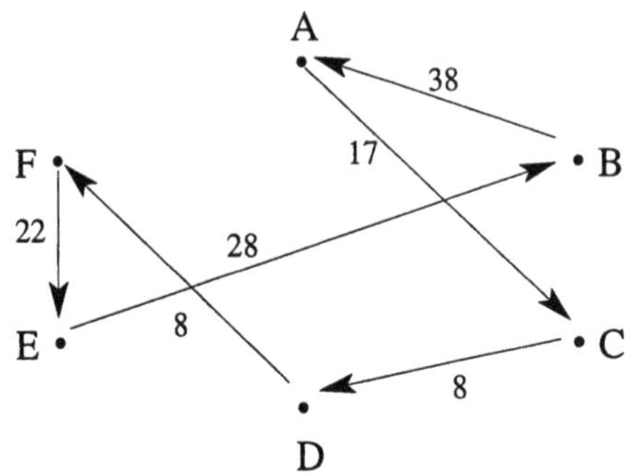

The sum of the costs per stage that formed the optimum circuit, supplied us, obviously, with the limit of the last vertex, that is 121 monetary units. In fact:

(A, C) : Reus → Grenoble	17	monetary units
(C, D) : Grenoble → Paris	8	" "
(D, F) : Paris → Brussels	8	" "
(F, E) : Brussels → Frankfurt	22	" "
(E, B) : Frankfurt → Rome	28	" "
(B, A) : Rome → Reus	38	" "
Total	121	monetary units

It was at least curious that the change of values on passing over from one extreme to the other provided in one case a graph and in the other case its opposite graph. To put this another way, when the costs were minimum it should be arranged in one direction and when the costs were maximum in the reverse direction. This was a special case. What is normal is that the graphs should be different, as a consequence of the graphs not being symmetrical for either of the two extremes (there was not the same cost for the outward trip as for the return, from one city to the next).

Jesús, with the concern that characterised him, took pencil and paper in order to make a few calculations which he then immediately abandoned on the table: with them he attempted to see what would have happened in the event of doing the same route by the upper extremes of the intervals as for the lower extremes. The result was conclusive:

Reus → Rome → Frankfurt → Brussels → Paris → Grenoble → Reus

35 + 29 + 21 + 10 + 10 + 19 = 124

There was no doubt whatsoever, that the figures we had arrived at would allow for following an optimum route. Theoretically another could exist which would give the same cost, but never a lower one.

From a general perspective, the treatment we had done had led us to a certain optimisation, limiting the uncertainty of the costs. In fact, we knew that these would never be less than 102 monetary units nor higher than 121 monetary units. Technically the result was expressed by means of the confidence interval [102, 121].

This was useful information, although it could never be accepted as complete and final. In fact, the costs could be located *between the two extremes* and when this occurred, which of the two circuits should be chosen?. All present guessed that it would be one of the two, or even perhaps neither of the two, but a further new one. All was dependent upon *the data* of the starting out matrix, that is to say from the costs of each possible route. But if this information were to be available, then the use of the algorithm would *simultaneously* provide the optimum circuit and its cost, which, incidentally, would be found *within* the confidence interval that was found (also considering the extremes). The great advantage of this algorithm is that it is easily programmable for computers, thanks to its simple automatisms. From this point of view it would be sufficient to introduce as data the elements of the matrix in order to immediately find the optimum circuit and its cost.

Those of you who have read these pages may well ask if it was worth doing all these calculation, certainly very long, for such a short circuit as the one we have described comprising six cities. It should be taken into account, nevertheless, that the process is valid for any finite number of the same. And, also, when the algorithm is computerised, it is easy to change cities, adding some and removing others. What we have then is an instrument of assistance for decision that is highly flexible and adaptable.

These short works had allowed us in conjunction with the famous ceramist Neus, to follow a pleasant path, along which her actual requirements were explained and treated by means of formal schemes, which up to not so long ago were reserved to the reduced circles of scientists. We have always felt that in the economic and management spheres the researcher should attempt that in any findings should always include the germ of its speedy application. Only in this way will we achieve the objective of our work not being useless and that the field of our contributions is not barren.

Neus continues her creative work, reaping many successes and opening windows through which art lovers all over the world can view the chromaticisms of multiform objects that only a privileged imagination is capable of conceiving. I am aware of the artistic exhibitions that she is now holding on different continents. What in the distance I am unaware of is if these modest tools have continued to be useful for her, if she continued to find support in them. I did not dare to ask her at our last meeting. If some day I find out, I can assure you that our readers will be the first to know.

REFERENCES

1 KAUFMANN, A., GIL ALUJA, J. and GIL LAFUENTE, A.M.: *La creatividad en la gestión de las empresas.* Publ. Pirámide, Madrid 1994 , pages. 18-19.

2 The notion of frequency in the alternative is due to Dr. Jaques Pezé.

3 GIL LAFUENTE. J.: *Marketing para el nuevo milenio.* Publ. Pirámide. Madrid 1997, pages. 349-352.

4 PICHAT, E.: *Algorithm for finding the maximal elements of a finite universal algebra.* Inform. Processing 68. Publ. North Holand 1969.

5 GIL ALUJA, J.: *Elements for a theory of decisión in uncertainty.* Kluwer Academic Publishers. Boston, London, Dordrecht. 1999, pages. 242.

6 GIL ALUJA, J.: *Interactive Management of the human ressources in uncertainty.* Kluwer Academic Publishers. Boston, London, Dordrecht, 1999, page. 18.

7 LITTLE, J.D.G. and others: *An algorithm for the travelling salesman problem I.O.R.S.A.* Vol.II nº 6, 1963, pages. 972 - 989.

8 KAUFMANN, A. and GIL-ALUJA, J.: *Técnicas operativas de gestión para el tratamiento de la incertidumbre.* Publ. Hispano-Europea. Barcelona 1987, pages. 168-170.

IX THE IMAGE OF A BUSINESS

Elements conforming the image

The outcome of events has substantially changed the surroundings in which organised groups in general and businesses and institutions, in particular, carry out their activity. We have repeated ad nauseam that the society in which it has been our good fortune to live is characterised by very rapid changes that affect, not only material objects but also mental objects, that is to say, ideas. Nevertheless it is very true to say that this evolutionary process, and even perhaps revolutionary, is not taking place without a certain amount of trauma, which can be detected more in the world of *ideas* than in the world of *things*. The human being, the last link, they say, in the evolutionary chain, is still "an animal of custom". And to break these customs requires a great effort which routine is reluctant to accept. Mental torpor is the *great sin* of our so-called civilisation.

Be this as it may, and with a different intensity according to the circles, in which the different groups that roam this world move, change ends up imposing itself. This change has affected aspects the incidence of which on social life can be considered as insignificant. But on the other hand it has affected others that are of vital importance for the coexistence of people. This is how a radical subversion has taken place of those "values" accepted as unchanging for many centuries. Where *patience, perseverance, sacrifice, work, family,...* all written in letters of gold are hidden within the thoughts of so many generations. Today there is a cult to the concepts such as *aggressiveness, success, the competitive spirit,* ..., and enthroned on the highest of altars *image*. We have entered into the *realm of the image* and any person or institution aspiring to occupy a place in the structure of our society must care for, spoil and strengthen the image. When new ideas are scarce, when the creative spirit becomes poor, when imagination gives way to automatisms, then vulgarity insinuates itself into souls and bodies, causing an alienating uniformity. In the absence of a differentiation of points of view, of a diversity of positions, of a variety of concepts, a plurality of approaches to life, refuge is sought in what is superfluous. Instead of delving into the depths of the oceans of thought, we are content to placidly contemplate the sterile foam of the ephemeral rolling of the waves.

Yet, nevertheless, it is possible to combine the profound with the superficial, the content and the form, *the contents and the image*. What is more, we feel it is necessary, not only for whom, person, business or institution, is the actor in the theatre of life, but also for those who, in some manner, are spectators, potential receivers of other people's activities. How many times have we seen shining in the firmament of culture, institutions the only merit of which has been knowing how to sell their teachings, empty of all scientific and technical content and filled with a dense cloud of smoke, whilst others go unnoticed because they dedicate their efforts to profound works and meritorious investigation. We have always rebelled against the ghosts that disappear at the very first breath of wind. Our poor efforts have not been sufficient for our shouts and complaints to make, those reduced groups of power that are capable of imposing, due to their position and strength, although not always their reason, aware of the "truth" of their interests. The fact of having protested for so many years only to receive meaningless replies, has not, however, reduced our enthusiasm, or our fighting spirit. We are proof against discouragement, nothing nor nobody will be able to prevent us, in our modest dwelling, from continuing working so that *the light* of science illuminates the spirits of those who want to contribute to the construction of a better world.

Over twenty years ago, in conjunction with Professor Kaufmann, we initiated certain studies that had the ambitious objective of *formalising* certain relations of incidence, or if you like, causality, in order to arrive at an exhaustive list, and therefore without error or omission, of the indirect effects, that is the second generation, which can occur when a set of causes give rise to a set of effects, some of these being related to each other. This was a complex problem, above all if the fact is taken into account that the different causality relations have *different levels*. It is relatively easy to present the levels of direct relations among the set of causes and the set of effects, but to find *all* the levels of the indirect relations was another kettle of fish altogether. The result of that research first came to light in a chapter of a joint book [1], and later in a book that became the support for many works from other universities and which gave rise to the so-called *theory of forgotten effects*[2].

Fifteen years further on, it is easy to find interesting applications of this formal scheme to many spheres of knowledge. Even in the spheres of economy and business management there are several authors who have become interested in the use of these models for seeking, and we believe finding, good solutions to problems that they had approached. In fields as wide apart as finance and commerce, for example, it is not difficult to find good examples of these[3]. In a recurring manner, we have seen the appearance of studies in which what was being sought after was the strengthening of the image of a politician, of a business or institution, by means of *the updating* in some aspects that constituted *the causes*, which had an *incidence* on a set of elements, *effects*, that went to make up *the image* of the person, business or institution. In all of these, without any exception whatsoever, the starting out point was a *limited* set of causes and a *limited* set of effects. And this was so as a consequence of the fact that each added cause or effect, multiplied the number of possible second generation relations, due to the complex interconnections existing between the possible causes, between the effects themselves and between causes and effects.

On the other hand, from a practical point of view, it was necessary to take into account that every *action* on any of the elements of the cause set incurs a cost. Financial means available for dedicating to the strengthening of *an image* were, are and will always be, limited. For this reason, it was fundamental that a determination be made on which *aspect or aspects* of the set *of causes* were the available financial means to be applied in order to arrive at better result in the elements making up the image, of which the person, business or institution was lacking.

We shared these thoughts with the President of the *Reial Academia de Doctors*, my friend and companion Josep, when the idea arose of attempting to use the instruments created for *transmitting an image* of this Royal Corporation to the different levels of society, in accordance with current times. The *Reial Academia de Doctors*, in its current status is a young institution, by the roots of which go way back to medieval times, when the *Consejo de los Sabios* (Council of Wise Men) had a value as a support for important decisions. The *permanent hundred members*, elected from all areas of knowledge, constitute, due to their knowledge and proven experience, a select nucleus of researchers, teachers and professionals who, in a certain manner, represent the intellectuality of this small country which is Catalonia. But as is wont to happen in this type of entity, all those works that either individually or collectively are done within the heart of the entity, are not known and therefore, not valued, by the different strata of society which, on the other hand, is the recipients of the benefit derived from the same. It is true that some scientific academies have become stagnated, anchored in the already declining past, which has been the reason why they have been classified as "elephant cemeteries". But this was not, and is not so today, the case of the *Reial Academia de Doctors*. The magnificent management of our President, and the indubitable merit of our member academicians, has equipped the institution with a dynamism, which has made it worthy of the very deepest respect from the entire scientific and technological world.

It should not come as a surprise then that the President should be concerned, better said very concerned, to attain, by means of *a strengthening of image*, the recognition from citizens of the important tasks that had been carried out and were being carried out. My friend did not hide the fact from me that following on recognition there would also be an increase in financing, with which to carry out the grand projects with which he had always dreamed.

In the case, which concerned us, the development of the work should be something different from those described before, since our President requested, and he had the right to do so, to become aware of the technical bases that were to be used. His participation would be, apart from anything else, very enriching. Therefore we commenced the task by starting out from the basic conceptual elements, to immediately, express the theoretical and technical elements that would serve as a support for the chosen model.

In the first place what we had to do was to define the set of "causes" or elements capable of modifying the image of the *Reial Academia*. In a symbolic manner these were expressed as:

$$A = \{a_1, a_2, ..., a_n\}$$

We then immediately moved on to define another set of elements that made up the image of the Royal Corporation, that is, those aspects by means of which society perceived this institution. These were symbolised as follows:

$$B = \{b_1, b_2, ..., b_m\}$$

Once we had this very valuable information available we proceeded *to establish a degree or level* of incidence *of each cause* on each effect, by assigning values in [0, 1] as was normal for us to do. The greater the incidence, the nearer to one would be the assigned value. The lesser the incidence, then the value would be closer to zero. The way of representing these incidences could be in the form of a matrix. And this is how we did it. Apart from this, with the object of explaining it better to our companions, whose specific interests was removed from our own, we represented it by means of the following example:

	b_1	b_2	b_3	…	b_m
a_1	.7	.5	.9	…	.2
a_2	.4	.8	0	…	.6
a_3	.1	.3	1	…	.4
…	…	…	…	…	…
a_n	.9	.2	.7	…	0

One reading, even superficially, of this matrix explains that the incidence of cause a_1 over effect b_1, is quite high 0.7; of a_1 over b_2 is medium (neither high nor low); a_1 over b_3 is very, very high 0.9; and so on successively.

Specialists are perfectly aware that these relations of incidences can also be represented by means of an *arrow form graph*. For this we place in the first column (to the left), *the causes*, one below the other, and in a second column (to the right), *the effects*, also one beneath the other. We draw an arc from the cause to the effect when an incidence existed, however small, and we only refrain from drawing an arc when the incidence is nil. Above each arc we place a degree or level of incidence. In our example the following was the result:

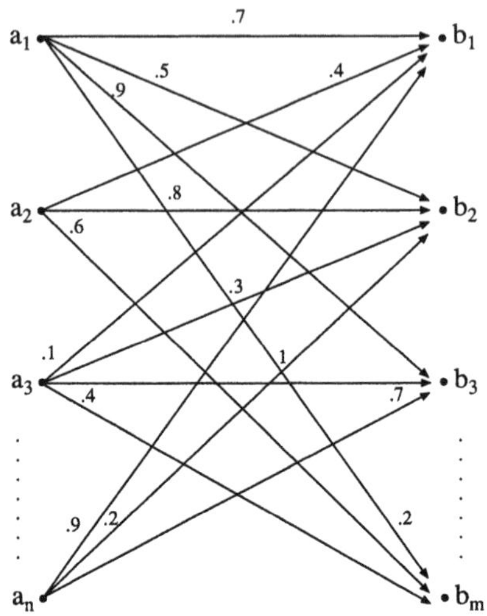

In this way we had covered the first phase that allowed us to formalise, that is to say theoretically "explain" the relations *of direct or first degree incidence*, by resorting to the two most normally used forms, the matrix form and the arrow form graph.

But as wisely suggested by our President, what actually acquired the greatest interest was to get to know the most possible overall *incidences*, or what amounted to the same, at least the *accumulated incidences* of first and second generation, It was common knowledge that, on many occasions the fact occurs that certain causes have an incidence on certain effects which, in turn, are causes of other effects. The question then was to what degree do *the primary causes* have an effect *on the last effects*. It was evident then that there was a need to expand the previous reasoning to the case of the existence of three sets: the first, A, would include *the primary causes*, the second, B, *the intermediary effects*, which in turn would be causes of a third, C, constituting *the final effects*. These we represented as follows:

$$A = \{a_1, a_2, a_3, ..., a_n\}$$

$$B = \{b_1, b_2, b_3, ..., b_m\}$$

$$C = \{c_1, c_2, c_3, ..., c_s\}$$

The necessary initial information (data of the problem) would now be; the degree of incidence of elements from A on those from B and the degree of incidence of elements from B on those from C. This could be presented in the same way we had done before: in the matrix form or the arrow form graph. We illustrate this with the following example, in which we have limited the elements of the three sets in order for greater clarity:

	b_1	b_2	b_3	b_4
a_1	.3	.7	0	.6
a_2	.8	.2	.1	0
a_3	0	.4	.5	.9

	c_1	c_2
b_1	.3	.8
b_2	.7	0
b_3	.5	.6
b_4	0	.9

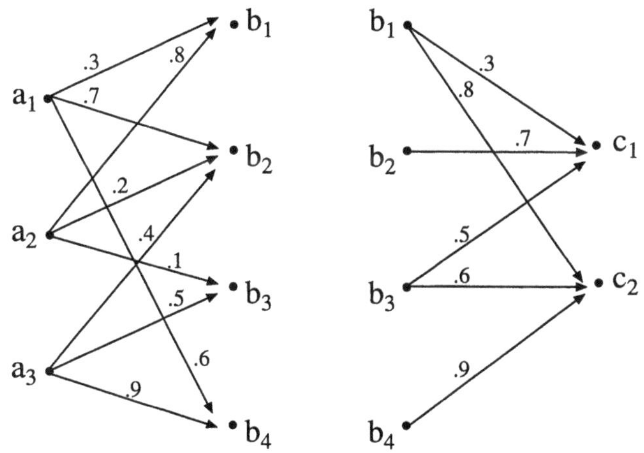

We continue the process from this second form (arrow form graph) in order later to pass over to the first (matrix form). For this we bring the two graphs "closer" until the *effects* of the first are superimposed on the *causes* of the second. The result was as follows:

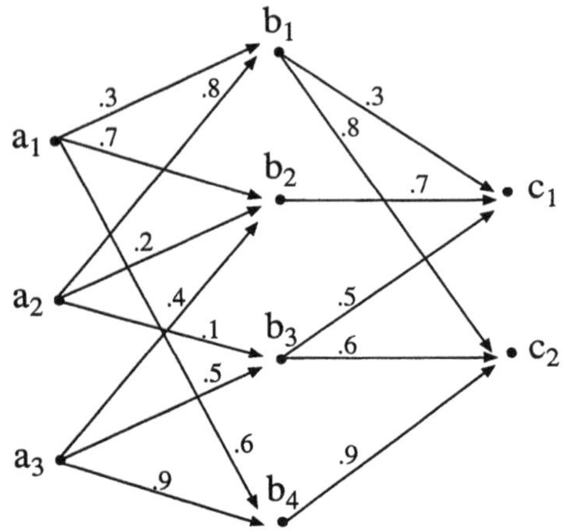

One simple glance of this graph would allow us to establish the *indirect* relations between the elements of set A and those of set C by the intermediation of those corresponding to set B. In this way we were able to determine the existence of an incidence from a_1 to c_1 by means of b_1 and b_2. This can be easily seen if we look at the part of the previous graph we have reproduced below:

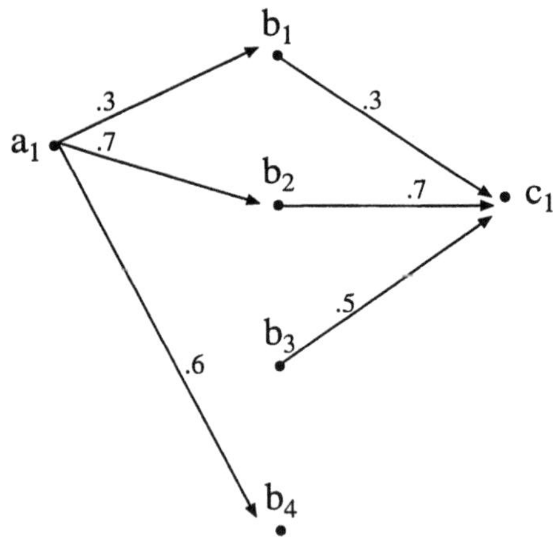

If, for teaching purposes, we consider that each arc is a channel along which passes a flow the volume of which is given by the noted valuation (level or degree of incidence), it is correct to think that the liquid that reached c_1, from a_1, by the intermediation of b_1, is given by the *minimum* of the flows *from a_1 to b_1 and from b_1 to c_1*. In our case there is no doubt whatsoever since $0.3 \wedge 0.3 = 0.3$. Neither is there any doubt when the intermediary is b_2, as $0.7 \wedge 0.7 = 0.7$. No flow arrives by means of intermediaries b_3 and b_4 as these are respectively $0 \wedge 0.5 = 0$ and $0.6 \wedge 0 = 0$. Of the four channels we choose the one that provides a *maximum* flow, that is $0.3 \vee 0.7 \vee 0. \vee 0 = 0.7$. We have in this way used an operator knows as "maxmin convolution" or "maxmin composition". The *maximum* of the *minimums* is taken, Jurists would say, in this case, to act with "the prudence of a good father" and mathematicians would write:

$$(0.3 \wedge 0.3) \vee (0.7 \wedge 0.7) \vee (0 \wedge 0.5) \vee (0.6 \wedge 0) = 0.7$$

We now show the indirect incidence of a_1 on c_2. We arrive at the following subgraph:

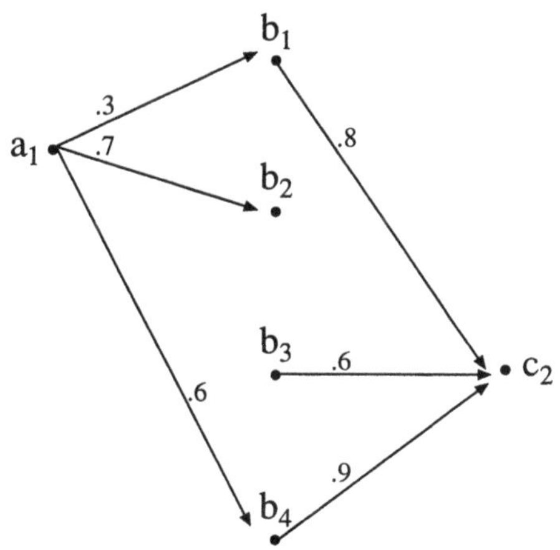

The flows that can reach c_2, from a_1 along the different channels are:

$$0.3 \wedge 0.8 = 0.3$$

$$0.7 \wedge 0 = 0$$

$$0 \wedge 0.6 = 0$$

$$0.6 \wedge 0.9 = 0.6$$

The maximum of these minimums would be 0.6. Therefore the selection would be along those channels with the intermediation of b_4.

Continuing with this process from each element of A to each element of C, we find the following graph:

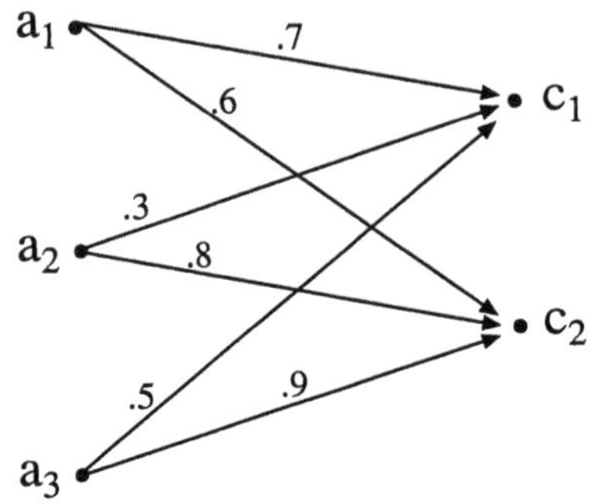

In this way we arrived at all the indirect incidences of the elements of set A on the elements of set C, by the intermediation of set B.

Now then, this "visual" procedure is very limiting, since it is only possible to use when the elements posses a few elements. For this reason it is recommendable to use the *matrix form* in order to represent relations of incidence and arrive at the maxmin convolution" of the matrices. In this case we write:

	b_1	b_2	b_3	b_4
a_1	.3	.7	0	.6
a_2	.8	.2	.1	0
a_3	0	.4	.5	.9

$\circ$

	c_1	c_2
b_1	.3	.8
b_2	.7	0
b_3	.5	.6
b_4	0	.9

where the symbol $\circ$ represents the operator of maxmin composition or convolution. The result is a matrix in the rows of which are the elements of set $A = \{a_1, a_2, a_3\}$ and in the columns those from set $C = \{c_1, c_2\}$. The operating process is as follows

Element (a_1, c_1): take row a_1 of the first matrix and column c_1 of the second matrix.

$$0.3 \wedge 0.3 = 0.3$$

$$0.7 \wedge 0.7 = 0.7$$

$$0 \wedge 0.5 = 0$$

$$0.6 \wedge 0 = 0$$

$$(0.3 \wedge 0.3) \vee (0.7 \wedge 0.7) \vee (0 \wedge 0.5) \vee (0.6 \wedge 0) = 0.7$$

Element (a_1, c_2): take row a_1 of the first matrix and column c_2 of the second matrix.

$$0.3 \wedge 0.8 = 0.3$$

$$0.7 \wedge 0 = 0$$

$$0 \wedge 0.6 = 0$$

$$0.6 \wedge 0.9 = 0.6$$

$$(0.3 \wedge 0.8) \vee (0.7 \wedge 0) \vee (0 \wedge 0.6) \vee (0.6 \wedge 0.9) = 0.6$$

Element (a_2, c_1): take row a_2 of the first matrix and column c_1 of the second matrix.

$$0.8 \wedge 0.3 = 0.3$$

$$0.2 \wedge 0.7 = 0.2$$

$$0.1 \wedge 0.5 = 0.1$$

$$0 \wedge 0 = 0$$

$$(0.8 \wedge 0.3) \vee (0.2 \wedge 0.7) \vee (0.1 \wedge 0.5) \vee (0 \wedge 0) = 0.3$$

Element (a_2, c_2): take row a_2 of the first matrix and column c_2 of the second matrix.

$$0.8 \wedge 0.8 = 0.8$$

$$0.2 \wedge 0 = 0$$

$$0.1 \wedge 0.6 = 0.1$$

$$0 \wedge 0.9 = 0$$

$$(0.8 \wedge 0.8) \vee (0.2 \wedge 0) \vee (0.1 \wedge 0.6) \vee (0 \wedge 0.9) = 0.8$$

Element (a_3, c_1): take row a_3 of the first matrix and column c_1 of the second matrix.

$$0 \wedge 0.3 = 0$$

$$0.4 \wedge 0.7 = 0.4$$

$$0.5 \wedge 0.5 = 0.5$$

$$0.9 \wedge 0 = 0$$

$$(0 \wedge 0.3) \vee (0.4 \wedge 0.7) \vee (0.5 \wedge 0.5) \vee (0.9 \wedge 0) = 0.5$$

Element (a_3, c_2): take row a_3 of the first matrix and column c_2 of the second matrix.

$$0 \wedge 0.8 = 0$$

$$0.4 \wedge 0 = 0$$

$$0.5 \wedge 0.6 = 0.5$$

$$0.9 \wedge 0.9 = 0.9$$

$$(0 \wedge 0.8) \vee (0.4 \wedge 0) \vee (0.5 \wedge 0.6) \vee (0.9 \wedge 0.9) = 0.9$$

In this way we completed the maxmin convolution, arriving at the following result:

$$
\begin{array}{c|c|c|c|c|}
 & b_1 & b_2 & b_3 & b_4 \\ \hline
a_1 & .3 & .7 & 0 & .6 \\ \hline
a_2 & .8 & .2 & .1 & 0 \\ \hline
a_3 & 0 & .4 & .5 & .9 \\ \hline
\end{array}
\circ
\begin{array}{c|c|c|}
 & c_1 & c_2 \\ \hline
b_1 & .3 & .8 \\ \hline
b_2 & .7 & 0 \\ \hline
b_3 & .5 & .6 \\ \hline
b_4 & 0 & .9 \\ \hline
\end{array}
=
\begin{array}{c|c|c|}
 & c_1 & c_2 \\ \hline
a_1 & .7 & .6 \\ \hline
a_2 & .3 & .8 \\ \hline
a_3 & .5 & .9 \\ \hline
\end{array}
$$

We have now detailed , as we did before, all the calculations, since we wanted to leave the mechanism of the maximin convolution as clear as possible, as this will be used very frequently. Now there were no doubts on which were the *degrees or levels* of incidence of a_1, a_2 a_3 on c_1, c_2 which were quite explicit in each one of the boxes of the resulting matrix. Likewise appearing in total clarity were the elements of $B = \{b_1. b_2, b_3, b_4\}$ which, in each case, acted as intermediaries. Therefore we had:

For	(a_1, c_1)	the intermediary was			b_2
"	(a_1, c_2)	"	"	"	b_4
"	(a_2, c_1)	"	"	"	b_1
"	(a_2, c_2)	"	"	"	b_1
"	(a_3, c_1)	"	"	"	b_3
"	(a_3, c_2)	"	"	"	b_4

The President of our Royal Corporation accepted this development of the process from the established premises as correct. Therefore the established scheme possessed internal coherence. What remained questionable, however, were the initial presumptions relative to the problem we had to tackle.

Indeed, up to that time the approach which we had resolved in that way consisted in determining the level or degree of incidence of a set of elements over another set of elements, which, in turn, had an incidence on a third set of elements. Our problem, nevertheless, "was reduced" to the incidence of one set over another set, that is, taking into consideration not only the direct effects but also the induced effects (first and second generation effects). Our objective on developing the previous model, was none other than the introduction, in a simple manner, of an operator, that of maxmin convolution, which, doing so on the central subject, could entail certain difficulties. On the other hand, once we had reached this point no insoluble obstacles should appear for us to carry out our aims with every success.

Arriving at the accumulated effects in a rectangular relation

When starting out from the existence of a set of causes that have an incidence on a set of elements and the elements that constitute the causes are different from the elements that constitute the effects, then the relations of causality are susceptible to being presented in the form of a matrix that will normally be rectangular, in the sense that the number of rows (causes) will be different from the number of columns (effects).

In order to illustrate what we have just stated, let us consider an example with a reduced number of causes and effects.

The set of causes were:

$$A = \{a_1, a_2, a_3\}$$

and the set of effects:

$$B = \{b_1, b_2, b_3, b_4\}$$

We started out from the presumption that the information relative to the level or degree of *direct incidence* of each cause on each effect was available, which we expressed as follows:

[M]	b_1	b_2	b_3	b_4
a_1	.4	0	.6	.7
a_2	.9	.5	.1	.8
a_3	.3	.2	.9	0

Now, the causes have an incidence on themselves, but can also have an incidence, to a greater or lesser degree, on the remaining causes. This required establishing a second matrix that showed these relations of causality. In our case it was a *square matrix* such as the following:

[A]	a_1	a_2	a_3
a_1	1	.6	.2
a_2	.8	1	.5
a_3	0	.9	1

In a like manner, the effects had an incidence on themselves, but also they could have an incidence , to a greater or lesser degree, on the remaining effects. Here we had further relations of causality, which also were susceptible of being represented in a new square matrix. In this case this was the one shown below:

[B]	b_1	b_2	b_3	b_4
b_1	1	.8	.6	.9
b_2	.2	1	0	1
b_3	.8	.5	1	.3
b_4	1	.2	.7	1

The valuation contained in these three matrices constituted the only essential information for providing an adequate solution to our problem. From the same, and with the correct use of the maxmin convolution operator, we would arrive at the sought after result. We would see this immediately. In fact, if we convoluted matrix [A] with matrix [M] we would find the incidences of the causes over the causes and the causes over the effects. In this way we would have completed the first step. The operation to be done was:

$$[\underset{\sim}{A}]\circ[\underset{\sim}{M}] = \begin{array}{c|c|c|c|} & a_1 & a_2 & a_3 \\ \hline a_1 & 1 & .6 & .2 \\ \hline a_2 & .8 & 1 & .5 \\ \hline a_3 & 0 & .9 & 1 \\ \hline \end{array} \circ \begin{array}{c|c|c|c|c|} & b_1 & b_2 & b_3 & b_4 \\ \hline a_1 & .4 & 0 & .6 & .7 \\ \hline a_2 & .9 & .5 & .1 & .8 \\ \hline a_3 & .3 & .2 & .9 & 0 \\ \hline \end{array} = \begin{array}{c|c|c|c|c|} & b_1 & b_2 & b_3 & b_4 \\ \hline a_1 & .6 & .5 & .6 & .7 \\ \hline a_2 & .9 & .5 & .6 & .8 \\ \hline a_3 & .9 & .5 & .9 & .8 \\ \hline \end{array}$$

The maxmin convolution on this occasion also consisted in taking successively the lowest values for each row and column choosing the greater of these. This is how this was done:

For element (a_1, b_1): row a_1 of matrix $[\underset{\sim}{A}]$ and column b_1 of matrix $[\underset{\sim}{M}]$.

$$1 \wedge 0.4 = 0.4$$

$$0.6 \wedge 0.9 = 0.6$$

$$0.2 \wedge 0.3 = 0.2$$

$$(1 \wedge 0.4) \vee (0.6 \wedge 0.9) \vee (0.2 \wedge 0.3) = 0.6$$

For element (a_1, b_2): row a_1 of matrix $[\underset{\sim}{A}]$ and column b_2 of matrix $[\underset{\sim}{M}]$.

$$1 \wedge 0 = 0$$

$$0.6 \wedge 0.5 = 0.5$$

$$0.2 \wedge 0.2 = 0.2$$

$$(1 \wedge 0) \vee (0.6 \wedge 0.5) \vee (0.2 \wedge 0.2) = 0.5$$

. . . .

. . . .

For element (a_3, b_4): row a_3 of matrix $[\underset{\sim}{A}]$ and column b_4 of matrix $[\underset{\sim}{M}]$.

$$0 \wedge 0.7 = 0$$

$$0.9 \wedge 0.8 = 0.8$$

$$1 \wedge 0 = 0$$

$$(0 \wedge 0.7) \vee (0.9 \wedge 0.8) \vee (1 \wedge 0) = 0.8$$

Now then, matrix $[\underset{\sim}{A}] \circ [\underset{\sim}{M}]$ only represents a part of the accumulated effects. We were missing the relations of causality that the effects caused on the effects, which we added to those already arrived at. This was possible if we managed to convolve matrix $[\underset{\sim}{A}] \circ [\underset{\sim}{M}]$ with matrix $[\underset{\sim}{B}]$. We did this as follows:

$$[\underset{\sim}{A}] \circ [\underset{\sim}{M}] \circ [\underset{\sim}{B}] = \begin{array}{c|c|c|c|c|} & b_1 & b_2 & b_3 & b_4 \\ \hline a_1 & .6 & .5 & .6 & .7 \\ \hline a_2 & .9 & .5 & .6 & .8 \\ \hline a_3 & .9 & .5 & .9 & .8 \\ \hline \end{array} \circ \begin{array}{c|c|c|c|c|} & b_1 & b_2 & b_3 & b_4 \\ \hline b_1 & 1 & .8 & .6 & .9 \\ \hline b_2 & .2 & 1 & 0 & 1 \\ \hline b_3 & .8 & .5 & 1 & .3 \\ \hline b_4 & 1 & .2 & .7 & 1 \\ \hline \end{array} = \begin{array}{c|c|c|c|c|} & b_1 & b_2 & b_3 & b_4 \\ \hline a_1 & .7 & .6 & .7 & .7 \\ \hline a_2 & .9 & .8 & .7 & .9 \\ \hline a_3 & .9 & .8 & .9 & .9 \\ \hline \end{array}$$

The convolution mechanism or maxmin composition acted, once again in the same way.

For element (a_1, b_1): row a_1 of matrix $[\underset{\sim}{A}] \circ [\underset{\sim}{M}]$ and column b_1 of matrix $[\underset{\sim}{B}]$.

$$0.6 \wedge 1 = 0.6$$

$$0.5 \wedge 0.2 = 0.2$$

$$0.6 \wedge 0.8 = 0.6$$

$$0.7 \wedge 1 = 0.7$$

$$(0.6 \wedge 1) \vee (0.5 \wedge 0.2) \vee (0.6 \wedge 0.8) \vee (0.7 \wedge 1) = 0.7$$

For element (a_1, b_2): row a_1 of matrix $[\underset{\sim}{A}] \circ [\underset{\sim}{M}]$ and column b_2 of matrix $[\underset{\sim}{B}]$.

$$0.6 \wedge 0.8 = 0.6$$

$$0.5 \wedge 1 = 0.5$$

$$0.6 \wedge 0.5 = 0.5$$

$$0.7 \wedge 0.2 = 0.2$$

$$(0.6 \wedge 0.8) \vee (0.5 \wedge 1) \vee (0.6 \wedge 0.5) \vee (0.7 \wedge 0.2) = 0.6$$

. . . .

. . . .

For element (a_3, b_4): row a_3 of matrix $[\underset{\sim}{A}] \circ [\underset{\sim}{M}]$ and column b_4 of matrix $[\underset{\sim}{B}]$.

$$0.9 \wedge 0.9 = 0.9$$

$$0.5 \wedge 1 = 0.5$$

$$0.9 \wedge 0.3 = 0.3$$

$$0.8 \wedge 1 = 0.8$$

$$(0.9 \wedge 0.9) \vee (0.5 \wedge 1) \vee (0.9 \wedge 0.3) \vee (0.8 \wedge 1) = 0.9$$

In this way we had arrived at the accumulated incidences of first and second generation of the elements constituting set $\underset{\sim}{A}$ over the elements forming set $\underset{\sim}{B}$ without any error or omission.

If we made a comparison between the accumulated effects shown in matrix $[\underset{\sim}{A}] \circ [\underset{\sim}{M}] \circ [\underset{\sim}{B}]$ and the direct effects considered initially by matrix $[\underset{\sim}{M}]$ we would see the importance, in some of the relations, of the second generation effects, or better said, the forgotten effects. A very easy way of finding these could be by arriving at the difference between the two matrices. This way we would have:

$[\underset{\sim}{A}] \circ [\underset{\sim}{M}] \circ [\underset{\sim}{B}] (-) [\underset{\sim}{M}] =$

	b_1	b_2	b_3	b_4
a_1	.7	.6	.7	.7
a_2	.9	.8	.7	.9
a_3	.9	.8	.9	.9

(-)

	b_1	b_2	b_3	b_4
a_1	.4	0	.6	.7
a_2	.9	.5	.1	.8
a_3	.3	.2	.9	0

=

	b_1	b_2	b_3	b_4
a_1	.3	.6	.1	0
a_2	0	.3	.6	.1
a_3	.6	.6	0	.9

One simple glance is sufficient for seeing the importance of "forgetting" (indirect effect) of the relation of incidence of a_3 over b_4. In the direct relation it was nil, in the accumulation the level reached 0.9. Other *indirect incidences* worthy of mention are (a_1, b_2), (a_2, b_3), (a_3, b_1), (a_3, b_2), all of them at level 0.6.

It now appeared that the technical and mechanical operating aspects of the model were ready to be used in the actual problem we were occupied with. We now had to put "names and surnames" to each one of sets $\underset{\sim}{A}$ and $\underset{\sim}{B}$, that is to say, to the set of causes (actions capable of moving the elements making up the image of the Reial Academia de Doctors) and the set of effects (elements constituting the image of our

Royal Corporation). In this way we had arrived at the "crunch of the matter".

Elements constituting the incidence matrices

The President of the Reial Academia, our friend Josep, was very sure that the renewal of this "institution" would not be done automatically, over the mere course of time. Inertia played a negative role in this respect. What was required therefore, were actions that would act on the "sensitive points" of the image of the Royal Corporation. But the question was raised on which were these aspects to be considered. They should be sufficient in order to cover the essential aspects, but not excessively so, taking into account the normal financial limitations to which we were submitted. During quite a few sessions we directed our efforts in attempting to list those elements that were capable of designing the image desired for the entity, which he presided. What it meant was an attempt to establish the elements, which *the effect* of the actions should receive, and which we had called "set B". We agreed to consider the following:

b_1: Improve headquarters.

b_2: Presence at scientific forums.

b_3: Good review in the communications media.

b_4: Correct consideration by public administrations.

b_5: Personal prestige of members.

b_6: Incorporation into international institutions.

b_7: Recognition by society as an active institution.

b_8: Feeling of public usefulness.

b_9: Public projection of academic activities.

We were conscious of the fact that this list did not take into account all, absolutely all the elements constituting the image of the Reial Academia de Doctors, but it did represent, in a certain fashion, those that, at that particular time were of most importance. On the other hand, our task was open to all suggestions emanating both from the academicians as from external experts. The importance was " to get up and walk" and create an instrument that was valid for the treatment of the specific problem. And this we were convinced that we had attained.

Having closed then in this manner the list of elements making up the image, we moved on to select the list *of actions to be carried* out in order to strengthen this image, which would give rise to the comprehensive *"set"* of causes. After the due selection process, we limited ourselves to the following:

a_1: Punctual publication of the records.

a_2: Solemnity in all acts.

a_3: Increase in sources of finance.

a_4: Rejuvenation of the members of the Reial Academia.

a_5: Stricter requirements in curricula for incorporation.

a_6: Presence at acts of very well known personalities.

a_7: Setting up and development of a press office.

a_8: Improve the Internet "web page".

a_9: Publication and divulging important works of academicians.

a_{10}: Promote incorporation of prestigious foreign academicians.

a_{11}: Strengthen the presence of academicians at citizen manifestations.

a_{12}: Setting up a prize of worldwide impact.

There was no doubt whatsoever on the vital importance of a wise decision in the selection, both of the "causes" and the "effects". But we were also conscious of the fact that the most delicate work was still to come, when we had to assign *degrees or levels* (valuations in [0, 1]), to all the relations between causes and effects. We felt it would be sensible, for this, to seek the co-operation of some of the members of the Royal Corporation, in this way, attempting to limit, even if only partially, the inevitable subjectivity of all valuations. We prepared a table in the form of a matrix with 12 rows and 9 columns of a notable size, which we placed on a wall in a large room where we installed all the "experts". We then read out one by one, all the relations of incidence, commencing, obviously, by the incidence of the *punctual publication of the records* on the *improvement of headquarters*. Here there was unanimity in assigning level zero. We then moved on to the incidence of the *punctual publication of records* on *the presence at scientific forums*. In this case opinions differed, although not excessively. This point was given valuations of between 0.5 and 0.7 so we accepted 0.6. And this we did successively until we had constructed the following fuzzy matrix:

[M] =

	b_1	b_2	b_3	b_4	b_5	b_6	b_7	b_8	b_9
a_1	0	.6	.7	.7	.5	.6	.7	.6	.4
a_2	0	.6	.9	.7	.3	.5	.7	.4	.9
a_3	1	.7	.1	.2	0	.7	0	0	.8
a_4	0	.8	.5	0	.5	.5	.7	0	.8
a_5	0	.9	.6	.8	1	.7	.5	.7	.5
a_6	.3	.1	.9	.8	.6	.5	1	.6	1
a_7	.4	.2	1	.7	.8	.5	.7	.8	1
a_8	0	.8	.7	.2	.1	.6	.8	.6	.8
a_9	.3	.9	.5	.7	1	.8	.7	.7	.7
a_{10}	0	.8	.8	.6	.7	.9	.7	.5	.6
a_{11}	.1	0	.7	.5	.9	0	.9	.4	1
a_{12}	0	.8	.9	.2	.5	.6	.9.	2	1

With more or less effort we had managed to terminate the task of "agreeing on" certain valuations that gave us a good idea of the *direct relations* between the actions to be tackled and the defining elements of the image of our institution. Although it was only this, we had before our eyes *a table* that put some order in our ideas and showed us a first approximation at the objective we were aiming for. But the importance was to find the degree of accumulated incidence, since only then would we have available the complete information. It was necessary then for us to have the information concerning the levels of incidences between causes and the levels of incidences between effects. This required the construction of square matrices $[\underset{\sim}{A}]$ and $[\underset{\sim}{B}]$. We followed the same procedure as that for arriving at matrix $[\underset{\sim}{M}]$ and the result was as follows:

$[\underset{\sim}{A}] =$

	a_1	a_2	a_3	a_4	a_5	a_6	a_7	a_8	a_9	a_{10}	a_{11}	a_{12}
a_1	1	0	.7	0	0	.4	0	.6	0	.5	0	0
a_2	.1	1	.7	0	.9	1	0	.6	.2	.5	.6	0
a_3	1	1	1	0	0	.7	.8	.8	.9	.6	.2	1
a_4	.5	0	0	1	0	0	.4	.1	.6	0	0	0
a_5	.2	.4	.5	0	1	.9	.2	.6	.6	0	.1	0
a_6	0	1	.7	0	.8	1	.5	.8	0	0	.8	0
a_7	0	.8	.5	0	0	.8	1	.9	.7	.3	.9	0
a_8	0	.6	.2	0	.6	.6	0	1	0	.6	.6	0
a_9	0	.4	0	0	.6	.7	0	.7	1	.6	.7	0
a_{10}	0	.8	0	0	.6	.8	0	.8	0	1	0	0
a_{11}	0	.3	.5	0	.2	.5	0	0	0	0	1	0
a_{12}	0	.4	0	0	.8	.7	.4	0	.9	.8	.2	1

$[\underset{\sim}{B}] =$

	b_1	b_2	b_3	b_4	b_5	b_6	b_7	b_8	b_9
b_1	1	.2	.6	.7	.7	.5	.7	.7	.9
b_2	0	1	.8	.9	.8	.8	.7	.6	.9
b_3	.3	.4	1	.8	.7	.6	.9	.8	1
b_4	.8	.3	.2	1	.4	.1	.2	.9	.5
b_5	.1	.9	.7	.8	1	.9	.6	.6	.7
b_6	.3	1	.7	.9	1	1	.8	.4	.8
b_7	.5	.2	.8	1	.9	.3	1	1	.7
b_8	.8	.6	.8	1	.7	.3	1	1	.7
b_9	.7	.3	.9	.7	.7	.4	.9	.8	1

Now we had available as much information as was necessary to commence the process that should lead us to the sought after objective, which was none other than to become aware of the degrees of accumulated incidence of the actions to be carried out, grouped in set A, on the elements constituting the image of the Reial Academia, joined together in set B. This would be attained if we could establish the incidence of each action on itself and on the remaining possible actions accumulated to the incidence of the actions over the elements that integrate the image, accumulating them, in turn, to the incidence of the elements representing the image with itself and over the remainder. Actually what this meant was two operations: the first would be resolved by means of the maxmin convolution of matrices $[\underset{\sim}{A}]$ and $[\underset{\sim}{M}]$; the second by convolving the result we had obtained, that is to say $[\underset{\sim}{A}] \circ [\underset{\sim}{M}]$ with matrix $[\underset{\sim}{B}]$. The result we arrived at, $[\underset{\sim}{A}] \circ [\underset{\sim}{M}] \circ [\underset{\sim}{B}]$ would contribute the desired totally accumulated relations of incidence.

Below we have reproduced the first convolution:

$[\underset{\sim}{A}] \circ [\underset{\sim}{M}] =$

↗	a_1	a_2	a_3	a_4	a_5	a_6	a_7	a_8	a_9	a_{10}	a_{11}	a_{12}
a_1	1	0	.7	0	0	.4	0	.6	0	.5	0	0
a_2	.1	1	.7	0	.9	1	0	.6	.2	.5	.6	0
a_3	1	1	1	0	0	.7	.8	.8	.9	.6	.2	1
a_4	.5	0	0	1	0	0	.4	.1	.6	0	0	0
a_5	.2	.4	.5	0	1	.9	.2	.6	.6	0	.1	0
a_6	0	1	.7	0	.8	1	.5	.8	0	0	.8	0
a_7	0	.8	.5	0	0	.8	1	.9	.7	.3	.9	0
a_8	0	.6	.2	0	.6	.6	0	1	0	.6	.6	0
a_9	0	.4	0	0	.6	.7	0	.7	1	.6	.7	0
a_{10}	0	.8	0	0	.6	.8	0	.8	0	1	0	0
a_{11}	0	.3	.5	0	.2	.5	0	0	0	0	1	0
a_{12}	0	.4	0	0	.8	.7	.4	0	.9	.8	.2	1

$\circ$

$[\underset{\sim}{A}]$

∘

	b_1	b_2	b_3	b_4	b_5	b_6	b_7	b_8	b_9
a_1	0	.6	.7	.7	.5	.6	.7	.6	.4
a_2	0	.6	.9	.7	.3	.5	.7	.4	.9
a_3	1	.7	.1	.2	0	.7	0	0	.8
a_4	0	.8	.5	0	.5	.5	.7	0	.8
a_5	0	.9	.6	.8	1	.7	.5	.7	.5
a_6	.3	.1	.9	.8	.6	.5	1	.6	1
a_7	.4	.2	1	.7	.8	.5	.7	.8	1
a_8	0	.8	.7	.2	.1	.6	.8	.6	.8
a_9	.3	.9	.5	.7	1	.8	.7	.7	.7
a_{10}	0	.8	.8	.6	.7	.9	.7	.5	.6
a_{11}	.1	0	.7	.5	.9	0	.9	.4	1
a_{12}	0	.8	.9	.2	.5	.6	.9	.2	1

=

[M]

=

	b_1	b_2	b_3	b_4	b_5	b_6	b_7	b_8	b_9
a_1	.7	.7	.7	.7	.5	.7	.7	.6	.7
a_2	.7	.9	.9	.8	.9	.7	1	.7	1
a_3	1	.9	.9	.7	.9	.8	.9	.8	1
a_4	.4	.8	.5	.6	.6	.6	.7	.6	.8
a_5	.5	.9	.9	.8	1	.7	.9	.7	.9
a_6	.7	.8	.9	.8	.8	.7	1	.7	1
a_7	.5	.8	1	.8	.9	.7	.9	.8	1
a_8	.3	.8	.7	.6	.6	.6	.8	.6	.8
a_9	.3	.9	.7	.7	1	.8	.7	.7	.7
a_{10}	.3	.8	.8	.8	.7	.9	.8	.6	.8
a_{11}	.5	.5	.7	.5	.9	.5	.9	.5	1
a_{12}	.4	.9	.9	.8	.9	.8	.9	.7	1

[A]∘[M]

Once we had finished this first stage of the convolution we proceeded to tackle the second and last, consisting in the composition of the result arrived at, [A]∘[M] with the fuzzy relation [B], that is to say ([A]∘[M])∘[B]. We finally arrived at:

[A]∘[M]∘[B]=

	b_1	b_2	b_3	b_4	b_5	b_6	b_7	b_8	b_9
a_1	.7	.7	.7	.7	.5	.7	.7	.6	.7
a_2	.7	.9	.9	.8	.9	.7	1	.7	1
a_3	1	.9	.9	.7	.9	.8	.9	.8	1
a_4	.4	.8	.5	.6	.6	.6	.7	.6	.8
a_5	.5	.9	.9	.8	1	.7	.9	.7	.9
a_6	.7	.8	.9	.8	.8	.7	1	.7	1
a_7	.5	.8	1	.8	.9	.7	.9	.8	1
a_8	.3	.8	.7	.6	.6	.6	.8	.6	.8
a_9	.3	.9	.7	.7	1	.8	.7	.7	.7
a_{10}	.3	.8	.8	.8	.7	.9	.8	.6	.8
a_{11}	.5	.5	.7	.5	.9	.5	.9	.5	1
a_{12}	.4	.9	.9	.8	.9	.8	.9	.7	1

∘

[A]∘[M]

∘

	b_1	b_2	b_3	b_4	b_5	b_6	b_7	b_8	b_9
b_1	1	.2	.6	.7	.7	.5	.7	.7	.9
b_2	0	1	.8	.9	.8	.8	.7	.6	.9
b_3	.3	.4	1	.8	.7	.6	.9	.8	1
b_4	.8	.3	.2	1	.4	.1	.2	.9	.5
b_5	.1	.9	.7	.8	1	.9	.6	.6	.7
b_6	.3	1	.7	.9	1	1	.8	.4	.8
b_7	.5	.2	.8	1	.9	.3	1	1	.7
b_8	.8	.6	.8	1	.7	.3	1	1	.7
b_9	.7	.3	.9	.7	.7	.4	.9	.8	1

=

[B]

	b_1	b_2	b_3	b_4	b_5	b_6	b_7	b_8	b_9
a_1	.7	.7	.7	.7	.7	.7	.7	.7	.7
a_2	.8	.9	.9	1	.9	.9	1	1	1
a_3	1	.9	.9	.9	.9	.9	.9	.9	1
a_4	.7	.8	.8	.8	.8	.8	.8	.8	.8
a_5	.8	.9	.9	.9	1	.9	.9	.9	.9
a_6	.8	.8	.9	1	.9	.8	1	1	1
a_7	.8	.9	1	.9	.9	.9	.9	.9	1
a_8	.7	.8	.8	.8	.8	.8	.8	.8	.8
a_9	.7	.9	.8	.9	1	.9	.8	.7	.9
a_{10}	.8	.9	.8	.9	.9	.9	.8	.8	.8
a_{11}	.7	.9	.9	.9	.9	.9	.8	.9	1
a_{12}	.8	.9	.9	.9	.9	.9	.9	.9	1

$= [\underset{\sim}{A}] \circ [\underset{\sim}{M}] \circ [\underset{\sim}{B}]$

With the use of this operator we had attained our objective. We now had a fuzzy matrix that was capable of bringing to light the *degree of incidence* of each action to be started over every element that went to make up the image of the Reial Academia. We should therefore move on from *study to action*. For this it would be necessary to establish an order of preference between the components of the image, according to how we felt they would have a greater importance, or on the contrary were deficient. Having done this task, we would simply have to see which activity or activities had a higher incidence. This or these would be the first we would have to tackle.

Analysis of the accumulated incidence matrix

On looking at the fuzzy relation $[\underset{\sim}{A}] \circ [\underset{\sim}{M}] \circ [\underset{\sim}{B}]$ we can observe an interesting fact: That its values or nearly all its values were high or very high. In fact none of the values was below 0.7. This indicated that, directly or indirectly, all the actions were good or very good for improving or up dating the image of the Reial

Academia. Now this general fact should not prevent us from seeking a better use for our resources and our efforts. Therefore, in conjunction with our President and friend Josep, we chose two of the constituting elements and established the following order of precedence:

b_1: Improve headquarters.

b_7: Recognition as an active institution by society.

This selection and ordering does not mean, to any extent, relegating the remaining elements, considering them as second category. We were certainly going to take them into consideration at a later stage.

With regard to the first of the elements. "improve headquarters", the action to be taken was obvious: "increase sources of finance". Others such as "solemnity in any acts carried out" and "setting up a Press office", for example, remained at a certain distance. The first action to be tackled left no doubt whatsoever, taking into account the direct incidence, brought to light right from the start in matrix [$\underset{\sim}{M}$]. But with regard to the other two, this was not quite so clear, above all if we thought that the relation of direct incidence had been established at level nil or low, respectively.

We showed, by means of the following figures, those *indirect relations* with an accumulated incidence of 0.8:

a) Incidence of "solemnity of acts carried out" over "improve institutional headquarters"

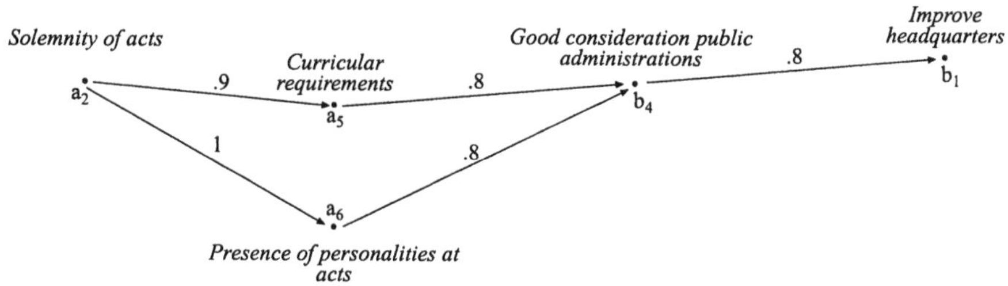

b) Incidence of "curricular requirements for incorporations" over "improvements in institutional headquarters":

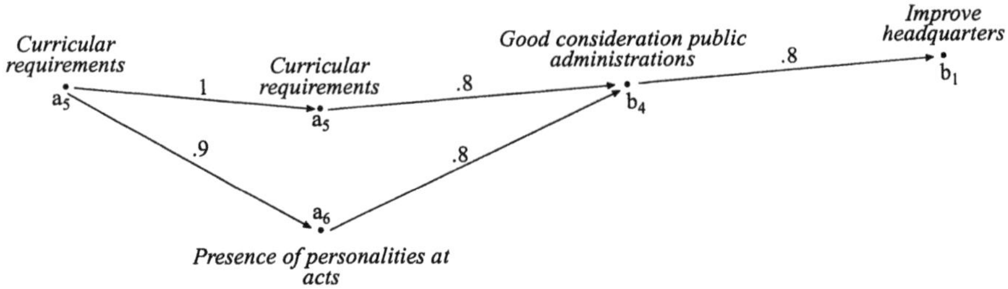

c) Incidence of "presence of personalities at acts" over "improvements in institutional headquarters":

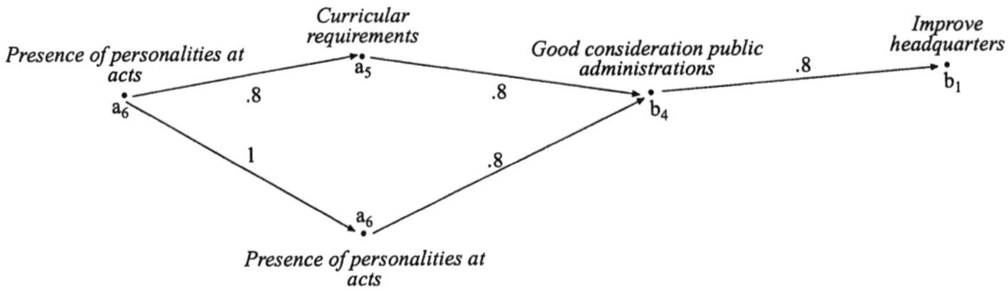

d) Incidence of "setting up a press office" over "improvements to institutional headquarters":

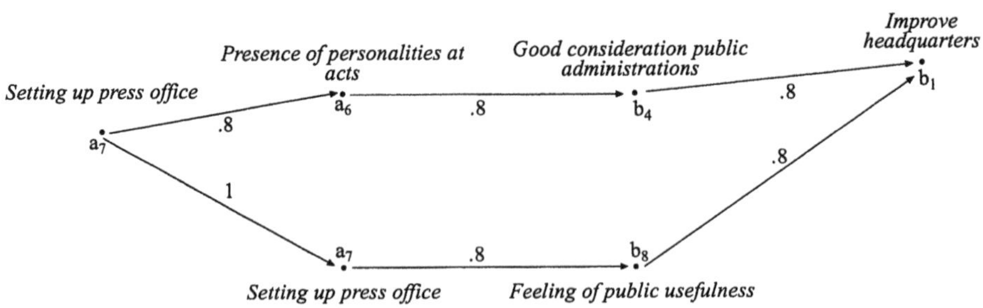

e) Incidence of "incorporation of prestigious foreign academicians" over "improvements to institutional headquarters":

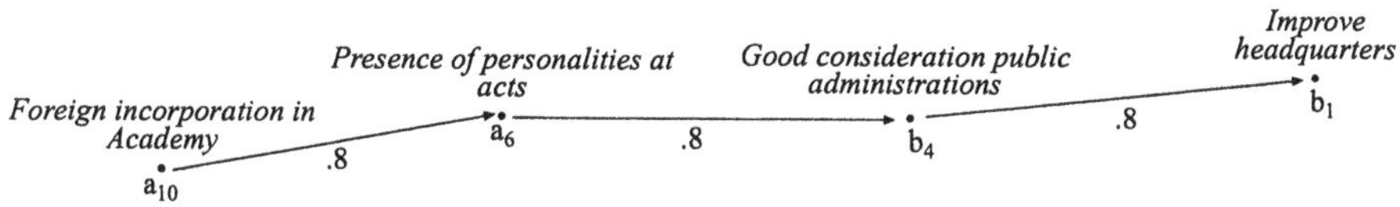

f) Incidence of "setting up a prize of international impact" over "improvements to institutional headquarters":

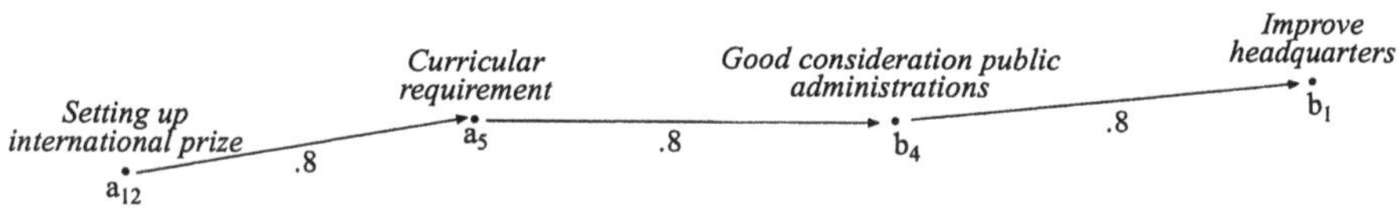

Each one of these figures brought to light the linking of the relations of incidence or causality at a level of 0.8, all of which allowed us to attain to a high degree, the objective of improving the *institutional headquarters*. We did not have to forget, however, that the most important action to be carried out in order to attain this end was to be found in the *direct relation*:

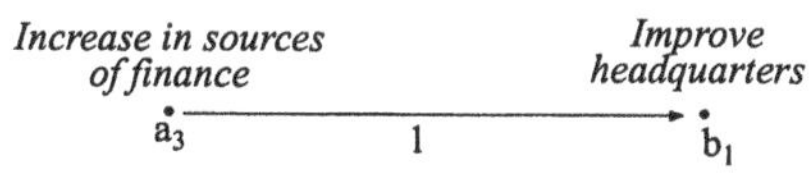

This is where we would have to concentrate all our efforts

But independently of this unquestionable truth we could also extract some important conclusions from the analysis of the six previous figures. In fact, in all the actions to be tackled: a_2, a_5, a_6, a_7, a_{10}, and a_{12}, there always appeared as intermediaries a_5 and/or a_6. This acquires a significant importance since as occurred in section f), the "curricular requirements" is the *effect* of the "setting up of a prize of world impact". Perhaps this *action* was not to be necessary, if we could "start out from" (because this depends on us and without any cost whatsoever) of the greater requirements in the "curricula" for candidates. The same does not happen with section e). Here the relation of causality is brought to light that: the "incorporation of prestigious foreign academicians" over the "presence of personalities at acts", which is very often essential as it is difficult to attract the attention of personalities when the acts that are carried out are outside the normal.

We could go on listing many others of the thoughts that the graphs presented suggested to us at the time. But we will not do so, so as not to tire our eventual readers, fully convinced that they will know how to make up for our modest explanations.

We now move on to the study of the causes that will allow us to strengthen the second of the elements that go to make up the image of the Reial Academia, which we had chosen as a first priority. This was "recognition as an active institution by society", symbolised by b_7. A simple glance at the matrix that resulted from the overall convolution $\underset{\sim}{A} \circ \underset{\sim}{M} \circ \underset{\sim}{B}$ allowed us to see as important accumulated incidences those from the causes (actions to be carried out), a_2, "solemnity in acts carried out" and a_6, "presence at acts of recognised personalities". In both, *the accumulated incidence* was total. If we also looked at the *direct incidence*, included in the original matrix $\underset{\sim}{M}$, we would have seen that this was already very high for the first, with a degree of 0.7, and total, for the second, with a degree of 1. Therefore we had no need to find the intermediary incidences as they were of little or no importance.

On analysing the accumulated incidences which followed them in importance, degree 0.9, we found as actions to be carried out, a_3, "increase sources of financing", a_5 "curricular requirements in incorporations", a_7, "setting up a press office", and a_{12}, "setting up a prize of world impact".

With regard to incidence a_{12}, "setting up a prize of world impact" there was no accumulation whatsoever of the indirect effects of incidence. Relative to a_7 over b_7, "setting up a press office" over "recognition as an active institution by society", the direct incidence was already 0.7, therefore the accumulated incidence only contributed a degree of 0.2. The indirect effects were more important in the accumulation of incidence a_5 "curricular requirements for incorporation", since this went from 0.5 to 0.9. Where the indirect incidences had a special prominence, was in those stemming from cause a_3, "increase in sources of finance", since in the original matrix no incidence whatsoever had been considered. It was then on these primary causes that we concentrated the greatest attention. Below we have reproduced the graphs in which the intermediary elements appear quite clearly.

a) Incidence of "increase in sources of finance" over "recognition as an active institution by society":

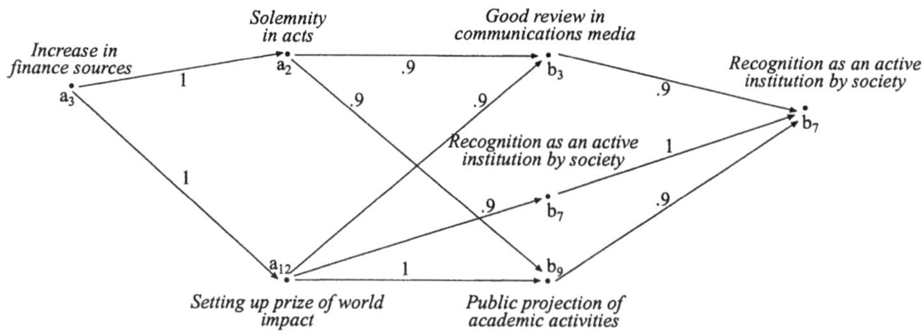

b) Incidence of "curricular requirement for incorporations" over "recognition as an active institution by society":

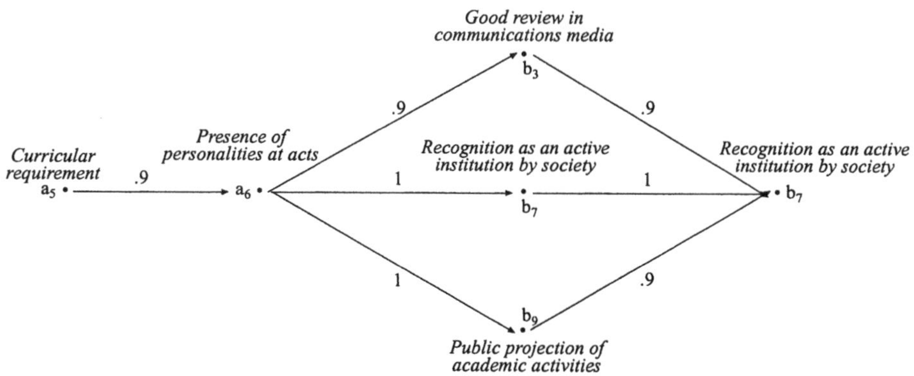

An examination, although even superficial, of the previous accumulated relations of incidence and of their intermediary elements, lead us to certain considerations which converge in the need of paying special attention to two actions which stand out as indisputable protagonists: "presence of personalities at acts" and "solemnity at acts carried out". In fact, their direct incidence is obvious in the objective of attaining "recognition as an active institution by society" with a degree of 1 and 0.7 respectively. But also, these possible actions act as intermediary elements in the "curricular requirements for incorporation", in the first of these, and in the "increase of finance sources" in the second. It for this reason, that we decided to temporarily abandon actions a_5, "curricular requirements for incorporation", and a_3, "increase in finance sources" (obviously, in this case for the objective we were seeking then "recognition as an active institution by society"). In this way fortunately the problem was reduced.

With regard to the relations with nil or reduced second-generation effects, the total incidences were shown in the following graphs:

c) Incidence of "setting up and development of a press office" over "recognition as an active institution by society"

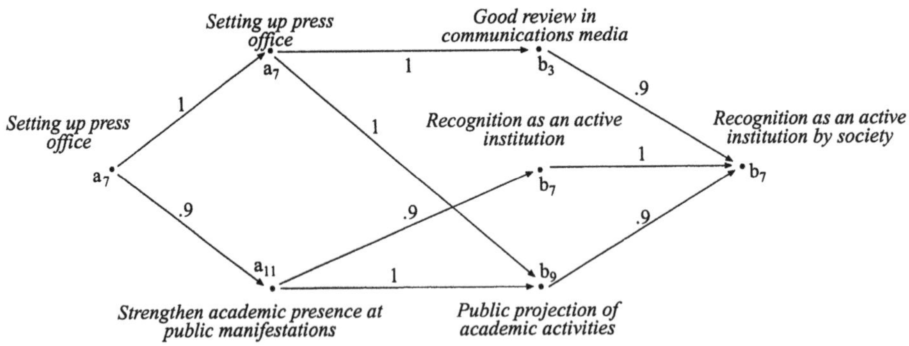

d) Incidence of "setting up prize of world impact" over "recognition as an active institution by society"

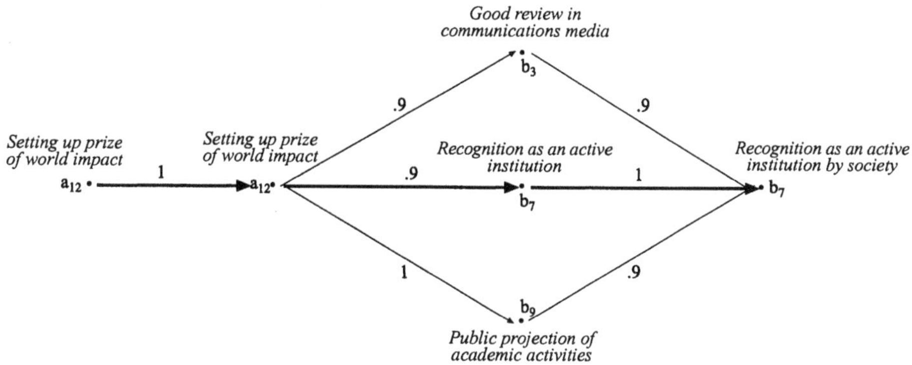

As it could be no other way, both in one case as in the other, the causes that act as intermediaries no longer exercise a basic influence. In fact, the "setting up and development of a press office" does not require any further action from other actions in order to attain the objective "recognition as an active institution by society", taking into account that "automatically" there would be "a good review in the communications media" and "public projection of academic activities" take place, both effects would have as an important effect (degree 0.9) the sought after "recognition as an active institution by society". Indeed it is true to say that the "setting up of a press office" would strengthen "the presence of academicians at public manifestations" (degree 0.9), which would have a direct incidence on "recognition as an active institution by society" and an indirect incidence on "public projection of the academic activity" it would also motivate the sought after recognition. Nevertheless, this indirect action does not add any degree whatsoever in the level of accumulated incidence over that already arrived at in a direct manner (with regard to actions, that is causes) by means of the effect caused in the *aforementioned good review and public projection.* We remain, therefore, with the direct effect caused by the "setting up and development of a press office".

Even clearer appears to be the *direct incidence* of the "setting up of a prize of world impact" over "recognition as an active institution by society". It is sufficient for this to see in the last graph the path of arcs $a_{12} \rightarrow a_{12} \rightarrow b_7 \rightarrow b_7$. The degree, 0.9 of direct incidence between a_{12} and b_7 also sets the accumulated incidence as it is not higher than the one arrived at along the other paths, which are none other than "good review from the communications media" and "public projection of academic activities".

Little by little we had been tackling all the effective possibilities of action (causes) necessary for attaining the objectives we had set (effects), which in a first approximation were limited to, let us recall:

Improve the institutional headquarters

Recognition as an active institution by society

We now had to make a summary of all the actions to be carried out, taking into account the always limited possibilities, both from a financial and a temporal point of view.

In order to "improve the institutional headquarters":

a) Increase financial sources (degree 1)

   This undoubtedly was the major action to be tackled on a priority basis.

b) Greater curricular requirements for incorporations (degree 0.8)

c) Presence at acts of very well known people (degree 0.8)

   These latter two causes, as they also appear as intermediaries, acquire a higher importance relative to those that follow.

d) Setting up and development of a press office (degree 0.8)

   This action had no need to act on the other causes therefore its interest stand out.

e) Solemnity in acts carried out (degree 0.8)

f) Promote the incorporation of prestigious foreign academicians (degree 0.8)

g) Setting up of a prize of world impact (degree 0.8)

For "recognition as an active institution by society":

a) Presence at acts of very well known people (degree 1)

This action stands out over the remainder due to its total direct incidence over the desired effect.

b) Solemnity at acts carried out (degree 1)

With a high level of direct incidence (0.7) its accumulation takes it to total incidence.

c) Setting up a press office (degree 0.9)

d) Setting up a prize of world impact (degree 0.9)

Interest here come from the fact that it does not require further action for

ttaining its accumulated incidence.

e) Increase in financing sources (degree 0.9)

f) Greater curricular requirements for incorporation (degree 0.9)

In this way we established, as we understood it, the order of the actions to be tackled in order to attain one or another objective. In the event of attempting to tackle the tasks in order to attain both at the same time, we had to take into account not only the order in each one of them, but also those actions that are repeated in each of them. This occurred in:

. Increase in sources of finance (1 + 0.9 = 1.9)

. Presence at acts of very well know people (0.8 + 1 = 1.8)

. Solemnity at acts carried out (0.8 + 1 = 1.8)

. Setting up and development of a press office (0.8 + 0.9 = 1.7)

. Greater curricular requirements for incorporation (0.8 + 0.9 = 1.7)

. Setting up a prize of world impact (0.8 + 0.9 = 1.7)

We are conscious of the "abuse" signified by the use of the sum of degree or levels as a criteria for ordering in a case such as we are occupied with, in which the addenda are not measurements but valuations, but we have granted ourselves this license as a concession to easier comprehension for those not requiring, in practice, highly pure method.

Some final considerations

We were commenting with the President of the Reial Academia on some of the possibilities of this methodology in order to take into account *all*, absolutely all, of the accumulated effects of first and second generation, which the human mind, at least in its current situation, is not capable of taking globally into account. We asked ourselves: *how many routes* would our thoughts have to take in order to cover all the direct and indirect relations? Just for curiosity, and only in this sense, we tackled the task of drawing graphically the mental effort required for carrying out "all" the necessary connections to be able to take into account, in their entirety, the relations of incidence or causality between a *single* cause and a *single* effect.

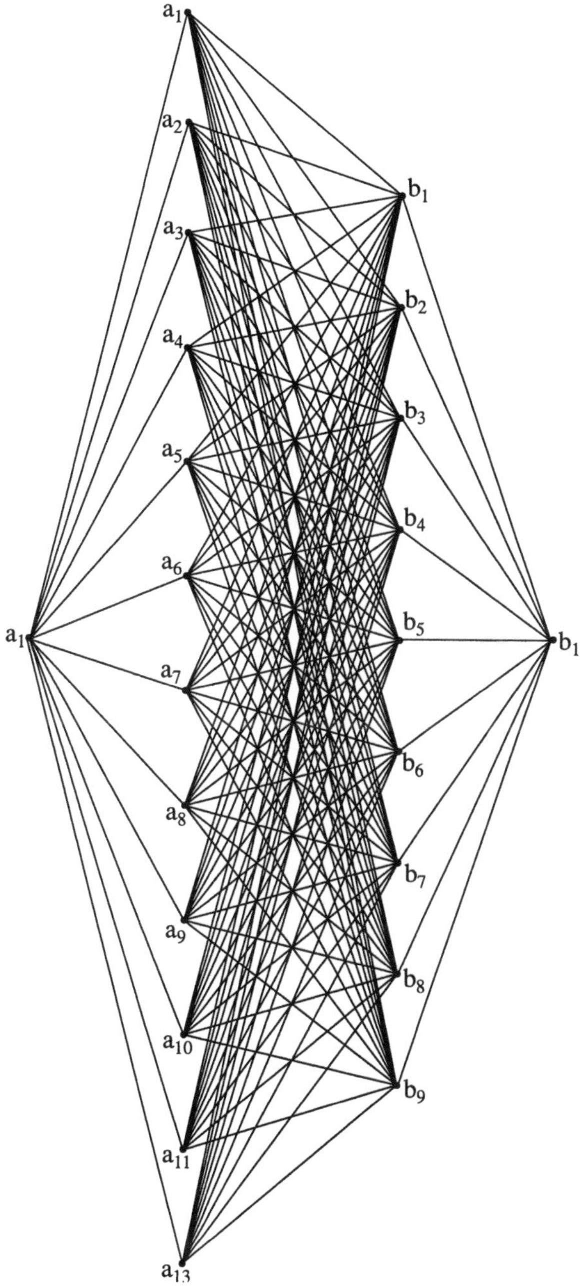
a1
a2
a3
a4
a5
a6
a7
a8
a9
a10
a11
a13
a1
b1
b2
b3
b4
b5
b6
b7
b8
b9
b1

The eloquence of the scheme we arrived at spoke for itself. And just to think that for a case as simple as ours, in which only *twelve* causes and *nine* effects were considered, the total connections would be the result of repeating the graph we had arrived at. 108 times! On the other hand we had arrived at our objective in an easy way and, what was even more important, without any possibility of error or omission.

This complexity, derived from the high numbers of connections that were necessary for covering all the accumulated relations of incidence, leads to, if the adequate technique is not used, the occurrence of forgotten effects. In our study some of these stand out (those that we considered as the most significant). But the method itself was sufficient for arriving at all of them, even with their "degree of forgotten effects". Apart from this, it was not difficult to attain, if we were satisfied with good approximations. It would be sufficient to find the difference between the matrix representing the accumulated effects $\underset{\sim}{A} \circ \underset{\sim}{M} \circ \underset{\sim}{B}$, and the original matrix (theoretically the matrix of direct effects) $\underset{\sim}{M}$, that is:

$$\underset{\sim}{A} \circ \underset{\sim}{M} \circ \underset{\sim}{B} \; (-) \; \underset{\sim}{M}$$

which for our problem was:

↗	b_1	b_2	b_3	b_4	b_5	b_6	b_7	b_8	b_9
a_1	.7	.7	.7	.7	.7	.7	.7	.7	.7
a_2	.8	.9	.9	1	.9	.9	1	1	1
a_3	1	.9	.9	.9	.9	.9	.9	.9	1
a_4	.7	.8	.8	.8	.8	.8	.8	.8	.8
a_5	.8	.9	.9	.9	1	.9	.9	.9	.9
a_6	.8	.8	.9	1	.9	.8	1	1	1
a_7	.8	.9	1	.9	.9	.9	.9	.9	1
a_8	.7	.8	.8	.8	.8	.8	.8	.8	.8
a_9	.7	.9	.8	.9	1	.9	.8	.7	.9
a_{10}	.8	.9	.8	.9	.9	.9	.8	.8	.8
a_{11}	.7	.9	.9	.9	.9	.9	.9	.9	1
a_{12}	.8	.9	.9	.9	.9	.9	.9	.9	1

(-)

$\underset{\sim}{A} \circ \underset{\sim}{M} \circ \underset{\sim}{B}$

(-)

	b_1	b_2	b_3	b_4	b_5	b_6	b_7	b_8	b_9
a_1	0	.6	.7	.7	.5	.6	.7	.6	.4
a_2	0	.6	.9	.7	.3	.5	.7	.4	.4
a_3	1	.7	.1	.2	0	.7	0	0	.8
a_4	0	.8	.5	0	.5	.5	.7	0	.8
a_5	0	.9	.6	.8	1	.7	.5	.7	.5
a_6	.3	.1	.9	.8	.6	.5	1	.6	1
a_7	.4	.2	1	.7	.8	.5	.7	.8	1
a_8	0	.8	.7	.2	.1	.6	.8	.6	.8
a_9	.3	.9	.5	.7	1	.8	.7	.7	.7
a_{10}	0	.8	.8	.6	.7	.9	.7	.5	.6
a_{11}	.1	0	.7	.5	.9	0	.9	.4	1
a_{12}	0	.8	.9	.2	.5	.6	.9	.2	1

$[\underset{\sim}{M}]$

=

=

	b_1	b_2	b_3	b_4	b_5	b_6	b_7	b_8	b_9
a_1	.7	.1	0	0	.2	.1	0	.1	.3
a_2	.8	.3	0	.3	.6	.4	.3	.6	.1
a_3	0	.2	.8	.7	(.9)	.2	(.9)	(.9)	.2
a_4	.7	0	.3	.8	.3	.3	.1	.8	0
a_5	.8	0	.3	.1	0	.2	.4	.2	.4
a_6	.5	.7	0	.2	.3	.3	0	.4	0
a_7	.4	.7	0	.2	.1	.4	.2	.1	0
a_8	.7	0	.1	.6	.7	.2	0	.2	0
a_9	.4	0	.3	.2	0	.1	.1	0	.2
a_{10}	.8	.1	0	.3	.2	0	.1	.3	.2
a_{11}	.6	(.9)	.2	.4	0	(.9)	0	.5	0
a_{12}	.8	.1	0	.7	.4	.3	0	.7	0

Forgotten effects

We have signalled with a circle those relations of incidence with the *highest degree o forgotten* effects, which in our case was 0.9. Tickled by curiosity we wanted to see which of the intermediary elements had caused such a high degree of incidence. One of these had already been analysed, that was $a_3 \rightarrow b_7$, that is, the incidence of "increase in the source of financing" over "recognition as an active institution by society ". The others were the following:

a) Incidence of "increase in the sources of financing" over "personal prestige of the members"

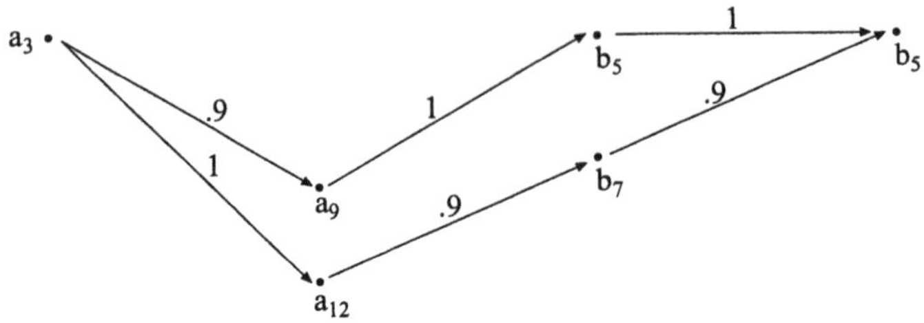

Initially a nil incidence had been considered. Later we became aware that, as a high incidence existed (degree 0.9) of the "increase in the sources of financing" over "publication and divulging the important works of the academicians" and total incidence (degree 1) of the latter action over "personal prestige of members", it was clear to see the incidence of "increase sources of financing" over "personal prestige of academy members".

By another path we also arrived at the same conclusion. In fact "the increase of sources of financing" had a total incidence (degree 1) over "setting up a prize of world impact" and this action very much affected (degree 0.9) "recognition as an active institution by society" which, in turn had an incidence to a greater degree (degree 0.9) over "personal prestige a academy members".

b) Incidence of "increase in sources of finance" over "feeling of public usefulness".

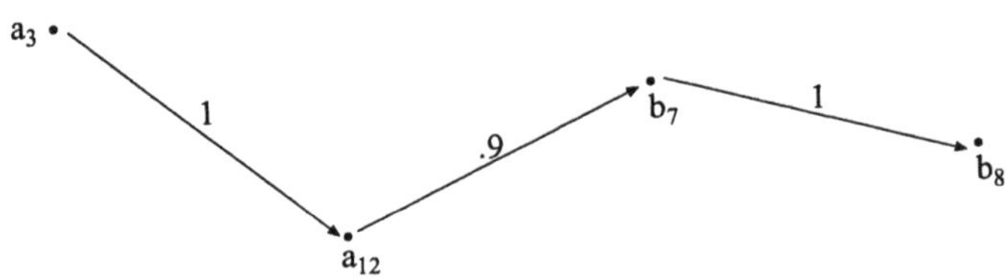

Initially, neither here did it appear that there existed any direct incidence whatsoever. After the study made, we could see that, as a consequence of the total relation (degree 1) between "increase of sources of financing" and "setting up a prize of world impact", of this action over "recognition as an active institution by society" (degree 0.9) and the incidence (degree 1) of the latter over "feeling of public usefulness" appeared the strong link existing that was detected in the corresponding box of the matrix of accumulations.

c) Incidence of "strengthen the presence of academicians at large public manifestations" over "presence in scientific forums".

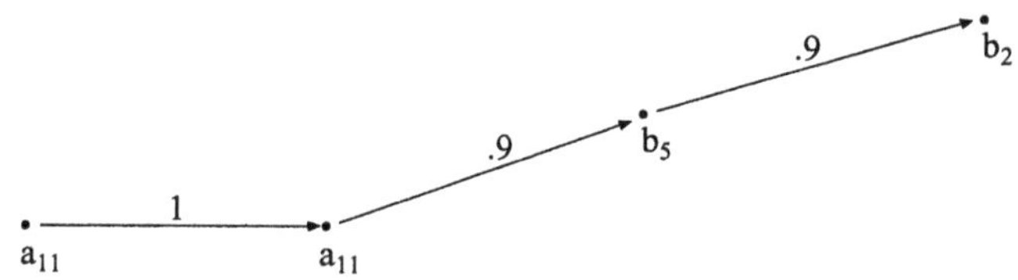

Once again was not easy, at a single glance, to establish the connection between this cause and this effect. But the theory of forgotten effects provided us with the path to follow. In fact, the incidence of "presence of academicians at large public manifestations" acted to a very high degree (degree 0.9) over "personal prestige of members" which made it very easy (degree 0.9) for "presence at scientific forums". There was no doubt that all this was included within the strictest rules of good reasoning.

d) Incidence of "strengthen the presence of academicians at large public manifestations" over "incorporation in international institutions".

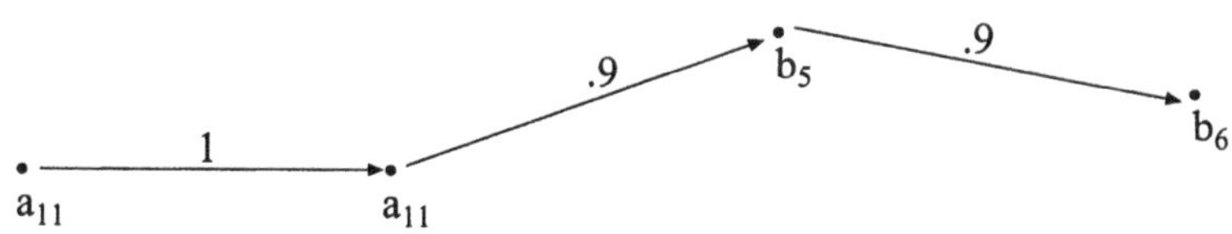

The initially difficult verification of the relation of direct causality was resolved by the same path as used in the previous section, that is, by the intermediation of "personal prestige of members" of the Reial Academia.

These four cases, that we detailed then, served as a clear example of the usefulness of the techniques employed. We were conscious of the fact, nevertheless, that the work we had carried out only had the value of opening up a new path, which we hoped would be followed by other academicians, who, taking advantage of the inter-disciplinary nature of our Royal Corporation, would be capable of greatly improving on it. We had thrown out a challenge. The love for our institution should serve as a stimulus for joining efforts in the interest of attaining our best theoretical fulfilment, but above all else for more efficient instruments that are suitable for raising, even more, the prestige of the Reial Academia to which our President had dedicated all his efforts for the best part of his life.

REFERENCES

[1] KAUFMANN, A. and GIL ALUJA, J.: *Técnicas operativas de gestión para el tratamiento de la incertidumbre*. Publ. Hispano-Europea. Barcelona 1987 pages. 133-151.

[2] KAUFMANN, A. and GIL ALUJA, J.: *Modelos para la investigación de efectos olvidados*. Publ. Milladoiro. Santiago de Compostela 1989.

[3] GIL LAFUENTE, J.: *Marketing para el nuevo milenio*. Publ. Pirámide. Madrid 1997, pages. 454-465

[4] GIL LAFUENTE, A.M.: *Nuevas estrategias para el análisis financiero en la empresa*. Publ. Ariel. Barcelona 2001, pages. 452-469.

EPILOGUE

Today we are members of a society that is lodged in uncertainty. The profound changes, often pulling in different directions, foresee a *new era* in which ideas, phenomena and things will never be the same. And, nevertheless, within these new surroundings adaptation will be necessary. The Reial Academia de Doctors has contributed and continues to contribute, for this situation , a very tiny grain of sand.

It is comforting to see how, for more than twenty year, certain researchers of our Royal Corporation have accepted the hard task of moving the foundations on which *Transmittable knowledge* is based. Little by little a *new scientific building* is being constructed, that is capable of housing Concepts, Methods and Techniques that are susceptible to providing instruments designed for redirecting social, economic and business activities towards a better world. A world in which justice and solidarity become the basis and the support of our coexistence.

It is in this sense that our Royal Corporation has attempted, over time, to radiate to the five continents the light of Science, and has wanted to transmit to the furthest corners the traces of our Culture.

In spite of this, it would not be correct to take pleasure in self-complacency when faced with a vision of a world shaken by so many tragedies. Faced with a possible rise in terror, it now corresponds to us to contribute, from our vantage point, *the task well done*, in the service of *progress*; *the spirit of tolerance in the service of concord and the expansion of new ideas in the service of the common well-being.*

This work intends to be a modest contribution for attaining these desires. For this we have chosen the difficult path of explaining things in a simple manner. Our readers will be the judges of whether we have achieved our objective. Each chapter has tackled a subject relative to business and institution management, by way of an event relative to actual life, of which the players have been friends from our childhood and, exceptionally, some happily found in our maturity. In a certain fashion this is a homage to friendship. Only a very timid homage, since there is no gift whatsoever that is capable or worthy of the prize represented by the fact of having a good friend. We believe that we have them. And this in spite of all the treachery and disappointments of many of those who approached with empty hands and went away full of vanity. But such is life. All forms a part of the context in which we wander, in which, the nature itself of human beings and things, supplies a multicoloured vision of our small universe, today subject to so many changes. We have become accustomed to the profound changes that take place both in the world of ideas as in the world of phenomena. Perhaps because of this it is difficult to imagine the context in which the citizens of this new century will carry on their activity in the future.

It would seem that there is total agreement on the fact that we are at the gates of a new era and that also that in it, it will be difficult to maintain the social and economic systems to which we have become accustomed. Activities that formed a part of daily events, routines followed by men of business and habits acquired by public managers, have given way to new ways of approaching the problems that arise on a daily basis, bringing to light the fact that the economic, financial and commercial scenario of businesses and institutions is changing with considerable speed. And, nevertheless, it is necessary to adapt to the new socio-economic context. Our prime objective is to live alongside these phenomena that keep appearing with insistent intermittency.

Those who wish to be actors in the new society that is just emerging, must prepare themselves, updating their skills and knowledge. It is convenient that we open the windows to the new ideas of this new world. Uncertainties, already today, form a consubstantial part of our lives. We may or may not like it, but this is a reality, which will, increasingly, show its fuzzy profiles, by changing its forms during the course of time.

Economic activity cannot avoid this new manner of conceiving a society in which the acceleration of time will impose itself as a fundamental element. Fortunately, scientists and technicians are working in order to contribute new concepts and new methods that are capable of successfully facing up to the challenges that arise. Today, it is possible to state that the advances within the realm of management and technology constitute veritable pillars on which the transformation of society is based.

Just a very few last words, sown very deeply, to manifest our profound gratitude to the Reial Academia de Doctors for the opportunity we have been offered to present a few ideas, which inserted in the form of techniques have been used for the resolution of old and new problems. We have been guided by the hope of contributing a very small grain of sand to the marvellous beach of Knowledge

BIBLIOGRAPHY

ABRAHAM, C. and THOMAS, A.: *Microéconomie: décisions optimales dans l'entreprise et dans la nation.* Publ. Dunod. Paris 1966, pages 325-327.

BARRE, Raymond : Prologue to the book by KAUFMANN, A. and GIL ALUJA, J.: *Técnicas operativas de gestión para el tratamiento de la incertidumbre.* Publ. Hispano Europea, Barcelona, 1987.

DALKEY, N., BROWN, B., COCHRAN, S.: *La prévision à long terme par la méthode Delphi.* Ed. Dunod. París 1972.

DESROUSSEAUX, J.: *L'evolution économique et le comportement industriel.* Publ. Dunod. París 1965.

GIL ALUJA, J.: *El estudio dinámico de la elección de las inversiones.* Revista Técnica Contable. Febrero 1967.

GIL ALUJA, J.: *Selección de personal:el problema de la polivalencia y el de la uniformidad.* Cuadernos Ceura. Madrid, 1987.

GIL ALUJA, J.: *FAMM (Fuzzy Adverse Minimum Method). Paper presented at the European Congress of AEDEM, published in the proceedings of the same.* Minsk, 20-23 Septiembre 1993.

GIL ALUJA, J.: *Lances y desventuras del nuevo paradigma de la teoría de la decisión.* Proceedings of the III SIGEF Congress. Buenos Aires, 10-13 Noviembre 1996.

GIL ALUJA, J.: *Elements for a theory of decision in uncertainty.* Kluwer Academic Publishers. Dordrecht, Boston, Londres, 1999.

GIL ALUJA, J.: *Interactive Management of the human ressources in uncertainty.* Kluwer Academic Publishers. Boston, Londres, Dordrecht, 1999.

GIL ALUJA, J.: *Investment in uncertainty.* Kluwer Academic Publishers. Londres, Boston, Dordrecht, 1999.

GIL LAFUENTE, A.M.: *El análisis financiero en la incertidumbre.* Publ. Ariel. Barcelona 1990.

GIL LAFUENTE, A.M.: *Fundamentos de análisis financieros.* Publ. Ariel. Barcelona 1993.

GIL LAFUENTE, A.M.: *Nuevas estrategias para la gestión de las empresas,* Publ. Ariel Barcelona 2001.

GIL LAFUENTE, J.: *Marketing para el nuevo milenio.* Publ. Pirámide. Madrid 1997.

GIL LAFUENTE, J.: *El "indice del máximo y mínimo nivel" en la optimización del fichaje de un deportista.* Proceedings of the AEDEM'01 Congress. Polytechnical University of Reggio 4-6 Septiembre 2001.

KAUFMANN, A.: *Méthodes et Modèles de la Recherche Opérationelle.* Tomo I. Dunod París 1970

KAUFMANN, A.: *Introduction à la théorie des sous-ensembles flous.* Vol I. Ed. Masson. París, 1973.

KAUFMANN, A.: *Modèles mathématiques pour la stimulation inventive*. Publ. Albin Michel. París 1979.

KAUFMANN, A. and GIL ALUJA, J.: *Introducción de la teoría de los subconjuntos borrosos a la gestión de las empresas*. Publ. Milladoiro. Santiago de Compostela 1986.

KAUFMANN, A. and GIL ALUJA, J.: *Técnicas operativas de gestión para el tratamiento de la incertidumbre*. Publ. Hispano-Europea. Barcelona 1987.

KAUFMANN, A. and GIL ALUJA, J.: *Modelos para la investigación de efectos olvidados*. Publ. Milladoiro. Santiago de Compostela 1989.

KAUFMANN, A. and GIL ALUJA, J.: *Las matemáticas del azar y la incertidumbre*, Publ. Ceura. Madrid 1990.

KAUFMANN, A. and GIL ALUJA, J.: *Nuevas técnicas para la dirección estratégica*. Publ. Universidad de Barcelona. Barcelona 1991.

KAUFMANN, A. and GIL ALUJA, J.: *Selection of affinities by means of fuzzy relations and Galois lattices*. Proceedings of the XI Euro O.R. Congress. Aachen, 16 - 19, julio 1991.

KAUFMANN, A. and GIL ALUJA, J.: *Técnicas de gestión de empresas. Previsiones, decisiones y estrategias*. Publ. Pirámide, Madrid 1992.

KAUFMANN, A. and GIL ALUJA, J.: *Técnicas especiales para la gestión de expertos*. Publ. Milladoiro, Santiago de Compostela 1993.

KAUFMANN, A., GIL ALUJA, J. and GIL LAFUENTE, A.M.: *La creatividad en la gestión de las empresas*. Publ. Pirámide, Madrid 1994.

KÖNIG, D.: *Théorie der endlichen und unendlichen graphen* (1916) later reprinted by Chelsea Publ. Cª. New York, 1950.

KUHN, H.W.: *"The hungarian method for the assignment problem". Naval Research Logistics Quartely Vol. 2* nos. 1-2, marzo-junio 1955.

LITTLE, J.D.G. and others: *An algorithm for the travelling salesman problem I.O.R.S.A.* Vol.II nº 6, 1963.

PICHAT, E.: *Algorithm for finding the maximal elements of a finite universal algebra*. Inform. Processing 68. Publ. North Holand 1969.

SCHNEIDER, E.: *Teoría de la inversión*. Publ. Ateneo. Buenos Aires 1956.

SCHNEIDER, E: *Economía política y económica de la empresa*. Publ. Sagitario Barcelona, 1968.

TERBORGH, G.: *Dynamic Equipment Policy*. Publ. Mc. Graw-Hill. New York, 1946.

TERBORGH, G.: *Business Investment Policy. Machinery and Allied Products Institute*. Washington 1958.

TERBORGH, G.: *Business Investment Management. Machinery and Allied Products Institute*. Washington 1967.

ZADEH, L.: *Fuzzy Sets. Information and Control*, 8, Junio 1965.

www.ingramcontent.com/pod-product-compliance
Ingram Content Group UK Ltd.
Pitfield, Milton Keynes, MK11 3LW, UK
UKHW020131290926
15471UKWH00003B/412

* 9 7 8 3 6 4 2 5 3 5 5 0 5 *